W9-ABH-784

An Introduction to GSM

The Artech House Mobile Communications Series

John Walker, *Series Editor*

Advanced Technology for Road Transport: IVHS and ATT, Ian Catling, editor

An Introduction to GSM, Siegmund H. Redl, Matthias K. Weber, Malcolm W. Oliphant

Cellular Radio: Analog and Digital Systems, Asha Mehrotra

Cellular Radio Performance Engineering, Asha Mehrotra

Mobile Communications in the U.S. and Europe: Regulation, Technology, and Markets, Michael Paetsch

Land-Mobile Radio System Engineering, Garry C. Hess

Mobile Antenna Systems Handbook, K. Fujimoto, J. R. James

Mobile Information Systems, John Walker, editor

Narrowband Land-Mobile Radio Networks, Jean-Paul Linnartz

Wireless Data Networking, Nathan J. Muller

For a complete listing of *The Artech House Telecommunications Library*, turn to the back of this book.

An Introduction to GSM

Siegmund H. Redl
Matthias K. Weber
Malcolm W. Oliphant

Artech House
Boston • London

Library of Congress Cataloging-in-Publication Data
Redl, Siegmund, H.
An introduction to GSM / Siegmund H. Redl, Matthias K. Weber, Malcolm W. Oliphant.
Includes bibliographical references and index.
ISBN 0-89006-785-6
1. Mobile communication systems. 2. Cellular radio. I. Weber, Matthias K. II. Oliphant, Malcolm W. III. Title.
TK6570.M6R43 1995 94-23893
621.3845'6–dc20 CIP

British Library Cataloguing in Publication Data
Redl, Siegmund H.
Introduction to GSM
I. Title
621.38456

ISBN 0-89006-785-6

Chapters 4 and 6–10 are largely adapted from and Chapters 3, 5 and 11 contain portions of text adapted from the German work *D-Netz-Technik und Meßpraxis* by Siegmund Redl and Matthias Weber, © 1993 FRANZIS-Verlag GmbH.

685 Canton Street
Norwood, MA 02062

International Standard Book Number: 0-89006-785-6
Library of Congress Catalog Card Number: 94-23893

10 9 8 7 6 5 4 3 2 1

CONTENTS

PREFACE

This book is about GSM, the global system for mobile communications, which is a digital cellular radio system. It contains a serious discussion of GSM 900 and DCS 1800 (digital cellular system on the 1,800-MHz band), which applies as well to their proposed derivative, PCS 1900 (personal communication system on the 1,900-MHz band, a reserved band in the United States). With this book, readers do not need to sort through all the details in the technical standards and recommendations, but are free to decide how much they want to learn about the different topics.

This book is for engineers and technicians who want a comprehensive introduction to GSM. Though the material is rigorous enough for a thorough first-pass description of the system, it is also general enough for project managers newly assigned to a GSM project who need to make analyses and decisions. It can also be an excellent accompaniment to a short, intensive seminar on the subject. This is not merely a theoretical work, but is a functional description of the system and its architecture, operation, and procedures.

Emphasis is placed on practical matters, particularly testing issues in all its forms. The authors have wide experience in cellular radio and have helped clients solve a variety of problems. They have seen many projects succeed, and they have also seen some fail.

The book also includes considerable descriptive details on important competitive digital cellular radio systems such as the North American digital cellular system (NADC or D-AMPS) and the code-division multiple access (CDMA) system.

Cellular Mobile Radio

Part I contains Chapters 1 and 2. The first chapter describes the cellular landscape before digital, discusses the diversity of analog cellular systems, as well as the commercial and technical forces that created them, and explains the need and reasons for the generation of new technology. The second chapter introduces some basic radio system engineering terms and concepts such as channel capacity and quality, as well as economic aspects, focusing on regions worldwide. This chapter also explores the effects of commercial and political forces on the search for a solution to the problems in analog systems.

Digital Cellular Mobile Radio—GSM

Part II contains Chapters 3 through 7, which describe the typical GSM and mention simpler and more intuitive analog systems where possible. Chapter 3 describes the history and development of the GSM standard, the services and security offered to the users of a GSM-based network, and the structures and architecture of the network. It also gives a detailed description of how a mobile terminal gains access to a GSM system. Chapter 4 briefly explains how GSM is embedded in the integrated services digital network (ISDN), describes the ISO/OSI (International Standards Organization/open systems interconnection) layer model, and explains how the lower three layers of this model function.

Chapter 5 explores in detail the lowest of the three layers, Layer 1, the physical layer, which is the air interface part of GSM: channel scheme, RF power, cell size, timing, time-division multiple access (TDMA), burst and frame structures, synchronization, the difference between traffic and signaling data, speech and channel coding, modulation, demodulation, and frequency hopping. Chapter 6 describes the next higher layer in the model, Layer 2, the data link layer, which contains the functions ensuring that signaling takes place in an orderly way between radios in GSM. The chapter descibes the structure and format of the fields and works out an example showing how Layer 2 orders and moves data between the air interface and Layer 3.

Chapter 7 describes the highest layer in the model, Layer 3, the network layer, which contains the functions that connect the radios to its users and networks. Like Chapter 6, Chapter 7 describes sublayers, fields, and messages. This chapter also explains timers associated with Layer 3 and works out an example.

New Testing Methods

Part III contains Chapters 8 through 11, which describe the testing methods used in GSM. This part is filled with practical examples and comparisons with appropriate analog functions where possible. Chapter 8 explores the salient differences between analog and digital radio testing, and uses the simpler and more intuitive analog examples as starting points for the digital cases. Chapter 8 also names some of the instruments that are necessary or available for GSM testing. Chapter 9 explains how the mobile stations are tested. Chapter 10 describes base station testing. Chapter 11 explains the five testing applications used in GSM: R&D, type approval, production, service, and network planning.

Competing Digital Cellular Standards

Part IV contains Chapters 12 and 13, which describe some important competing digital cellular systems from the GSM point of view: NADC, Japanese digital cellular/ personal digital cellular (JDC/PCD), and CDMA. The material is presented to enhance understanding of system functions and technical details of the GSM system, where the NADC and JDC/PDC systems are viewed as variations on GSM and the CDMA proposal is viewed as a completely different approach to digital cellular systems. These chapters clarify the place of GSM in the world of digital cellular systems.

Glossary and Appendix

The glossary tries to clear up some of the confusion surrounding acronyms, abbreviations, and specialized terms found in GSM literature. The appendix is a bibliography of applicable literature and GSM documents.

CHAPTER 1

The Global Scenario for Cellular Mobile Radio

Cellular mobile radio—the magic technology that enables everyone to communicate anywhere with anybody—has created an entire industry in mobile telecommunications, an industry that is rapidly growing and that has become a backbone for business success and efficiency and a part of modern lifestyles all over the world. This chapter looks at some aspects of the history of cellular mobile radio, as well as the present situation of cellular mobile communications markets, their commercial parameters and dynamics.

1.1 HISTORICAL BACKGROUND

Mobile radio has been used for almost 75 years. Even though the cellular concept, spread spectrum techniques, digital modulation, trunking techniques, and other modern radio technologies were known more than 50 years ago, mobile telephone service did not appear in useful forms until the early 1960s, and then only as elaborate adaptations of simple dispatching systems. The convenience of these early mobile phone systems was severely limited, and their maximum capacity was tiny by today's standards. Finally, analog, full FM duplex, trunked access, *cellular mobile phone* systems appeared in the 1980s. All the systems were popular and grew at rates tempered only by local regulatory anomalies, commercial considerations, and the available radio spectrum.

By the end of the 1980s, it became clear that the analog cellular systems would not be able to meet continuing demand into the next century unless something was done about four inherent limitations of these systems: (1) severely confined spectrum allocations; (2) a perception among more sophisticated users—the chief revenue generators—that the systems were limited in usefulness because of annoying sounds and interference as the mobiles moved about in a multipath fading environment, as well as their having only trivial access to a growing catalog of attractive network features; (3) inability to substantially lower the cost of mobile terminals and infrastructure, and (4) incompatibility among the various analog systems, especially in Europe, thus preventing the subscriber from using his or her phone abroad.

What is the solution? Spectrum space is probably the most limited and precious resource available in the industrialized world. Continued allocation of additional spectrum to meet the cellular services' growing demand is simply out of the question. The solution is to further multiplex traffic (e.g., in the time domain) into a radio system on top of the current and conventional frequency and spatial domains. Such time multiplexing requires that all the traffic and signaling functions be realized with *digital* techniques.

Digital radio hides the effects of fading and interference from the user. Moreover, digital modulation and all its logical extensions make access to network features much easier. Half-duplex digital radios are theoretically cheaper to produce than full-duplex analog radios. These practical considerations, together with political pressure in Europe, created the GSM system in the late 1980s in which specific features were included to allow international roaming and a wide variety of auxiliary services.

The history of cellular mobile radio in the forms we see today began only recently. The first real commercial cellular system in the United States (in Chicago) went into service in 1983 after an earlier system in Japan. This was followed by some halting years and some changing marketing scenes, finally leading to eventual acceptance by the mobile customer. Still, the cellular radio market has to be seen as a fast-changing one, subject to changing consumer moods and significantly influenced by the available services. The available services are, in turn, derivatives of whatever new technologies are employed and the hardware offered for sale. Even though there were the unavoidable little problems and hiccups when introducing new technologies and services, cellular phone service has been accepted worldwide as a comfortable, useful, and indispensable accessory to daily life and as an expression of the modern lifestyle. The growth in the number of subscribers always was, and still is, immense. Cellular's success can only be compared to the introduction of the radio, the television, and the telephone [1].

The cellular phone has clearly proven itself and has justified its usefulness in a world full of alternative means of wireless communications, including pagers, trunked private radio, and telepoint services (cordless phones for inhouse and city use). These alternatives enjoy healthy growth in their own right and attest to the attractiveness of mobile communications in any form. Cellular will hold its position when the *personal communications networks* (PCN) and the U.S. *personal communications systems* (PCS) are introduced. In fact, PCNs and PCSs are simply microcellular systems that emphasize low-cost service, virtual customer networks, and hand-portable phones with a long battery life.

1.2 SITUATION OF GLOBAL CELLULAR MOBILE RADIO BEFORE DIGITAL

1.2.1 Different Standards Worldwide

The cellular services available throughout the world have matured beyond infancy. Especially in the industrial countries, the mobile phone has established a strong position among business executives and private users. The cellular services offered up to the year 1992 used exclusively analog radio technology. Analog cellular networks still exist in several countries, and the vast majority of them are based on only a few different standards. The most important standards, with some variations for national interests and technical considerations, are:

- *Advanced mobile phone service* (AMPS), U.S. standard EIA-553 (Electronics Industries Association), on the 800-MHz band;
- *Total access communications system* (TACS), U.K. standard, based on the AMPS protocol with some adaptations in bandwidth and allocated spectrum on the 900-MHz band;
- *Nordic mobile telephone system* (NMT), Scandinavian standard on the 450- and 900-MHz bands. Though similar to each other, there is considerable difference in the details between the NMT-450 and the NMT-900 analog systems.

Table 1.1 contains partial lists of countries using each of the three analog cellular standards mentioned above.

Table 1.1
Analog Cellular Standards and Their Use Worldwide

Standard	*Some of the Countries Using Standard*
AMPS	United States, Canada, Mexico, Australia, New Zealand, Taiwan, South Korea, Singapore, Hong Kong, Thailand, Brazil, Argentina
TACS	United Kingdom, Ireland, Spain, Italy, Austria, United Arab Emirates, Bahrain, Kuwait
NMT 450/NMT 900	Denmark, Finland, Norway, Sweden, Belgium, Austria, France, Hungary, Netherlands, Spain, Turkey, Switzerland

Various other standards with less significance and usage are restricted to single countries. Though most of them have remote origins in one of the three systems in Table 1.1, some were developed particularly for use in only a few countries, such as the C-Netz standard in Germany (then used also in Portugal and South Africa) and the Radiocomm 2000 system in France.

To sum up, we can say that the world is covered with a quilt of analog cellular radio systems on different frequency bands and with different signaling protocols. Especially when considering European countries, it becomes clear that the different and therefore incompatible standards will not allow *international roaming* with a single cellular phone (i.e., using the same phone in more than one country). One exception is Scandinavia, where the NMT system provides for this roaming feature. The economic and social consequences of this continued segregation of mobile phone services is a huge subject by itself, beyond the scope of this book.

Another important point is that the various cellular standards largely describe a radio interface, or a common air interface (CAI), which has been attached to or adapted to a fixed telephone network. The standards do not describe standalone radio communications systems.

1.2.2 Commercial Aspects of Cellular Communications

The explosive growth in the number of subscribers to cellular services in the past decade resulted in the fast and steady expansion of the cellular networks. The installations and expansions increased both coverage and capacity. This constant building and expansion will not be finished for years to come. *Coverage* can be expressed as the percentage of a country's area covered by cellular service or the portion of a region where cellular mobile service is available. *Capacity* refers to the number of calls that can be handled in a certain area within a certain precise period of time. Capacity can also refer to the probability that users will be denied access to a system due to the simple unavailability of radio channels.

Today, the analog cellular networks still cover their respective countries and regions and provide adequate coverage and capacity. These systems will not disappear; they will run and expand well into the next decade. Some of the systems still need to recover the initial investments made to build them up. Still other systems will continue to operate profitably, because there is no reason for the end users to change to another—eventually digital—system. The most important figure, which directly influences the required capacity of a cellular network, is the number of subscribers it has to serve, especially in densely populated areas at the peak usage times. This figure is influenced by a host of parameters, including the following:

- *Cost of ownership:* The price (sale or lease) of mobile terminals, subscription fees, taxes, and air time fees;

- *Quality of service:* The quality of speech transmission, the ability to access services, the incidences of dropped or blocked calls, crosstalk, interference, coverage, security, and auxiliary services (e.g., mailbox, call forwarding, data and fax transmission);
- *Demography and social status of a population or society:* The number of residents who can afford a cellular phone, the number of residents who are likely to need cellular service for business, and the manner in which business is conducted within a society;
- *Whatever symbolic status applies to cellular service:* Attractiveness of owning and using a cellular phone, and the influence of marketing campaigns and advertisements.

There are more interdependencies among all the above parameters than this book can ever hope to cover. There are, however, some salient additional parameters worth mentioning that serve to prove how dynamic cellular communication is for business and the culture in which it is installed.

- *Low cost of ownership*—tariffs and terminal costs—will lower the boundary of financial acceptance and lead to higher subscriber numbers. This encourages investment in additional capacity. In some countries, the phone terminals are merely given away to anyone who agrees to buy service for a certain period of time.
- *High cost of ownership* keeps subscriber numbers low and the revenue per subscriber—for those who can afford the service—relatively high. This situation generally does not stress capacity.
- *The status symbol of having a cellular phone* loses its value when it is available for lower cost to a larger number of people. On the other hand, more people will be able to possess this status symbol when the phone and the service is cheap. Marketing experts know how to balance these competing situations for their maximum effect.
- *Cost per subscriber* is another parameter that influences still others; the investment an operator of a cellular mobile telephone network has to make to provide both coverage and capacity of a certain quality is not at all trivial.

Other measurable parameters are the cost per channel or voice path, the cost per cell site (equipment and rent), the price of insurance, and the price of leased phone lines or microwave channels between the sites. In addition, there are still other costs that need to be considered and that need to be recovered by the cellular network operator: cost of network and cell site planning, maintenance, billing, test equipment, additional network features (value-added services), transportation, and weather (e.g., the possibility of a base station site being knocked out of service by lightning).

High quality of service is indispensable for mobile customer satisfaction and is an important prerequisite for healthy revenue. Low capacity always yields a diminished quality of service. As the number of dropped or blocked calls, interferences, crosstalk, and poor reception conditions increase, the casual mobile customer will not be satisfied with the service and will stop using it. The less casual users or users who truly depend on mobile phone service (e.g., the mobile trades and outside sales people) have a relatively high tolerance to marginal service conditions.

Competition in providing cellular service where, for example, two or more operators are allowed in the same country or region leads to very dynamic scenarios of tariffs and services. End users always benefit from competition.

There is no point in giving sample figures here. The dynamics of the parameters above are complex and are considered by the vast numbers of experts responsible for the financial health of cellular service providers. The industry is so dynamic that the figures, when available, become obsolete within increasingly shorter periods of time. An important and familiar parameter is the price of mobile terminals: decreases in price, add-on features, and give-away policies can take on stunning forms.

The trends of some remaining parameters are in the direction of (1) better quality of service, (2) higher coverage and capacity, (3) low cost of ownership, (4) lower costs per subscriber, (5) lower costs per voice channel, and (6) lower operational costs. All of these increase the number of subscribers.

REFERENCE

[1] Calhoun, G., "Introduction," in *Digital Cellular Radio*, Boston: Artech House, 1988, pp. 10–12.

CHAPTER 2

▼▼▼

The Transition From Analog to Digital Cellular Radio—Why?

Given the expense and accompanying confusion, it is fair to ask why anyone should bother participating in the transition from analog to digital cellular. How can an industry as successful as cellular radio possibly be in need of a major rework? There are, it turns out, a number of reasons why a move from conventional analog cellular radio to digital cellular radio is taking place. The transition to digital radio also takes on different forms in the various defining standards and the variety of technologies involved. Some of the reasons for the transition and some of the forms it takes, particularly in the new standards as contrasted with the myriad modifications to older ones, are controversial and remain the subject of intense discussion. In this chapter, we explore the reasons for the digital transition, together with some of the forms the transition takes. It will become clear that the motivations for the move to digital technology are regionally oriented and have different commercial, technical, and political origins.

2.1 THE CAPACITY SYNDROME

Especially in Europe and the United States, the continuing growth in demand for cellular services was only marginally influenced by the traditional factors of cost, tariffs, and quality. Demands left unmet often led to backlogs in cellular capacities.

The most critical regions of unmet demand were in metropolitan areas with high subscriber density and high traffic during peak hours.

Restricted spectrum and even the most clever attempts at the reuse of spectrum—a frequency channel can be reused only within some minimum distance—soon presented limitations to the operators of cellular networks [1]. Operators asked for more spectrum to cover their needs. They also called for new techniques to enhance capacity without sacrificing quality.

Due to the limited availability of additional spectrum, new technologies were proposed rather than strong-arm methods with government regulators. A cooperative atmosphere prevailed. The regulators, the operators, and the industry worked hard to achieve the utmost in network capacity. Among the solutions were (1) narrowband radios (less bandwidth per channel), (2) microcells, (3) sectored cells, and (4) time-multiplexing digital radio techniques. The coverage of dead spots (where service was not available at all) and hot spots (where service was often not available due to high traffic) was also a relevant topic. Eventually, some analog techniques (particularly narrowband radio and microcells) were able to cope with the problems to a certain degree, and they were in place even before digital equipment appeared. The capacity issue, then, became somewhat obscured by the dust of the rush to digital.

The cost of adding new capacity finally became an important factor. In some countries, a new spectrum was set aside or designated for new cellular services. Since spectrum remained a precious resource, the issue of which technology to choose for the best spectral efficiency was raised again. It must be noted that digital radio is not, and will not be, the expected or even the predicted perfect overall solution to *all* capacity problems.

2.2 ASPECTS OF QUALITY

Analog radio has the effect of directly passing physical influences or disturbances in radio transmission links into the audio path of the receiver, to users' occasional dismay. Such disturbances can be fades, interferences, spurious signals, or multipath reception. The results are static, hums, hisses, crackling sounds, crosstalk, and fadeouts, all of which occasionally annoy cellular users. Digital radio hides these effects. With digital radio techniques, the audio signal is not transmitted as such, but is transformed into digital data patterns we call *bits*, or *bit patterns*. Digital coding and error correction mechanisms support the reconstruction of the transmitted data in the receiver after it is sent over the hostile radio channel.

A comparison of speech quality will show that both analog and digital cellular radios perform very well under ideal reception conditions. If such were not the case, there would not be a healthy cellular industry today. Under critical and more difficult reception conditions, such as low received levels, multipath, and fades, digital radios perform significantly better than analog radios.

The quality of service can also be rated by any supplementary network features offered to users. Digital is expected to facilitate the introduction of value-added services, thanks to some technical (digital or data) compatibility with the established digital wire-line services. The integrated services digital network (ISDN) is an example of this easy compatibility in the terrestrial network. Services beyond traditional speech (e.g., data transmissions offered by ISDN) will eventually start to be accepted by end users [2]. A prerequisite to the initial acceptance of these auxiliary services is that operators of cellular networks offer such services and are able to attractively market their benefits to users.

2.3 SOME POLITICAL AND ECONOMIC ASPECTS

When looking at the issues surrounding the steady growth in the cellular industry with its demands (spectrum) and benefits (happy customers and revenue), there are also some topics that must be sorted out by regulatory bodies, such as national and international telecommunication associations and standardization bodies. There are some other players in the industry, which exert some peripheral influence on the decisions taken by those who regulate the mobile communications markets. There are the operators, national post and telecommunication companies, and the privately owned consortia, whose main interests are to serve the maximum number of subscribers with maximum quality at minimum investment and operational costs.

Then we have the manufacturers of mobile terminals and infrastructure equipment who depend on the designers and developers of the indispensable semiconductor devices within the terminals and infrastructure. The manufacturer's aim is to develop and manufacture products for large markets based on proven standards and technologies.

Most important are the end users, the mobile customers, who pay all the bills. They want to have adequate services at reasonable cost. As products and their service features improve, end users will continuously increase their demands. What was a brand new feature yesterday will be seen as standard and absolutely necessary tomorrow.

The combination of all these facts and factors, parameters and restrictions, demands and compromises, as well as the players and the markets, led to the development and introduction of digital cellular mobile standards. There were, however, some differences in the ways this process was tackled in various regions and countries.

Generally speaking, there were three different forms of modernizing cellular services in different areas in the world.

The U.S. Way

In the United States, the attitude toward modernization is to let the industry think about something, let it prove that it works, and then make a standard out of it.

which was finally agreed to in 1982 by the European Posts and Telecommunications Conference (CEPT, Conférence Européenne des Administrations des Postes et des Télécommunications);
- Capacity limitations on the existing (analog) cellular networks in many European countries.

The cellular networks, which were opened in the early 1980s, have experienced astronomic growths in the number of subscribers. For instance, the Nordic countries and the United Kingdom barely coped with the demand by continuously installing their NMT-450/900 and TACS networks, respectively. They even had to borrow frequencies from the spectrum allocated to a future pan-European cellular system. Due to their outstanding performance, these networks have been accepted eagerly by end users and will continue to meet cellular demands until the mid 1990s, when GSM is expected to gain significant momentum in these countries.

In Germany, the existing C-Netz system (opened in 1985 on the 450-MHz band) was running out of capacity at approximately 900,000 subscribers, even though the number of subscribers was regulated by the high cost of the terminals and service. As another extreme, the introduction of the GSM standard was immediately accepted by almost 1,000,000 end users within 18 months.

Some other factors influenced the face and parameters of GSM. There were different, and even divergent, requirements for the technical layout of the system. GSM should be able to economically cover (1) vast areas, as well as (2) densely populated urban and suburban territory. It should work reliably at (3) high speeds (in fast-moving cars or trains), (4) in urban jungles in the hands of slowly moving pedestrians among lots of tall buildings, and (5) within large buildings, parking structures, airports, and train stations.

It should be noted that it was not an initial demand that the system should be digital, as it eventually turned out to be. This decision was only made in the course of developing the standard, and the digital character of GSM came about because of the complementary services offered by the new digitally fixed ISDN wire-line standard. The digital character of GSM brings easier implementation of both technology and services in the digital domain.

2.4.3 The Cellular World Outside Western Europe and North America

2.4.3.1 Central and Eastern Europe

The opening up of the central and eastern European countries, which were blank spots on the map of cellular radio, have led to the rapid introduction of cellular services there. The need for cellular services was, and remains, driven by the absence of sufficient fixed wire-line telephone systems in central and eastern Europe. Most countries in central and eastern Europe now have cellular licenses and operational systems. Today, we can see the quick installation of some analog systems to cover increasing demand rather than the expenditure of huge sums of money and time to install wire-line infrastructure. This demand is mainly fueled by the needs of business-oriented people. The systems are mostly NMT-450, due to spectrum availability in the 450-MHz band. Some countries, such as Russia, Hungary, and the Baltic states, have already decided in favor of the GSM standard as the basis of future installations. These decisions follow from the relatively low cost of a new system and the easy resolution of roaming issues [3]. Another motivation for using GSM in central and eastern Europe is its compatibility with modern digital wire-line services and equipment (ISDN), which are to be installed instead of repairing the antiquated infrastructure.

Some risks remain: spectrum clearance (military use of 900-MHz spectrum), investment incentives, costs, export restrictions, and acceptance of the installation of a new system instead of improving fixed wire-line services and existing analog cellular systems. Basic telecommunications is still under development in central and eastern Europe, and this region will definitely be a huge marketplace for both manufacturers and operators. Since many Western network operators have valuable experience in the design, installation, and operation of cellular networks, we can see some joint ventures between these experts and local telecommunication authorities. This trend will continue.

The introduction of telecommunication services, especially modern digital mobile cellular service, is vital to the economic recovery of this area and will create new markets for suppliers of equipment and services. The GSM standard is most likely to be the big winner in this region.

2.4.3.2 *The Asian and Pacific Regions*

This region is expected to become increasingly important in the economic world picture. Not only is this region a major supplier of raw materials for Western industry and mass products for the world, it is also a major market for Western industry and technologies. There are sharp contrasts in the availability of cellular services in this region. Many countries already have working analog cellular networks, while some other countries are now installing their first new cellular networks. In those countries and areas with installed networks, the systems are based largely on the AMPS technology, but there are also some TACS and NMT systems. Measures have already been taken to increase capacity in some Asian and Pacific areas. In many discussions and decision processes, which will continue throughout the next few years, the GSM standard often wins out over the U.S. digital IS-54 and IS-95 standards. Countries or network operators looking into their first installations of real cellular networks also tend to use GSM equipment [4]. Again, the term *international roaming* is often heard during discussions on this subject.

2.4.3.3 Middle Eastern, Gulf Region, and North African Countries

Most of the Islamic countries have operational mobile networks. Different conventional analog systems, mostly TACS, NMT, and some others, contribute to a situation similar to the one in Europe, where different incompatible analog standards separate neighboring countries. The incentives for introducing a single new digital standard in a pan-Islamic cellular system are to be found in roaming capabilities. The incentives and intentions are (1) agreements with operators worldwide, (2) improved quality of service, (3) infrastructure cost, and (4) economic stimulation. In support of developing a new regional standard, the GSM standard has been recommended and will be chosen as the future basis for national, regional, and *international* cellular interconnection.

2.4.3.4 South America and Africa

Most South American and some African countries rely on the AMPS system. Not many attempts to introduce digital cellular services can be seen, except in Cameroon and South Africa, which intend to introduce the GSM standard.

REFERENCES

[1] Calhoun, G., "Cellular Realities," in *Digital Cellular Radio*, Boston: Artech House, 1988, pp. 115–120.
[2] Shosteck, H., "Digital/Next Generation Cellular Technologies: A Competitive Analysis and Forecast," Silver Spring, Dec. 1993, pp. 90–94.
[3] "Regional Focus: Eastern Europe," *Mobile Communications International Magazine*, Issue 14, Summer 1993, p. 24.
[4] "Regional Focus: Asia Pacific," *Mobile Communications International Magazine*, Issue 14, Summer 1993, p. 35.

PART II

Digital Cellular Mobile Radio—GSM

CHAPTER 3

GSM—A Digital Cellular Mobile Radio System

Different requirements and the dedication to meet them led to the development of the GSM standard. An unprecedented effort has been taken by telecommunication authorities, network operators, and industry sectors to establish and maintain a state-of-the-art cellular standard for the benefit of the entire industry and all its customers.

In this chapter, we confine our focus to GSM. We start with a brief review of its history and then follow with the evolution of the GSM standard and its worldwide adoption. We include a description of some of the services offered by a GSM public land mobile network (PLMN), and introduce some basic terms and the general architecture of a cellular network. Also introduced are the new terms and the components of a GSM system, and which come from the structure and operation of analog networks, which may already be known to the reader. We focus especially on security features inherent in a GSM network.

A brief note concerning the term GSM is in order now. Since GSM stands for *global system for mobile communications*, it is redundant to say "GSM system," for this would mean we said "global system for mobile communications system." However, since terms such as *GSM system* are widely used in the industry, it is common to be redundant when using system abbreviations (e.g., ISDN network), since the redundancy in such terms has some clarifying effect. This book will use both the redundant and nonredundant forms with reckless abandon.

3.1 DEVELOPMENT AND INTRODUCTION OF THE GSM STANDARD

In Table 3.1, the milestones in the course of establishing the GSM system specifications and the spread of GSM over the globe are listed [1].

When looking only at the adoption of the GSM standard in Europe, it becomes clear that the unification of cellular mobile radio will finally become a reality. Figure 3.1 illustrates this.

Words cannot easily express the tireless efforts expended to propel the development of the GSM standard, design a network architecture, test and verify technical parameters, prove functionalities, promote the system itself, and design and manufacture the necessary equipment. What we see today is the result of this work.

With teamwork to an extent never before seen in Europe in the whole industry, in administrations and their national or international institutes, and among operators and manufacturers, a rich and detailed standard for a promising, future-proof mobile communications system has been developed.

The new standard has given new momentum to the economy and has created new markets. A common standard for a market whose customers number in the tens of millions leads to minimized costs for the manufacturers of appropriate equipment. They can produce larger numbers of terminals for a large market, which drives down the cost to end users. Together with deregulation, the provision of cellular service means competition, again to the benefit of end users. Tariffs, products, and services become subject to the higher dynamics imposed by this competition.

New services and features, especially the *roaming* and *security* features, as well as the digital advantages, such as reduced power consumption (state-of-the-art semiconductor devices, TDMA technology) and improved speech quality are the keys that convince network operators and potential subscribers to choose GSM. Other attractive features and services, which have not yet appeared in any cellular network, are waiting in line (see following paragraphs).

The standardization work is not yet finished. It will be continued for years to come. The GSM standard can be regarded as an evolving standard. Today, networks are already operational, but to what extent does the equipment and the provision of cellular services comply with the standard? When it became obvious that the whole standardization process could not be completed before an actual launch of the services—necessary because of economic factors—a phased approach to rolling out the specifications and the networks was adopted. This meant that a *subset* of network features was to be introduced. The reduced features were initially designed to be upwardly compatible *add-ons* of services and functions. The subset was called *GSM Phase 1*. The additional supplements to full implementation of all the planned services and network features were called *GSM Phase 2*. Now, there is even thought given to features beyond Phase 2. These future implementations are known by the term *Phase 2+*. The networks, which are operational now, feature GSM Phase 1, whose specifications were frozen as early as 1991. GSM Phase 2 specifications and

Table 3.1
Development and Spread of the GSM Standard

Year	*Events/Decisions/Achievements*
1982	CEPT decides to establish a *Groupe Spéciale Mobile* (the initial origin of the term *GSM*) to develop a set of common standards for a future pan-European cellular mobile network.
1984	Establishment of three *Working Parties* (WP1-3) to define and describe the services offered in a GSM PLMN, the radio interface, transmission, signaling protocols, interfaces, and network architecture.
1985	Discussion and adoption of a list of recommendations to be generated by the group >100 recommendations in a series of 12 volumes).
1986	A so-called *permanent nucleus* is established to continuously coordinate the work, which is intensely supported by industry delegates.
1987	Initial *Memorandum of Understanding* (MoU) signed by telecommunication network operator organizations (representing 12 countries). Major objectives: • Coordinating the introduction of the standard and time scales • Planning of service introduction • Routing, billing, and tariff coordination
1988	Validations and trials (particularly the radio interface) show that GSM will work.
1988/89 to 1991/1992	With the establishment of the *European Telecommunications Standards Institute* (ETSI), the specification work was moved to this international body. GSM becomes a *technical committee* within ETSI and splits up into GSM groups 1–4, later called *Special Mobile Groups* (SMG) 1–4, which are technical subcommittees. Recommendations and Technical Specifications eventually become *Interim European Telecommunications Standards* (I-ETS), and after approval will become *European Telecommunications Standards* (ETS) some time in the future. GSM finally stands for *Global System for Mobile Communications.*
1990	The GSM specifications for the 900-MHz band are also applied to a *Digital Cellular System* on the 1,800-MHz band (DCS 1800), a PCN application initiated in the United Kingdom.
1991	The GSM Recommendations comprise more than 130 single documents including more than 5,000 pages.
1991	July: Planned commercial launch of GSM service in Europe (MoU plan) delayed to 1992 because of nonavailability of type-approved terminals (GSM then stands for *God Send Mobiles*).
1992	Official commercial launch of GSM service in Europe (*God Has Sent Mobiles*).
1993	The GSM-MoU has 62 members (*signatories*) in 39 countries worldwide. In addition 32 potential members (*observers/applicants*) in 19 other countries [2].
1993	The end of 1993 shows one million subscribers to GSM networks; however, more than 80% of them are to be found in Germany alone.

Table 3.1
Development and Spread of the GSM Standard (*continued*)

1993	First commercial services also start outside Europe: Australia, Hong Kong, New Zealand. GSM networks are operational in Denmark, Finland, France, Greece, Ireland, Italy, Luxembourg, Norway, Portugal, Sweden, Switzerland, United Kingdom [2].
1994 onwards	GSM services can be expected in the following additional countries: Andorra, Austria, Belgium, Brunei, Cameroon, Cyprus, Estonia, Iceland, Iran, Kuwait, Latvia, Malaysia, Netherlands, Pakistan, Qatar, Singapore, South Africa, Spain, Syria, Thailand, Turkey, United Arab Emirates, and many others [2].

Figure 3.1 The GSM standard in Europe.

products should be available in 1994. Phase 2+ addendums are intended to be updated on a regular basis according to market needs and the availability of specifications.

An appropriate comment on this chapter comes from Jonas Twingler, GSM coordinator of ETSI [3]:

Originally, GSM was seen as a Pan-European "only" system, in one single version and a fixed step towards the future. It turned out, for several reasons, that it would be more beneficial for all parties, to launch an interim version of the standard at an early point in time—Phase 1—and to produce the full version of GSM—Phase 2—adapted to the "latest" news, situations and experience gained in Phase 1.

By this, the GSM platform was created, a platform which is full of hooks, mechanisms and not at least potential to continue to build on and to provide mobile communication in all its possible forms and varieties. Even before the Phase 2 standard has been completed, GSM has grown far beyond its original geographical "limitations" and the Global System for Mobile communication really starts to deserve its name. With Phase 2, and in particular with Phase 2+, GSM will also expand far beyond its originally intended functional boundaries and open up for new applications, new access methods, new technologies and thus altogether for new categories of markets, needs and users.

It looks promising.

3.2 SERVICES OFFERED IN A GSM SYSTEM

The features and benefits expected in the new system were (1) superior speech quality (equal to or better than the existing analog cellular technology), (2) low terminal, operational, and service costs, (3) a high level of security (confidentiality and fraud prevention), (4) international roaming (under one subscriber directory number), (5) support of low-power hand-portable terminals, and (6) a variety of new services and network facilities. This section explores the services that are offered in a GSM PLMN.

It was only a logical consequence of the prevailing reality that a measure of interworking compatibility with the services offered by other existing telecommunication networks was sought. In particular, the basis for the services in the GSM standard can be found in the ISDN concept.

We can distinguish three categories of services: (1) *teleservices*, (2) *bearer services, and* (3) *supplementary services.* As already mentioned above, the phased approach to introducing the services led to a subset of these services being included in Phase 1, and another set of services added later. In the following, these services are listed according to their phases (1 and 2), and the features supplied in the individual phases are discussed. Complete lists of the services defined for a GSM PLMN can be found in [4].

3.2.1 GSM Phase 1 Services

The Phase 1 technical specifications, valid for the networks in operation since 1992, provide the definitions for the set of services and features listed in Table 3.2 [3]. It

Table 3.2
GSM Phase 1 Services

Service Category	*Service*	*Comment*
Teleservices	Telephony (speech)	So-called full rate, 13 kbps
	Emergency calls (speech)	
	Short-message services: point-to-point and point-to-multipoint (cell broadcast)	Alphanumeric information: user-to-user and network to all users
	Telefax	Group 3
Bearer services	Asynchronous data	300–9,600 bps, 1,200/75 bps
	Synchronous data	300–9,600 bps
	Asynchronous PAD (packet-switched, packet assembler/disassembler) access	300–9,600 bps
	Alternate speech and data	300–9,600 bps
Supplementary services	Call forwarding	For example, subscriber busy, not reachable or does not answer
	Call barring	For example, all calls, international calls, incoming calls

must be noted that bearer services are restricted to a maximum of 9,600 bps for technical reasons. ISDN networks use rates of up to 64 kbps. Adaptations are necessary.

These services are made available, or *can* be made available, by the operator of a Phase 1 network (their implementation is optional for the operators). Of course, the manufacturers and developers of infrastructure equipment and mobile terminals have to cope with the specifications of these services, since they have to provide the specific functions in their products.

3.2.2 GSM Phase 2 Services

Following the availability of the technical specifications for GSM Phase 2, there will be additional services defined that can be made available to end users. There are a number of supplementary services defined for Phase 2. Table 3.3 gives an overview of these [3].

GSM Phase 2 also has many enhancements made possible through the experience with operational Phase 1 networks, through new ideas, and through the dedication of the involved parties to steadily improve the system and its services.

Table 3.3
Services Added Through GSM Phase 2

Service Category	*Service*	*Comment*
Teleservices	Telephony (speech)	Half rate, 6.5 kbps
	Short-message services	General improvements
Bearer services	Synchronous dedicated packed data access	2,400–9,600 bps
Supplementary services	Calling/connected line identity presentation	Displays calling party's directory number before/after call connection
	Calling/connected line identity restriction	Restricts the display of the calling party's number at called party's side before/after call connection
	Call waiting	Informs the user about a second (incoming) call and allows to answer it
	Call hold	Puts an active call on hold in order to answer or originate another (second) call
	Multiparty communication	Conference calls
	Closed user group	Establishment of groups with limited access
	Advice of charge	Online charge information
	Unstructured supplementary services data	Offers an open communications link for use between network and user for operator-defined services
	Operator-determined barring	Restriction of different services, call types by the operator

3.3 WHAT IS A CELLULAR NETWORK?

Now that we know where GSM stands in the world, how it got to where it is today, and what it is supposed to do for its customers, we take a first look at how it works. We start at the very top, at the network level, to get an initial glimpse of how a GSM system works. First, however, a note on language. From now on, this book does lots of explaining of how radios work in cellular systems. In these explanations, the radios are referred to with language and terms usually reserved for living beings, such as "when the base station 'discovers' that . . .," or "the mobile 'replies' with. . . ." Base stations do not really "discover" anything, and only people "reply" to commands. Radios in cellular systems do many things, and it helps if one separates a radio's signaling tasks from everything else it does (carry voice traffic). To do this, the authors refer to radios as if they were alive and have personalities and habits, both good and bad, when describing signaling tasks. Users do not care about all the signaling tasks cellular radios perform; they only want their voices heard at the other end of the channel and to hear what is being said to them. In order for reliable communications to occur in a digital cellular system, radios have to do lots of channel maintenance tasks in the background, and they do all these tasks—invisibly and silently—with signaling routines as if they were, in fact, alive and carrying on their own little private conversations with each other, united in the task of moving the user's voice from one place to another. Think of radios as efficient and busy servants, and you will have much less trouble making sense out of all the details. You may also enjoy this book more.

3.3.1 A Little Bit of History

It was some time since the days of Heinrich Rudolf Hertz until the first real achievements of cellular radio. In the years 1887 and 1888, Hertz discovered that invisible waves which originated from an electric spark were able to transport *influence* or, as we call it today, *information* through the air. Only a few years later, this phenomenon was further investigated and developed until it was possible to actually transmit and receive signals over a distance of several kilometers. Guglielmo Marconi performed a dramatic demonstration of this several years later.

These early experiments formed the basis not only of cellular radio, but also of many types of transmissions. One merely has to think of early radio broadcasting, which was introduced in the early 1920s in the United States and Europe, to see how far these first experiments have taken society. Later applications for radio found quick and numerous paths to mass markets, even though the quality of the early AM transmissions were not very good by today's standards. The introduction of FM by Edwin H. Armstrong in 1929 was a breakthrough for quality of reception, and it became the standard for the remainder of the century. The current analog cellular networks are still based on Armstrong's FM.

Mobile radio applications took a longer and more halting path to their markets. In the days when the first transmitters started broadcasting, people were trying to make use of this technique for mobile applications, but they had a problem in that the transmitters were still very large. In the first applications for mobile radio, only the receiving system was mobile, similar to the paging systems which are so popular today. There were experiments by police departments, which used only one high-power transmitter to cover a whole city. The called police officer had to get out of his car at the next public telephone to report back to his office for further instructions. This awkward procedure and the limited ability of the receiver to withstand the problems of propagation and road hazards were limiting factors for mobile radio [5].

When FM was introduced, the quality of received information increased a great deal, but the applications were still limited by transmitter size and the huge amounts of power consumed at the mobile end of the communications links by those early transmitters.

Commercial mobile phone service had to wait for the perfection of the public dispatch systems, such as police and other public safety applications. The breakthroughs were (1) small, low-power transmitters (run from motor generators in the vehicle), and (2) the move to higher operating frequencies (above 30 MHz) to further decrease the size and weight of mobile transmitters. An initial step toward viable mobile phone service appeared with the radio common carrier (RCC) and mobile telephone service (MTS) systems. These were simply conventional land mobile radios fitted with a special control panel, called a *control head*, which were suitable for commercial use by people who were unfamiliar with operating two-way radios. The RCC and MTS systems could direct calls from a single transmitter to a particular mobile, but remained, after all, simple dispatch systems in which the users set up all their calls through a mobile operator. Later, some additional inband tone signaling was added to the MTS system to make the newer improved mobile telephone service (IMTS), which automated to a considerable extent the interface between the mobile customer and the fixed telephone network. The mobile operator almost disappeared from the mobile phone landscape when IMTS was introduced. Cellular radio became popular only when carefully designed, engineered, and thoroughly tested systems like AMPS and TACS started to work.

3.3.2 Cellular Structure

In the beginning of radio, engineers were happy to achieve a simple dedicated link between a transmitter and a receiver. As we saw in the previous section, these first links were not even two-way ones, but remained one-way dispatch links; that is, the people who called the mobiles did not get a response right away and did not even get a confirmation that their calls had reached the mobile addressees. The next step was to establish a two-way transmission link that allowed an immediate response. This came with mobile transmitters, but the structure of the network was simplistic and awkward to use. Service was limited to a certain area that could be reached with one transmitter or a small collection of transmitters on different channels at a single base site (Figure 3.2). We call the coverage area a *cell*. The cell or network size was determined by the transmitter's power. It was not possible to have a link between two different cells, or coverage areas, since an orderly means of directing traffic (voice audio) between transmitter sites and moving mobiles was missing. It was important to select the frequency of the transmitter and receiver in the cell carefully so that there was no interference from other systems, perhaps in the next town, which would interfere with the system's local operation.

The disadvantage of this is obvious to everyone from today's perspective. A small set of frequencies was used for a huge area. The transmitters were so powerful

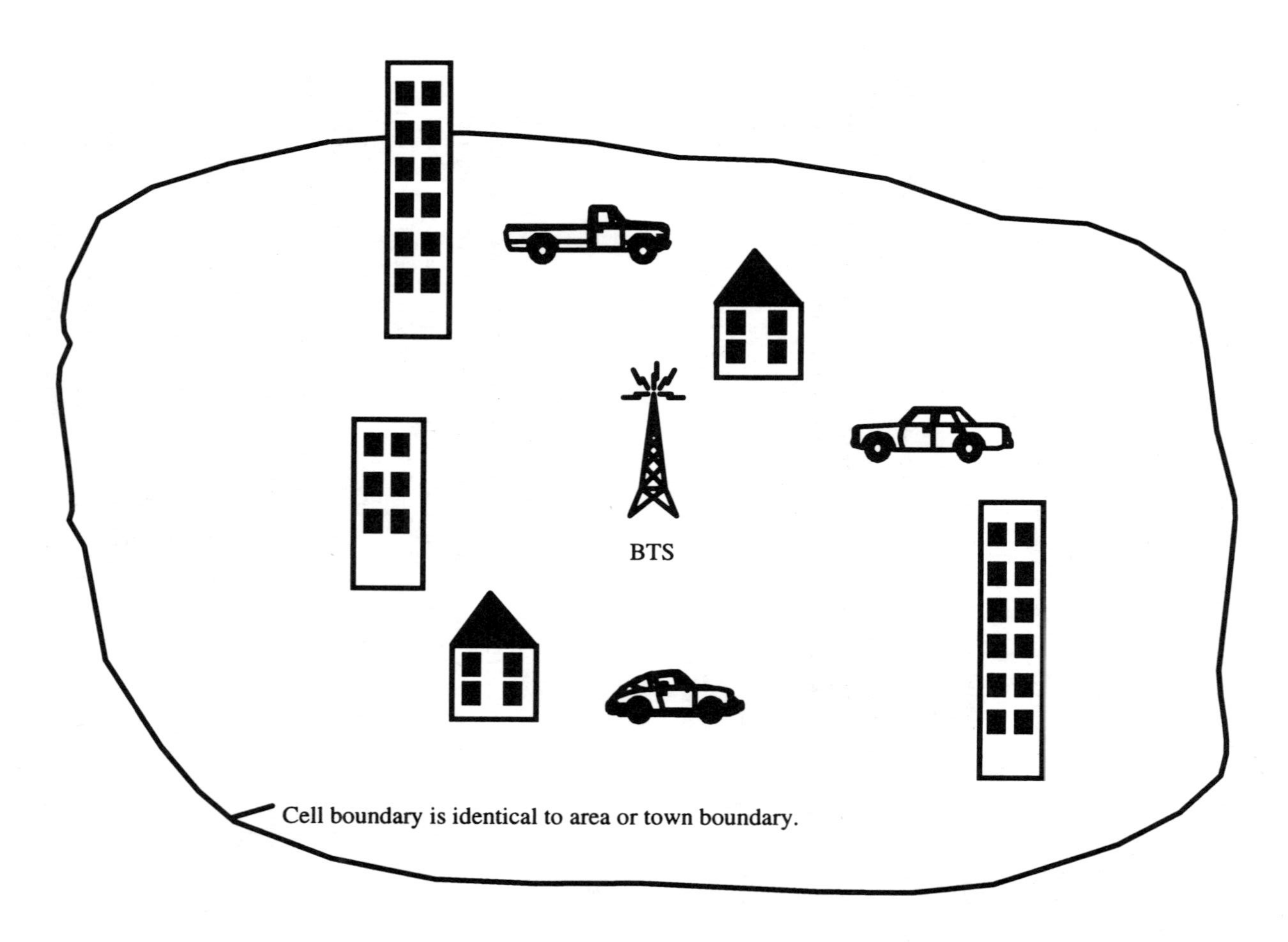

Figure 3.2 Single-cell structure.

that their operating frequencies could not be reused for hundreds of kilometers. This was a major limitation to the capacity of the system; once a channel was in use, the channel was tied up over the whole coverage area, even though the need for a mobile communications channel was confined to a small part of the network's service area. One could argue that capacity was not an issue in those days, for the mobile radios were expensive enough to limit the need for capacity below that of the technical limits of the system. Eventually the price of the mobile equipment dropped so low that the artificial capacity threshold was broken, and long waiting lists were common in the 1960s and 1970s for even rudimentary mobile phone service. A search for a solution continued in many countries. The possibility of allocating more frequency space was not a viable one. Other institutions and agencies needed spectrum too.

An idea was proposed to split the frequency band allocated to one cell among many cells, and have several cells coexist next to each other (Figure 3.3). The cellular structure was born. In order for this scheme to work properly, some restrictions had to be applied:

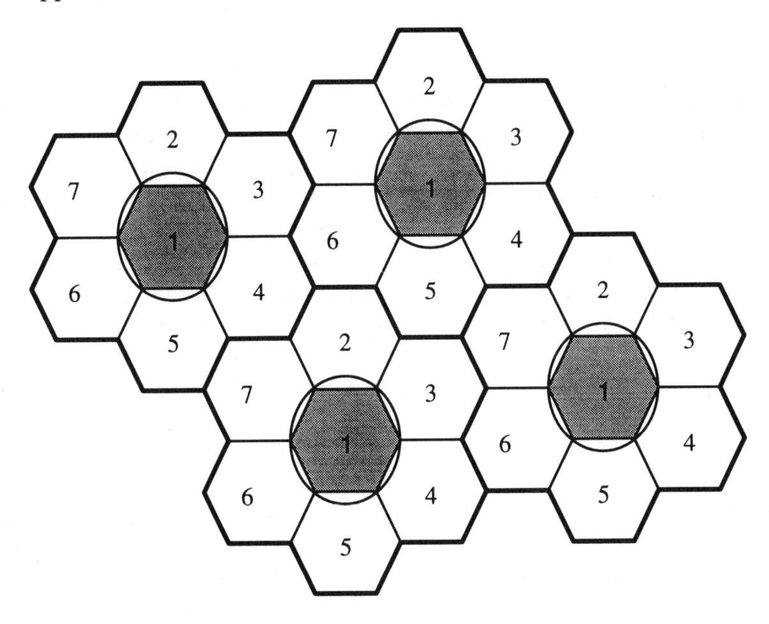

Transmission range of cell No.1

Border of a cell

Cluster of cells using different frequencies

Figure 3.3 Cellular structure.

- The frequencies had to be reused only within a certain pattern in order to reduce the interference between two different stations using the same channel. Neighboring cells could not share the same channels.
- The power levels used within the single cells had to be carefully limited, again to reduce the interference between the different stations.
- Receiver filters had to be improved.

The pattern used for early systems was the *seven-cell reuse pattern*, which was a result of the distance required between cells using the same frequencies, yet again to preclude excessive interference. Interference had to be limited to some level that could be handled by the input filters of all the receivers in all the cells. Typically, a distance of about 2.5 to 3 times the diameter of an average cell had to be reserved between base site transmitters to guarantee that interference would not render the system useless. Calculations and experiments dictated this reuse pattern. If the systems were carefully designed and installed, more users could be accommodated as the same frequencies were used more and more times in the system.

In the early systems, it was not possible to *roam* among the cells. This meant that a user was not able to travel freely between cells while engaged in a single phone call. It was also not possible to place a call from the fixed network to a particular mobile station (MS) without knowing the exact position of the mobile. Each area had its own code, and a mobile station within this area had to be called with this code in a manner similar to the use of area codes in the fixed public network. The only difference is that in the fixed public network, the phones are fixed and the area codes do not change. The introduction of much more intelligence into the network, together with additional audio routing equipment, made roaming possible. Registers were installed in the network, which traced all the mobiles and stored their positions in order to route calls to them. These registers could be queried so that the audio in a call could be passed from cell to cell as needed. A single incidence of this process is called a *handoff* or *handover*, and a host of details within both the network and the mobile station itself need to be carefully coordinated for this process to happen reliably. Mobile stations, for example, have to be equipped with synthesized transmitters which can change operating frequencies quickly. The network has to have sufficient equipment and signaling to make sure the handover or handoff is directed to the correct cell site. We will investigate how all this is done in the coming sections and chapters. For now, we can leave the mobile caller content to stay on the phone for a nearly unlimited time and still travel anywhere within the network's cell site coverage areas.

3.3.3 Network Planning

If one thinks of a country as varied in population density as the United States, it is easy to understand that it does not make sense to apply the same size to each cell.

It makes a difference if an operator has to supply a big and densely populated city, such as New York, with a network, or a remote and sparsely populated area, such as the island of Hawaii. Different possibilities of network planning and cell planning have been developed:

Cell-splitting or microcell applications. As the number of subscribers grew larger, the density within these networks also became higher. The operators and radio engineers had to look for new capacity funds. A rather basic idea was to split the existing space into smaller portions, thus multiplying the number of channels available (Figure 3.4). Along with this simple scheme, the power levels used in these cells decreased, making it possible to reduce the size of batteries required for mobile stations. With the decreased power required for mobiles came decreased size and weight. This made the networks more attractive to new users.

Selective cells. It does not always make sense to have circular cells. Radio engineers designed cells with a wide variety of shapes, together with the required antennas, which are able to confine transmitted power within a particular area and exclude power from adjacent areas. The most common of these selective coverage schemes is the sectored cell, where coverage is confined to individual 120-deg sectors rather than the typical full 360-deg coverage (Figure 3.5). Such antennas may be located

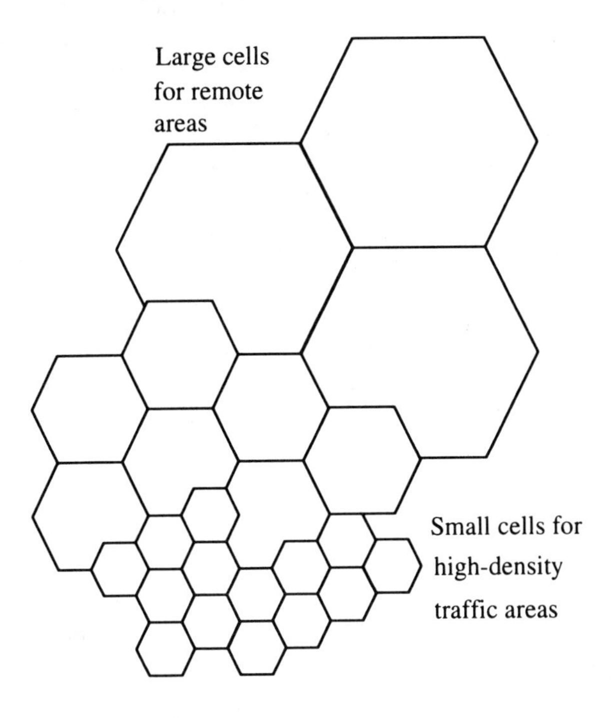

Figure 3.4 Cell splitting and microcells.

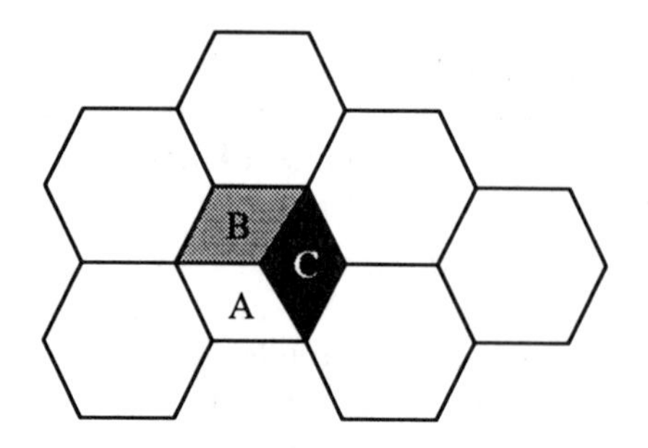

Figure 3.5 Selective cells.

at the entrances of tunnels, on the edge of a valley, or at the ends of streets among skyscrapers.

Umbrella cells. When the cell-splitting technique was first applied, the operators realized that a freeway crossing within very small cells caused a large number of handovers among the different small cells. Since each handover requires additional work by the network, it is not particularly desirable to increase the number of such events. This is particularly true on European freeways, where the average speed is very high. The time a mobile on such a European freeway would stay in one cell decreases with increasing speed. Umbrella cells were introduced (Figure 3.6) to address this problem. In an umbrella cell, power is transmitted at a higher power level than it is within the underlying microcells and at a different frequency. This means that when a mobile that is traveling at a high speed is detected as a fast mover, it can be handed off to the umbrella cell rather than tie up the network with a fast series of handoffs. Such a mobile can be detected from its propagation characteristics or distinguished by its excessive handoff demands. In this cell, the mobile can stay for a longer period of time, thus reducing the workload for the network.

3.4 SYSTEM ARCHITECTURE

It is difficult for typical wire-line phone users to understand and appreciate the overhead necessary to process a call to another city or country. It is even more difficult for cellular subscribers to understand that there is a little bit more in a cellular network outside their phones. To supply cellular service to subscribers, a network operator has to install a complete and separate network, which, at a certain

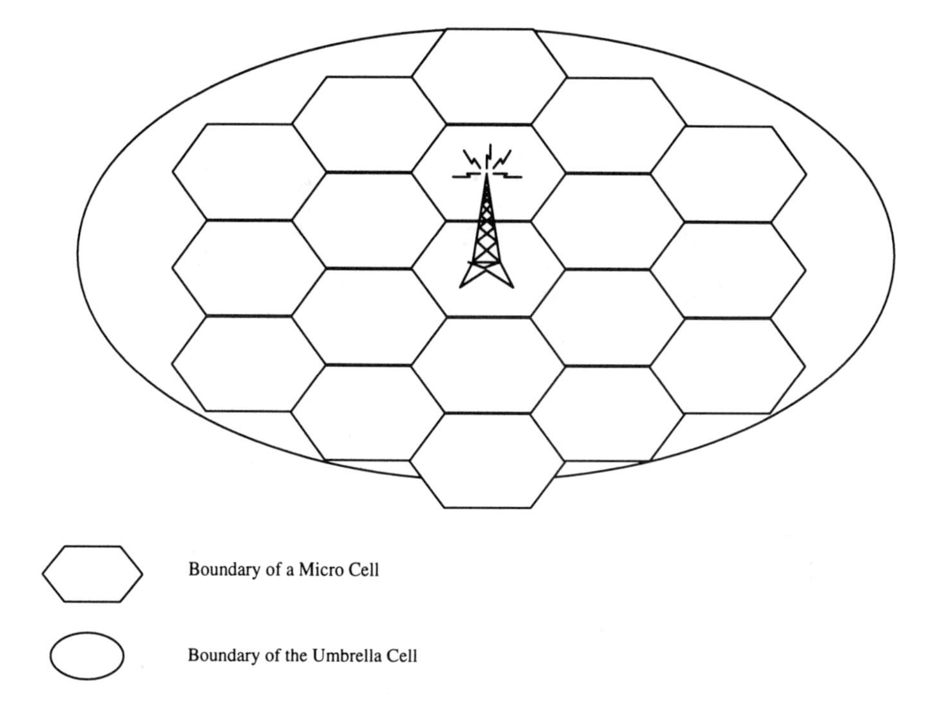

Figure 3.6 Umbrella cells.

point, has to interface to the *public switched telephone network* (PSTN). In addition to the standard national roaming feature, which applies to the current analog systems, the new GSM system was also designed to allow international roaming. This means that users can enjoy the option of taking their phone abroad and using it in foreign GSM systems. Furthermore, users can still be reached under their own subscriber number in their home country, independent of their location, as if they had never left town.

A description of the different entities in the GSM system follows. Most of these entities are also used in analog networks. The recommendations and the specifications for GSM networks do not merely specify the air interface and the message flow between mobile stations and the cellular network on that air interface. They also describe the whole infrastructure and all the other parts of the system that are mentioned and described here (Figure 3.7).

3.4.1 Mobile Station Terminal Equipment

The best known part of the cellular network is certainly the mobile station. Different types of stations are distinguished by power and application. *Fixed* mobile stations

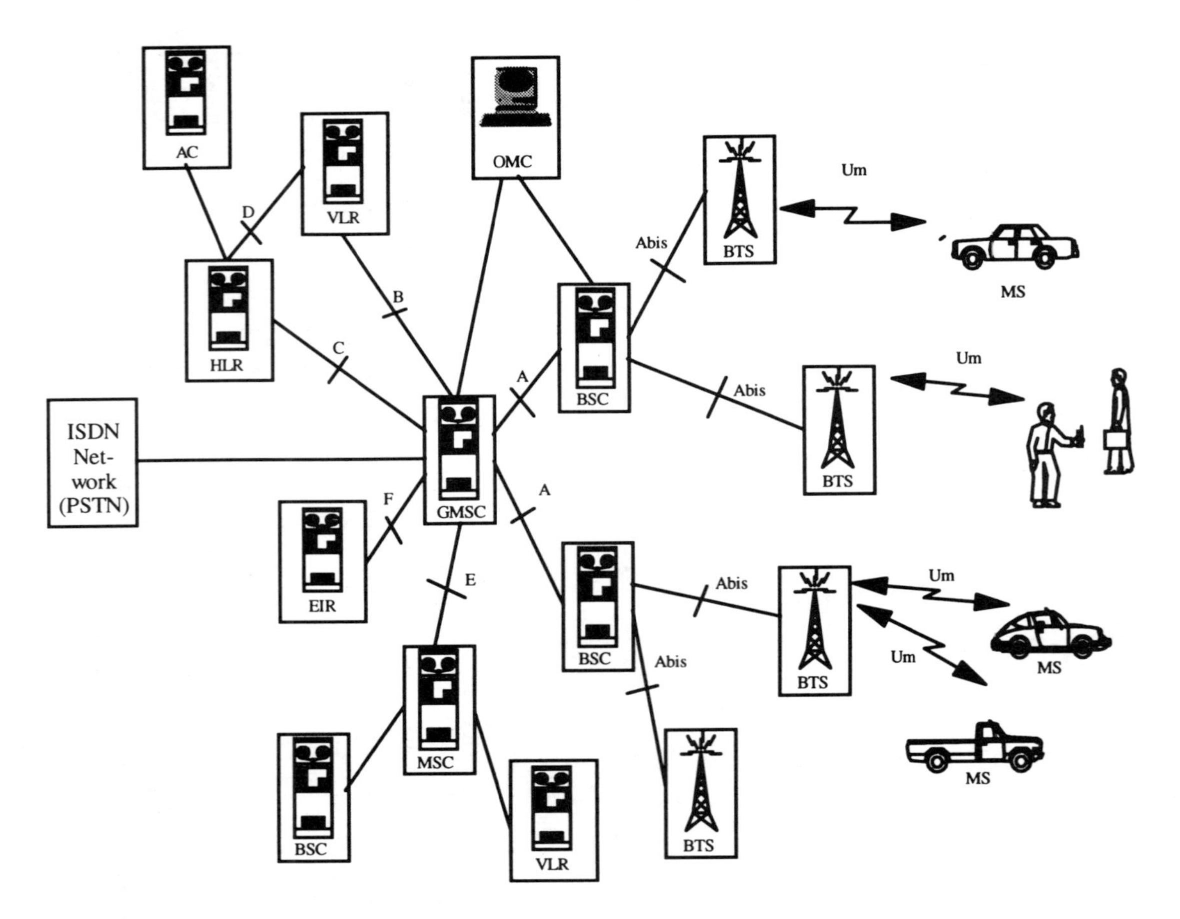

Figure 3.7 GSM system architecture.

are permanently installed in a car and may have a maximum allowed RF output power of up to 20W. Portable units (bag phones) can emit up to 8W and hand-portable units up to 2W. With second-generation mobiles (on the market since 1993), the GSM system is becoming more and more attractive. Hand-portable units are becoming much smaller and are now not much larger than analog units. This is giving the system a boost in popularity, especially in those markets with a particular demand for small mobiles, such as in the Asian and Pacific areas.

3.4.2 Subscriber Identity Module

The *subscriber identity module* (SIM) provides mobile equipment with an identity. Without a SIM, a mobile is not operational (except for emergency calls). The SIM is a smart card and has a computer and memory chip permanently installed in a plastic card the size of a credit card. The SIM has to be inserted into a *reader* in a mobile station before the mobile terminal can be used for its intended routine purposes. For very small hand-portable phones, the credit-card type is too large. There is, therefore, a small version of the SIM, called the *plug-in SIM*.

Certain subscriber parameters are stored on the SIM card, together with personal data used by the subscriber, such as personal phone numbers. The SIM card identifies the subscriber to the network. Since only the SIM can personalize a phone, it is possible to travel abroad, taking only the SIM card, rent a mobile phone at the destination, and then use the phone (with the SIM card inserted) just as if it were a personal mobile phone at home. Anyone may reach a subscriber using the subscriber's home number. Every phone call, from wherever it is placed, is billed to the subscriber's home account.

Short messages received from the network may also be stored on the card. The recent introduction of larger memories and better microprocessors will make the SIM card even more flexible and powerful in the future, combining it with different services, such as credit and service cards.

To protect the SIM card from improper use, a security feature is built in. Before they can use the mobile, users have to enter a four-digit *personal identification number* (PIN). The PIN is stored on the card. If the wrong PIN is entered three times in a row, the card blocks itself, and may only be unblocked with an eight-digit personal unblocking key (PUK), which is also stored on the card.

We have explored only the salient aspects of the SIM card, which are related directly to the GSM system. With the increasing number of services and applications of smart cards, additional auxiliary functions and features may also be introduced on the SIM cards intended for use in the GSM systems. One example might be different access priorities for good customers, or restricted usage in certain areas.

3.4.3 Base Station or Base Transceiver Station

The counterpart to a mobile station within a cellular network is the *base transceiver station* (BTS), which is the mobile's interface to the network. A BTS is usually located in the center of a cell. The transmitting power of the BTS determines the absolute cell size. A base station has between one and sixteen transceivers, each of which represents a separate RF channel. Some of the intelligence, which was incorporated into analog base stations and the host network, such as measurements on the radio channels as criterion for handover, is now shifted to the mobile stations (see Section 3.7). Dumping some of the work on the mobile's "desk" makes the GSM infrastructure cheaper than that of some analog systems. The result is that in some less wealthy countries, digital cellular systems (such as AMPS, NMT, or TACS) are installed instead of analog ones.

3.4.4 Base Station Controller

The *base station controller* (BSC) monitors and controls several base stations, the number of which depends on the manufacturer and can be between several tens and several hundreds of stations. The chief tasks of the BSC are frequency administration, the control of a BTS, and exchange functions. The hardware of the BSC may be located at the same site as the BTS, at its own standalone site, or at the site of the mobile switching center (MSC). BSC and BTS together form a functional entity sometimes referred to as the *base station subsystem* (BSS).

3.4.5 Gateway Mobile Services Switching Center

The *gateway mobile services switching center* (GMSC) is the interface of the cellular network to the PSTN. It is a complete exchange, and with all its registers it is capable of routing calls from the fixed network—via the BSC and the BTS—to an individual mobile station. The GMSC also provides the network with specific data about individual mobile stations. Depending on the network size, an operator might use several interfaces to the fixed network, thus using several GMSCs or only one. If the traffic within the cellular network requires more exchange capacity than the GMSCs can provide, additional *mobile services switching centers* (MSC) might coexist with no access to the fixed network. If not otherwise explicitly distinguished from each other, the capabilities of the GMSC and the MSC are the same. A major difference between the two is that the MSC has no related home location register (HLR).

3.4.6 Operation and Maintenance Center

The *operation and maintenance center* (OMC) has access to both the (G)MSC and the BSC, handles error messages coming from the network, and controls the traffic load of the BSC and the BTS. The OMC configures the BTS via the BSC and allows the operator to check the attached components of the system. As the cells become smaller and the number of base stations increases, it will not be possible in the future to check the individual stations on a regular basis for transceiver quality. Therefore, it is important to put remote control of the maintenance in place to save costs, but still maintain the quality of the system. This is supported by better self-test functions in the BTS. The distribution of maintenance tasks is treated differently by different manufacturers.

3.4.7 Home Location Register

The HLR stores the identity and user data of all the subscribers belonging to the area of the related GMSC. These are permanent data such as the *international mobile subscriber number* (IMSI) of an individual user, the user's phone number from the public network (which is *not* the same as the IMSI), the *authentication key* (see Section 3.8.1), the subscriber's permitted supplementary services, and some temporary data. Temporary data on the SIM include such entries as (1) the address of the current *visitor location register* (VLR), which currently administers the mobile station (see Section 3.4.8), (2) the number to which the calls must be forwarded (if the subscriber selects call forwarding), and (3) some transient parameters for authentication and ciphering.

The IMSI is permanently stored on the SIM card. The IMSI is one of the pieces of important information used to identify a subscriber within the GSM system. The first three digits of the IMSI identify the *mobile country code* (MCC) and the next two digits are the *mobile network code* (MNC, see Table 3.4). Up to ten additional digits of the *mobile subscriber identification number* (MSIC) complete the IMSI. The following IMSI:

262 02 454 275 1010

identifies a subscriber from Germany (MCC = 262), who is paying his or her monthly bill to the private operator D2 privat (MNC = 02). The subscriber's network identity number (MSIC) is 454 275 1010. The number with which the subscriber may be reached from the public network is totally different from the IMSI, and starts with an area code of 0172, followed by a seven-digit subscriber number. The first digits of this subscriber number identify the subscriber's related HLR. The number of digits used for this purpose is dependent on both the network size and the number of HLRs in the network. The IMSI is only used for internal network purposes.

3.4.8 Visitor Location Register

The VLR contains the relevant data of all mobiles currently located in a serving (G)MSC. The permanent data are the same as data found in the HLR; the temporary

Table 3.4
List of Different Country Codes Among the GSM Systems

MCC	*Country*	*MNC*	*Network*
505	Australia	01	Telecom Australia
505	Australia	02	Optus Communication
505	Australia	03	Vodafone
232	Austria	01	A E-NETZ
206	Belgium	01	BEL MOB-3
238	Denmark	01	DK TDK-MOBIL
244	Finland	91	SF Tele Fin
208	France	01	F France Telekom
208	France	02	F SFR
262	Germany	01	D1-Telekom
262	Germany	02	D2 privat
222	Italy	01	I SIP
204	Netherlands	08	NL PTT
530	New Zealand	01	Bell South
228	Switzerland	01	CH Natel D
240	Sweden	01	S TELE RADIO
240	Sweden	07	S COMVIQ
240	Sweden	08	S NORDICTEL
234	United Kingdom (UK)	10	UK CELLNET
234	United Kingdom (UK)	15	UK VODAFONE

data differ slightly. For example, the VLR contains the *temporary mobile subscriber identity* (TMSI), which is used for limited periods of time to prevent the transmission of the IMSI via the air interface. The substitution of the TMSI for the IMSI serves to protect the subscriber from high-technology intruders and helps point to the location of the mobile station through the cell identity.

The VLR has to support the (G)MSC during a *call establishment* and an *authentication procedure* as it furnishes data specific to the subscriber.[1] Locating subscriber data in the VLR, as well as in the HLR, reduces the data traffic to the HLR, because it is not necessary to ask for these data every time they are needed. Another reason for storing nearly identical data at two different locations (in the HLR and the VLR) is that each serves a different purpose. The HLR has to provide the GMSC with the necessary subscriber data when a call is coming from the public network. The VLR, on the other hand, serves the opposite function, providing the host (G)MSC with the necessary subscriber data when a call is coming from a mobile station (e.g., during authentication).

[1]The VLR receives the data for a particular subscriber from that subscriber's HLR.

If a mobile station is located in its own (G)MSC, it still uses the two different registers, even though the VLR, in this particular case, seems redundant. The consistency simplifies the underlying procedures.

3.4.9 Authentication Center

The *authentication center* (AC) is related to the HLR. It provides the HLR with different sets of parameters to complete the authentication of a mobile station. The AC knows exactly which algorithm it has to use for a specific subscriber in order to calculate input values and issue the required results (see Section 3.8.1). Since all the algorithms for the authentication procedures are stored within the AC, they are protected against abuse. The SIM card issued in an area assigned to an AC contains the same algorithms for authentication as the AC does. If the AC provides input and output parameters for these algorithms to either the HLR or the VLR, either location register can verify (authenticate) the mobile station.

3.4.10 Equipment Identity Register

The *equipment identity register* (EIR) is an option that is up to the network operator to make use of. The implementation of the EIR is a relatively new security feature of the GSM system. Within the EIR we find all the serial numbers of mobile equipment that is either stolen or, due to some defect in their hardware, may not be used in a network. The *international mobile equipment identity* (IMEI) is not only the serial number of a certain mobile station, but it also reveals the manufacturer, the country of production, and type approval. The idea is to check the identity at each registration or call setup of any mobile station, and then, depending on its IMEI, admit or bar access of the mobile station to the system.

An example of a forbidden mobile might be that the RF quality of a certain mobile station from a certain manufacturer is not as good as the recommendations specify; for example, it produces spurious emissions or disturbs other radio services in the area. The operator may check for this specific mobile station (since the model number and manufacturer are part of the IMEI) and reject it from its network.

Figure 3.7 not only shows the different parts of the network we have listed here, but also the names of the different interfaces between them. This book concentrates on the characteristics of the *Um interface*. The Um interface is sometimes called *air interface*. We also include a description of the Abis interface between the BTS and the BSC. An understanding of the Abis interface is required for testing a base station. The other interfaces are just named for further information, since they have nothing to do with the cellular quality or the radio character of GSM. How the different radio and network entities work together is shown in the next two sections about registration and call establishment.

3.5 REGISTRATION

After a mobile station is switched on, it scans the whole GSM frequency band with a certain scanning algorithm in order to detect the presence of a network in the least amount of time. When the network is detected, the mobile station reads the system information on the *forward* or, as it is called in GSM, the *base* channel. With this information, the mobile station is able to determine its current position within the network. If the current location is not the same as it was when the mobile was last switched off, a *registration procedure* takes place. Figure 3.8 describes the required actions in a registration procedure and their relationships to the different entities within the network.

First, the mobile station requests a channel from the network, which will be assigned by the base station. Before the channel is actually assigned to the Um interface, the BSC has to activate a channel on the BTS, which has to acknowledge the activation to the BSC in return. Now that the mobile station is connected to the infrastructure, the mobile station tells the system that it wants to perform a *location update*. This wish is passed on, via the BSC, to the (G)MSC, which, being very stubborn and bureaucratic, requires an authentication of the mobile station before

MS	BTS	BSC	(G)MSC	VLR	HLR	Action
→	→					Channel Request
	←					Channel activation command
	→					Channel activation acknowledge
←	←					Channel Assignment
→	→	→				Location Update Request
←	←	←				Authentication Request
→	→	→				Authentication Response
			↔			Comparison of the Authentication parameters
←	←	←				Assignment of the new area and the temporary identity (TMSI)
→	→	→				Acknowledgment of the new area and the temporary identity (TMSI)
			↔ ↔	 ↔		Entry of the new area and identity into the VLR and HLR
←	←					Channel Release

Figure 3.8 Registration in the network.

taking any further action. Upon the receipt of the correct parameters, the (G)MSC accepts the mobile, via the BSC and the BTS, into its new location, and—if this option is used in the network—assigns a *temporary identity* (TMSI), which the mobile station also has to acknowledge. When this procedure is finished, the channel is released from the BSC via the BTS.

The registration could as well be accomplished by the network. This is the case if the system always wants to know exactly which mobiles are currently available on the system. (Mobile stations would notify the network when they are switched on or off.) The registration procedure is a means to limit the message flow within the network and still give the network virtual control. The network always knows the contents of the HLR. Whatever the HLR knows the GMSC also knows, and whether or not a mobile is switched on or off becomes common knowledge within the network. If someone wants to call a switched-off mobile station, the GMSC can immediately signal a message to the caller indicating that the specific mobile is not available, rather than try to route the call to the area where the mobile station was last heard from.

3.6 CALL ESTABLISHMENT

Before a call can be established, the mobile station must be switched on and registered into the system. There are two different procedures. One is for the *mobile-originated call* (MOC), and the other is for the *mobile-terminated call* (MTC). Of these two, only the MOC is described here, in order to give an impression of the signaling overhead (message exchanges) used in the GSM system. Be careful! Fourteen different messages are exchanged between the mobile station and the network before an actual call (i.e., exchange of user data) starts.

In a manner similar to the location update procedure, the mobile starts with a channel request, which is answered by the system with a channel assignment. The mobile station informs the system why it wants a channel (e.g., it wants to establish a call). Before the procedure is allowed to continue, the mobile again has to authenticate itself. To protect any further signaling messages from eavesdroppers, the network may now, with the next message, tell the mobile station to start ciphering its data. *Ciphering* means that the messages are transmitted in a scrambled way that only the mobile station and the BTS understand. In the Setup message, the mobile transmits the number it wants to call. While the call is proceeding, the BSC (via the BTS) assigns a traffic channel on which the exchange of user data is performed. Different types of messages and user data move on different types of channels. The different channel types will be explained in detail in Chapter 5. For the moment, it should be enough to know that there are channels that only support the message exchanges, and other channels on which the user data are handled. If the called party is not busy, the mobile alerts and the connection is established when the called phone is brought off the hook (Figure 3.9).

MS		BTS	Action
	————→		Channel Request
	←————		Channel Assignment
	————→		Call Establishment Request
	←————		Authentication Request
	————→		Authentication Response
	←————		Ciphering command
	————→		Ciphering complete, from now on ciphering is in place
	————→		Setup message, indicating the desired number
	←————		Call Proceeding, the network routes the call to the desired number
	←————		Assignment of a traffic channel for the designated use "exchange of user data"
	————→		Assignment complete, from now on all messages are exchanged on the traffic channel
	←————		Alerting, the called number is not busy and the phone is ringing
	←————		Connect, the called party accepted the call
	————→		Connect acknowledge, now the call is active and both parties can talk to each other
	←———→		Exchange of speech data

Figure 3.9 Mobile-originated call establishment.

The opposite MTC procedure is almost identical to the MOC procedure and is too redundant to be included in our discussion. We will, however, see many other examples of all kinds of message exchanges throughout this book.

3.7 HANDOVER/HANDOFF

The *handover* or *handoff* procedure is a means to continue a call even when a mobile station crosses the border of one cell into another. As mentioned earlier, the handover or handoff technique, from one cell to another, finally made the mobile station really mobile. Before the introduction of this feature, a call was simply dropped when the cell border was crossed or when the distance between the mobile station and one particular base station became too large.

In a cellular network, one cell has a set of neighboring cells. The system, therefore, has to determine which cell the mobile station should be passed to. The method used to determine the next cell to use differs in analog and digital systems. The difference in the procedure can be determined from the different names. The handoff comes from the analog world, whereas handover was introduced by GSM. The term *handover* will be used when talking about the GSM system, and the term *handoff* will be used when talking about analog systems.

In analog systems, the base station monitors the quality of the link between a mobile station and itself. When the base station realizes that the quality of the link has degraded and the distance to the mobile station has become too large, it requests the adjacent cells to report the power level they see for the mobile back to the network. It is reasonable that the strongest reported power level for the mobile comes from the closest cell to the mobile station. The network then decides which frequency channel the base station should use in the new cell and which corresponding frequency the mobile should tune to. Eventually, the mobile station is commanded to perform a channel change.

The mobile station is the passive participant in the handoff process. All the measurements and subsequent work are done in the base stations and the network. Cell sites are equipped with a *measuring receiver* used to measure the power level of the different mobile stations on the various frequency channels in use. For those readers interested in analog cellular systems, the distance measurements within cells is sometimes determined from the relative phase of the *supervisory audio tones* (SAT) that the mobiles transpond back to the base stations. The distance is half the time of the phase shift multiplied by the propagation speed of the signal.

The situation in the GSM system is different. The mobile station must continuously monitor the neighboring cell's perceived power levels. To do this, the base station gives the mobile a list of base stations (channels) on which to perform power measurements. The list is transmitted on the base channel (again, system information), which is the first channel a mobile tunes to when it is turned on. The mobile station performs continuous measurements on the quality and the power level of the serving cell, and of the power levels of the adjacent cells. The measurement results are put into a *measurement report*, which are periodically sent back to the base station. The base station itself may also be performing measurements on the quality and power of the link to the mobile station. If these measurements indicate the necessity for a handover, such can be performed without delay, as the appropriate base station for a handover is already known. The measurements are coming in constantly, and they reflect the mobile's point of view. It is up to the operator to act upon different quality or power levels, and the handover constraints or thresholds can be adjusted in accordance with changing environment and operating conditions.

The GSM system distinguishes different types of handovers. Depending on what type of cell border the mobile station is crossing, a different entity may have to control the handover to ensure that a channel is available in the new cell. If a handover has to be performed within the area of a BSC, it can be handled by the

BSC without consulting the MSC, which, in any case, must at least be notified. This type of handover is called a simple handover between BTSs (Figure 3.10).

If, instead, a mobile station is crossing the border of a BSC (rather than a BTS), then the MSC has to control the procedure in order to ensure the smooth transition of the conversation. This can be continued for a handover between two MSCs (Figure 3.11). The only difference, in this latter case, is that even though the mobile is eventually handled by the second MSC, the first MSC still has to maintain control of the call management.

In theory, it is possible to perform a handover at a political border between two countries. There are no technical restrictions to this feature. Due to the different roaming agreements, however, it is not possible to start a phone call, let us say, in Germany, and cross the border to Switzerland and still continue the call. The call will be dropped, and subscribers have to register themselves in the new foreign network.

3.8 SECURITY PARAMETERS

In the previous sections and chapters some *security parameters* have already been mentioned. Now it is time to summarize these features and describe their operations.

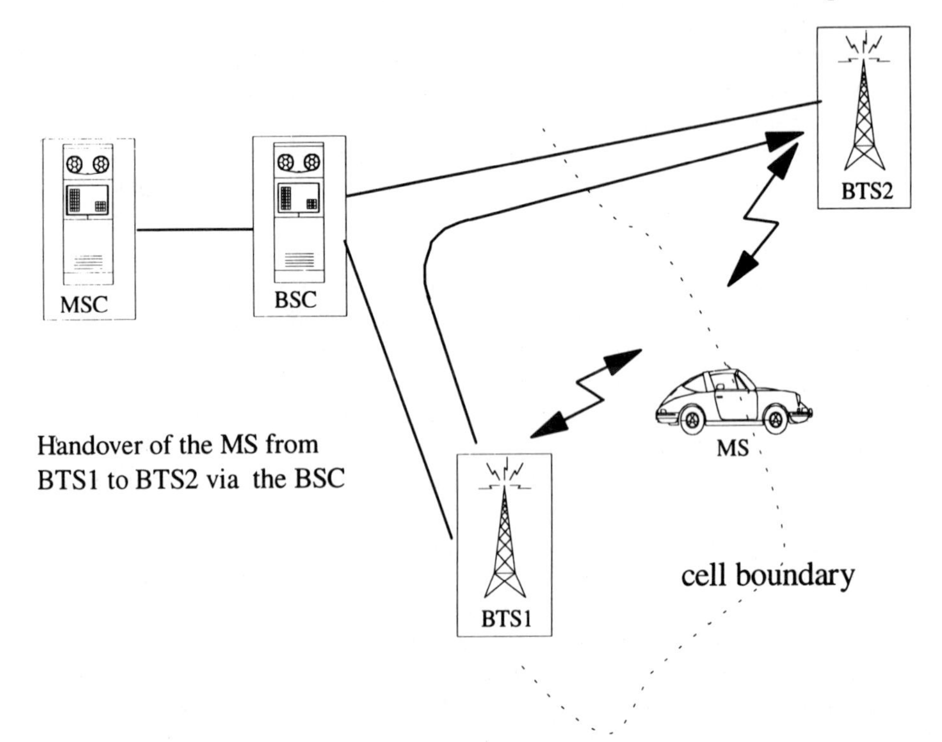

Figure 3.10 Handover between BTSs.

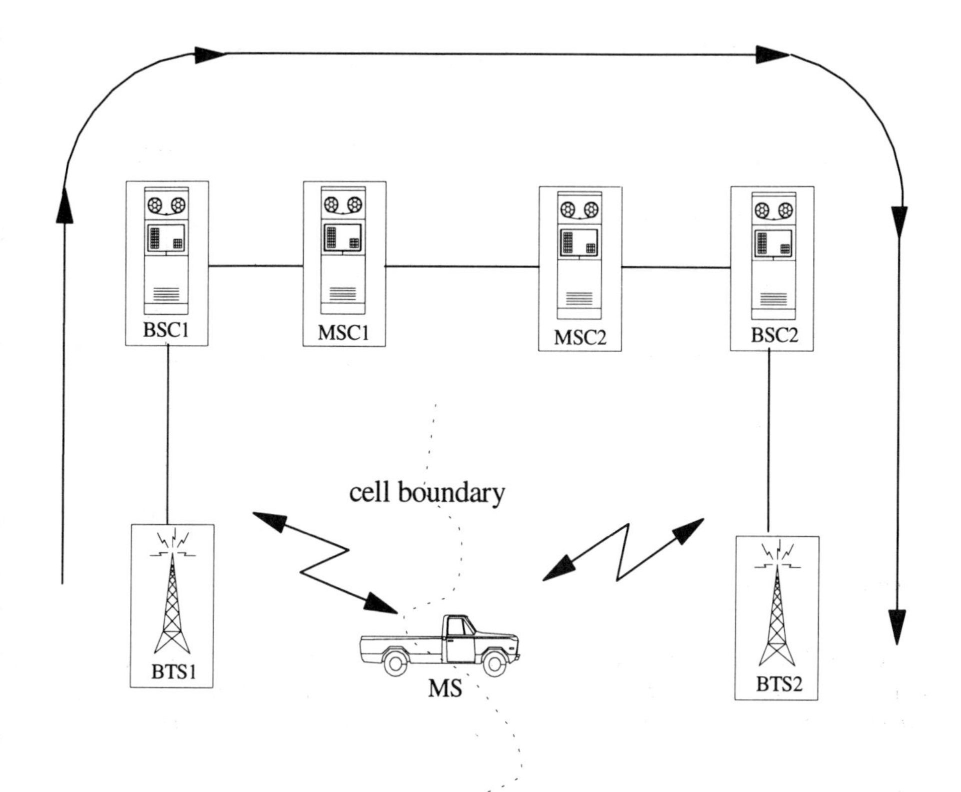

Figure 3.11 Handover between MSCs.

3.8.1 Authentication

The *authentication* procedure (Figure 3.12) checks the validity of subscribers' SIM cards, and whether they are permitted in a particular network. The authentication is based on the *authentication algorithm*, A3, which is stored on the SIM card and in the AC.

The A3 algorithm uses two input parameters: one is the authentication key, Ki, which is stored only on the SIM card and in the network. The second value, the randomly generated number (RAND), is transmitted to the mobile station on the Um interface. The mobile station passes the RAND to the SIM card where it is used as an input value for the A3 algorithm. The result, SRES, is returned—via the Um interface from the mobile station—to the network where the value of SRES is compared with the calculated value from the AC. A set of authentication parameters (RAND and SRES) is stored in the HLR and VLR for use by the AC. Usually, a number of sets of these parameters are stored there because a different set is used for each call setup or registration and are discarded after each use. If the HLR or

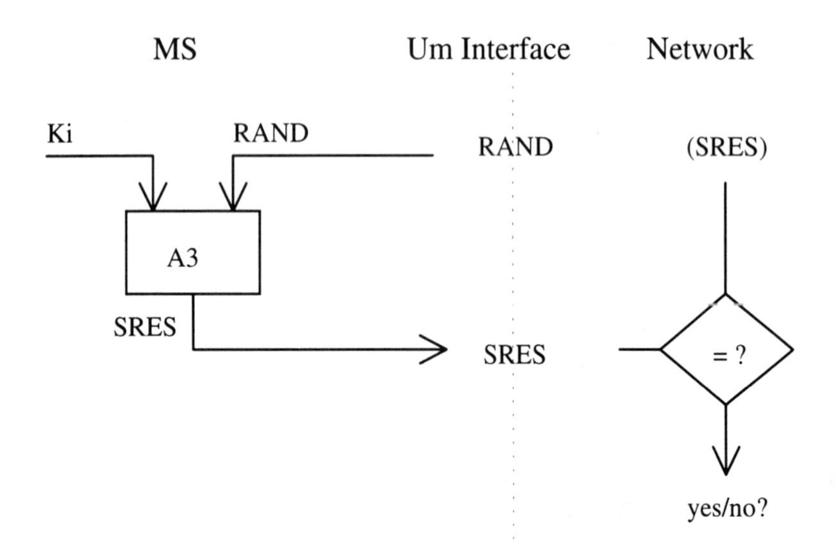

Figure 3.12 Principle of authentication.

VLR runs low on parameter sets, some new ones are requested from the AC. One important point of this security feature is that the relevant parameters (A3 and Ki) are stored in secure places and are never transmitted on the Um interface.

3.8.2 Ciphering

Digital transmission is suitable for the ciphering of data, because bit streams merely have to be scrambled with a certain method known only to both sides of the air interface. The GSM system uses such a *ciphering* method to protect signaling and user data. In order to ensure that the ciphered data from one side can be deciphered on the other side, a reversible algorithm is used. This means that if the *ciphering algorithm*, A5, is used to encipher a data stream, the same algorithm is used to decipher this stream and get back the original data stream.

One can easily understand that it is important that both entities use the same ciphering algorithm. In the current system only one algorithm is used, which is called A5/1. This is a special, protected algorithm, and the protection afforded the algorithm makes it is difficult to export the GSM system, with this ciphering capability, into countries other than the COCOM states (the Eastern European states restricted from access to certain Western technologies). Typically, it does not matter whether the ciphering algorithm is disabled or not, as long as it is part of the software. It is becoming even more difficult, if not impossible, to export the GSM system (with the ciphering algorithm) into former non-COCOM countries such as Russia, the other former USSR states, and China. Although COCOM does not exist anymore, the basic rules for exporting systems using the elaborate A5/1 ciphering still apply (i.e., A5/1 is not allowed to be exported). There are even new assemblies dealing with this kind of question. In a way, it might even be said that a company willing to export cellular equipment has to apply for individual export licenses.

To make exporting easier, the ETSI developed a new, simpler algorithm called A5/2, which is used for these former non-COCOM countries. The ciphering with this new algorithm is as secure as it is with the old one. This technique for achieving security does not use as much mathematics as the old one uses. Both algorithms can coexist in a network, and measures are employed to make sure that a mobile coming from a country using only the A5/2 algorithm has access to the European systems where the A5/1 algorithm is used. This is the reason why western European networks support both algorithms.

The designers of the ciphering aspects claim that this algorithm is so well protected against eavesdropping that even if someone knows the complete specifications, he or she would not be able to listen in on the data. This, of course, means that the security services, which in some countries listened in on private mobile phone conversations in the past, are no longer able to do so. This situation led to the delay of the acceptance of GSM in some countries.

Having only two algorithms for ciphering could make the life of professional eavesdroppers relatively easy. The algorithms, therefore, require a specific key, Kc. This key is calculated from a random number, RAND, delivered from the network. This is the same number that was used for the authentication procedure. The only difference is that a different algorithm, A8, is used to produce the *ciphering key* (Figure 3.13). The A8 algorithm is stored on the SIM card. The mobile equipment does not know anything about the security-related algorithms A3 and A8. This Kc key issued by the A8 algorithm is then used with the ciphering algorithm, A5/1 or

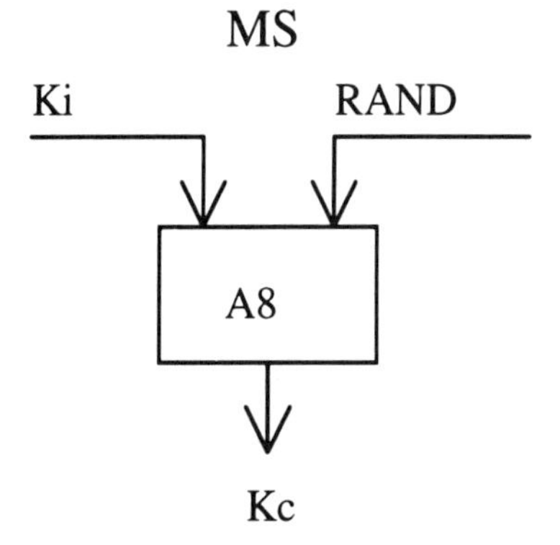

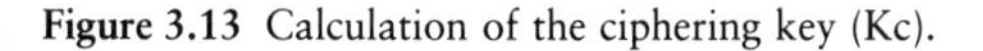
Figure 3.13 Calculation of the ciphering key (Kc).

A5/2, to encipher or decipher the data. The A5 algorithm is implemented in the mobile station whether it is A5/1 or A5/2.

To start the ciphering procedure, the network commands the mobile station to start ciphering with a specific ciphering sequence. From this time onward, the mobile station transmits ciphered data, where even the acknowledgment that ciphering is being used is already transmitted with enciphered data (Figure 3.14).

3.8.3 Temporary Mobile Subscriber Identity

To prevent a possible intruder from identifying GSM users by their IMSI, which is a permanently assigned number, a *temporary identity* is assigned to all subscribers while they are using the network. This identity is stored, along with the real identity, in the network. The temporary identity is assigned during the location updating procedure, and is used as long as a subscribers remain active in the network. The mobile station uses this temporary number when it reports to the network or originates a call. Similarly, the network uses the temporary number to page the mobile station. The assignment, administration, and updating of the TMSI is performed by the VLR. When it is switched off, the mobile station stores its TMSI on the SIM card to make sure it is available when it is switched on again.

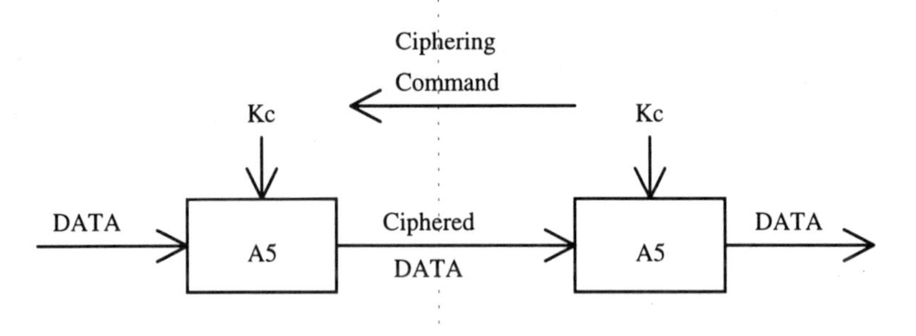

Figure 3.14 Start and execution of ciphering.

REFERENCES

[1] Mouly, M., and B. Pautet, *The GSM System for Mobile Communications*, Palaiseau, 1992, pp. 28–32.

[2] Schmitt, G., "GSM, a Success in Europe, an Opportunity for the Middle East and the Arab World," *Proc. IBC Middle East and Gulf Mobile Communications Conf.*, Dubai, Nov. 1993, IBC Technical Services Ltd., London.

[3] Twingler, J., "The GSM Standard in the Present and in the Future," *Proc. IBC Middle East and Gulf Mobile Communications Conf.*, Dubai, Nov. 1993, IBC Technical Services Ltd., London.

[4] GSM Technical Specifications, ETSI, Sophia Antipolis, Vols. 02.01 to 02.05.

[5] Calhoun, G., Chap. 2 of *Digital Cellular Radio*, Boston: Artech House, 1988.

CHAPTER 4

▼▼▼

LAYERS AND SIGNALING PRINCIPLES IN THE GSM SYSTEM

A telecommunications system cannot work without a minimum number of signaling functions to organize the interworking of its network entities and the interworking with other networks. Signaling is required to establish, maintain, and terminate connections or communication links. Signaling is necessary to make sure that the provision of services is taking place by the use of defined procedures. Measures have to be taken for all cases of service and system usage and in case of problems or malfunctions. Compared to the means of signaling used in the old, conventional telephone networks (e.g., dial tone, ring-down, dial pulses) and taking into account the higher complexity of a mobile network, which provides uncounted service features, a considerably higher signaling overhead can be expected for a system like GSM. The signaling overhead is great, although transparent (i.e., not obvious to or recognized by the end user).

This chapter describes the architecture of the complex signaling system used in GSM. An introduction to the layered structure of the ISO/OSI model is given, with an example to illustrate its functionality. Finally, an overview of the functional entities in the signaling system is given. Because this book emphasizes signaling protocols between a GSM mobile station and the fixed network (MSC/BSC/BTS), particular and relevant aspects are introduced.

4.1 SIGNALING IN THE GSM NETWORK

In order to cope with the task of having to organize a complex network system, the creators of the GSM specifications have chosen to rely on some already defined procedures and interface descriptions, and to adapt them for the mobile environment. The GSM system makes use of the so-called *signaling system no.* 7 (SSN7), which ISDN signaling is also based on. The SSN7 has been described by CCITT (International Telegraph and Telephone Consultative Committee), which originally defined it for fixed-network signaling and exchange switching. For use in a mobile telephone system, it had to be adapted and expanded by adding, for example, a so-called *mobile application part* (MAP), which contains additional procedures for coping with the mobility of the user. SSN7 provides a number of signaling channels and procedures for the communication (signaling transport) between the single entities in a GSM PLMN (e.g., the MSC, BSC, BTS, VLR, HLR, AC, and MS).

One means of describing the architecture and the interworking of the abovementioned signaling system is the layered model according to the International Standards Organization (ISO) with the open systems interconnection (OSI) presentation.

4.2 THE ISO/OSI LAYER STRUCTURE

The ISO/OSI layer model contains seven layers, each of which represents a certain entity with its particular functions and tasks.

4.2.1 What Is a Layer?

A separate layer can be regarded as a logical block in a communications entity, such as a mobile phone or a telephone exchange (switch). A logical block has certain functions, tasks, or assignments. It also has its own tools for executing its tasks. These tools consist of protocols, comparable to a language, that allow the layer (e.g., in a mobile station) to contact and communicate with its *peer* layer, which is a layer on the same level within another communications entity (e.g., in a switching center). The protocol used by two peer layers to communicate with each other is called *peer-to-peer protocol.*

Such a protocol also contains functions and conventions that describe how to use its own functions and those provided by the next lower layer to (1) execute the tasks in its own layer and (2) provide functions to the next higher (the upper) layer.

We can also find *notifications*, which are exchanged between one layer and its upper or lower layer in the same entity; for example, to inform the upper layer that a message (information) has been received or to instruct the lower layer to pass on a message (information). Such notifications are called *primitives*. They are part of the layer's protocol to organize the interworking with the neighboring layers (upper and lower). Figure 4.1 illustrates the functions and interconnection of layers.

4.2.2 An Example of the Layer Model

An example of communication over three layers can help with understanding the functionality and the dynamics of the layer model and its use in a communications system.

In this example, we watch the captains of two large ships, one coming from Brazil and the other one from Germany. They happen to meet, as they did very often before, at sea, far out on the ocean. Both captains are fond of soccer and they always like to exchange the latest results from their home country's first division. They also discuss the latest weather forecasts for the regions they are sailing to. Unfortunately, the Brazilian captain only speaks Portuguese and the German captain only knows his mother tongue. To get around this little problem, they employ their cooks as interpreters. The cooks both speak fluent French—what else, for a cook? Still, both sides also need the help of their radio operators, who have the task of "physically" connecting the two cooks; that is, to transmit their voices or dictated text between each other. Figure 4.2 illustrates this arrangement.

The two captains can be seen as part of the uppermost layer; in this example, it is Layer 3. They need the help (functions) of the lower layers in order to be able to communicate with each other. The two cooks, the interpreters, are part of Layer

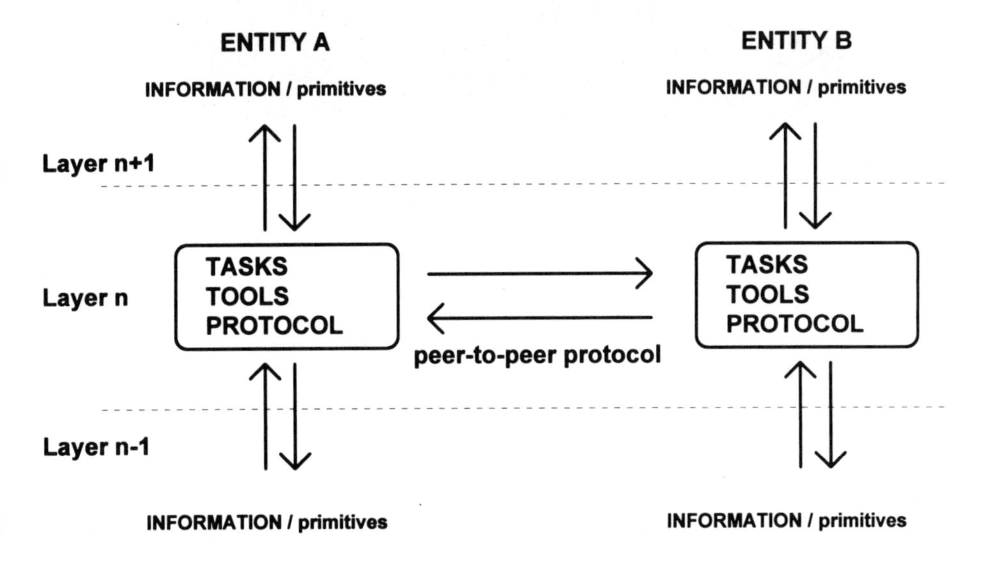

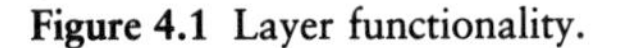
Figure 4.1 Layer functionality.

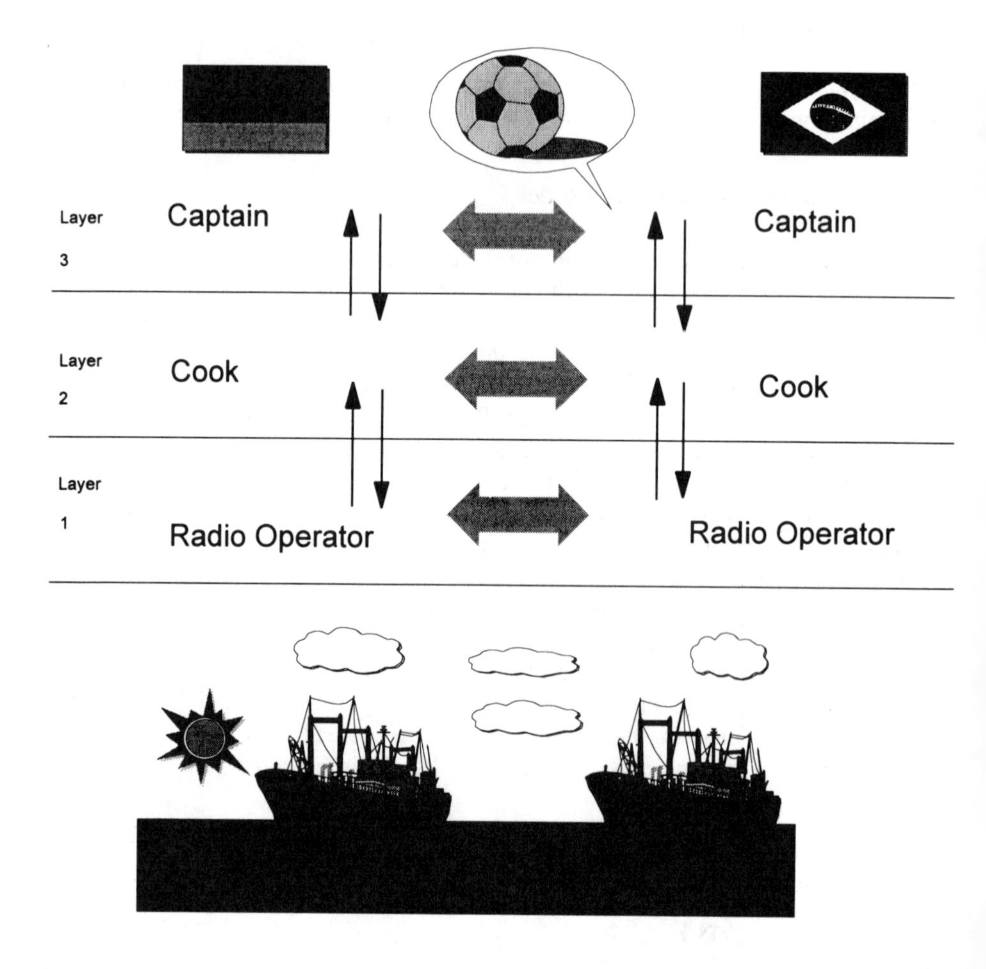

Figure 4.2 Example of a communication over three layers.

2 and the radio operators are located in Layer 1, the lowest layer. With their own tools and their lower layers' functions, the three layers can now start to communicate with each other.

Even though the two captains seem to pass and receive their information in a vertical direction, the communication is *virtually horizontal*, and so is the conversation between the cooks and between the radio operators.

Each layer keeps to a common protocol in order to successfully communicate with its peer without contention and in order to make use of functions or supply functions to the layers below or above. Such a protocol may consist of the following procedures.

Layer 1, the radio operators:

- Transmit messages, or information, they receive from their cooks on request;
- Notify their respective cooks when a message is received and pass it on when each cook is ready;
- Agree on a form of radio transmission, such as which frequency channel and whether to transmit speech or Morse signals;
- Comply with the international rules for radio transmission and communication;
- Acknowledge messages, ask whether the transmissions were readable, and notify the other party if something is not understood;
- Change the channel or increase the transmitted radio power in the case of bad reception.

Layer 2, the cooks:

- Translate the received information from their mother tongues into French and vice versa;
- Inform their captains every time there is a new message, such as another result of a soccer match;
- Partition the flow of speech from their captains into short and understandable blocks;
- Inform their counterpart when a block of information is not understood and may ask for a repetition;
- Count the information blocks (e.g., each soccer result), and should the need arise, can tell by the number which block has not been received.

Layer 3, the captains:

- Introduce themselves to each other and tell their counterpart what they would like to know;
- Exchange the final results of last weekend's soccer matches;
- Discuss the latest weather forecast;
- Pass their information to the cooks and instruct them to translate a message and pass it on for transmission.

In each of the virtually horizontal connections between peer layers, changes in the protocol of the other layers are not relevant and have no influence. Theoretically, the following could happen.

- The captains change the subject and start to chat about their last climbing tours in the mountains, since both are fond of this activity in their leisure time. What kind of information (text) they translate or transmit is irrelevant to the cooks and radio operators, respectively.
- Both cooks find out by chance that both of them study the Chinese language (Mandarin) for a better understanding of the Chinese kitchen. In order to

exercise and improve their skills, they decide to use Mandarin instead of French. Theoretically, presuming a certain level of mastery, this change will have no influence on the conversation between the captains or on the radio transmissions. A channel change or the use of Morse instead of voice radio transmission in Layer 1 will have no virtual influence on the other two upper layers.

4.3 THE SEVEN LAYERS OF THE ISO/OSI MODEL

The ISO/OSI model for an open communication system defines seven layers. Table 4.1 shows these layers and their meanings according to the model.

The signaling between all the interfaces from a GSM mobile station to the MSC takes place in the lower three layers (i.e., Layers 1 to 3).

Their functionality, with emphasis on the signaling taking place between a GSM mobile station and the fixed network (Um or radio interface), will be dealt with in the following chapters.

It should be noted here that layer 1, the physical layer, can be regarded as a means for transporting signaling data as well as user data. For the transmission of user data (e.g., speech data) the layer model does not apply. Due to the fact that in GSM (as in all communication systems) user data is transmitted over basically the same physical channel, with differences only in the logical organization and in the coding, our description of Layer 1 will also include, in the following chapters, the handling and transmission of user data.

Table 4.1
The Seven Layers of the ISO/OSI Model

Layer 7	APPLICATION	Application protocols, user-oriented provision of communication media
Layer 6	PRESENTATION	Application-specific format transfer
Layer 5	SESSION	Connection of application processes, billing
Layer 4	TRANSPORT	Flow control for point-to-point connections
Layer 3	NETWORK	Connection and switching of communication links
Layer 2	DATA LINK	Control of signaling links, block transfer of signaling data
Layer 1	PHYSICAL	Physical transmission, coding, error correction, modulation, etc.

CHAPTER 5

The Physical Layer—Layer 1

Radios move information from one place to another over channels, and a radio channel is an extraordinarily hostile medium on which to establish and maintain reliable communications. The channel is particularly messy and unruly between mobile radios. All the schemes and mechanisms we use to make communications possible on the mobile radio channel with some measure of reliability between a mobile and its base station are called the *physical layer*, or the *Layer 1* procedures. These mechanisms include modulation, power control, coding, timing, and a host of other details that manage the establishment and maintenance of the channel. In this chapter, we examine in considerable detail how mobile radios can gain access to a channel in a trunked radio system. Then we look at the characteristics of a typical radio channel. With the knowledge of how radio channels behave and the methods a mobile radio can use to gain access to their use, we turn our attention to the specific means GSM uses to acquire and maintain a channel for reliable communications.

Our approach, therefore, is the same as that a visitor to a strange city would take when discovering how to use the local tram system. First, he or she stands at the tram station and looks at the posted schedules to learn how to access the transportation services. Once boarded and settled in the tram he or she has a look around to become familiar with how to comfortably use the tram system during his or her stay, to learn the tram's characteristics: When is the best time to use the system? What is the most convenient route? Where is the quietest place to sit? What is done with the ticket? Finally, if our visitor decides to seek employment as a mechanic on the tram system, he or she would have to learn all the details of how the tram cars and the network of tracks are constructed, run, and maintained.

5.1 ACCESS TO TRUNKING SYSTEMS

Unless the reader who is familiar with analog radio systems also has considerable experience with fixed wire-line telephone systems, the notion of access will be unfamiliar and confusing. It is for this reason, therefore, that we explore various forms of access to trunked radio resources in an intuitive way.

If the number of channels available for all the users of a radio system is less than the number of all possible users, then such a system is called a *trunked* radio system. Trunking is the process whereby users share a limited number of channels in some orderly way. Sharing channels, or providing fewer channels than there are users who need them, works because we can be sure that the likelihood that everyone will want a channel at the same time is very low. The method of sharing channels is termed *access*; users ask for and are granted access to the trunked resource. A cellular phone system is a trunked radio system, because there are fewer channels than there are subscribers who could possibly want to use the system at the same time. Access is granted to multiple users of the system by dividing the system into one or more of its operating domains: frequency, time, space, or code.

5.1.1 Frequency-Division Multiple Access

Frequency-division multiple access (FDMA) is the most common form of trunked access and is what most radio people think of when they hear the words *access* or *channel*. With FDMA, users are assigned a channel from a limited set of channels ordered in the frequency domain. If the number of available channels is greater than about 15, then the best trunking efficiency is realized if the initial assignments to channels are made from a common control channel, to which all radios tune for instructions when they first try to use the system. There is a bewildering variety of signaling schemes used in FDMA systems. Radios designed to be used in such systems seek out and then tune to the designated control channel when they are turned on. On this control channel, the radios find signaling data that contain instructions on which frequency new entrants should tune their transmitters and receivers to when they want to pass user traffic, which is often voice. The signaling in these systems works something like a directory of businesses one may see on the lobby wall of a large office building. The larger the number of users assigned to an FDMA system, the greater the catalog of available frequencies must be. Frequency channels are precious and are assigned to systems by government regulatory bodies in accordance with the common needs of a society. When there are more users than the supply of frequency channels can support, users are *blocked* from access to the system. With more available frequencies comes more users, and more users means more signaling information has to pass on the control channel. Very large FDMA systems often have more than one control channel to handle all the access control tasks. An important characteristic of FDMA systems is that once a frequency is assigned to a user, the assigned frequency is used exclusively by the user (even during short periods of silence) until he no longer needs the resource.

There are many small trunked radio systems in the world with fewer than 15 frequencies available for use. Such small systems, which are seldom cellular, usually do not have a dedicated control channel. Instead, a simple signaling routine (tones or data) is included on each and every channel, which tells radios where to tune to for an idle traffic channel. The signaling works something like a "no vacancy" sign in a motel window.

Channels in an FDMA system are defined by their assigned number (e.g., 3), their center frequencies (e.g., 890.600 MHz), and their widths (e.g., 200 kHz).

5.1.2 Time-Division Multiple Access

TDMA is common in fixed telephone networks. The enabling technology in radio systems is speech encoding and data compression, which remove redundancy and silent periods and decrease the time it takes to represent any period of speech. Users are assigned access to a channel in accordance with a schedule. Though there is no strict technical requirement for it, cellular systems, which employ TDMA techniques, always overlay TDMA on top of an FDMA structure. A pure TDMA radio system would have only one physical operating frequency, not a particularly useful system. TDMA is a rather old concept in radio systems, too. The short-wave broadcast services on the HF band (3 to 30 MHz) are assigned channels (operating frequencies) on a worldwide schedule so that they do not interfere with each other during certain propagation conditions.

In modern digital cellular systems, TDMA implies the use of digital voice compression techniques, which allow multiple users to share a common channel on a schedule. Modern voice encoding greatly shortens the time it takes to transmit voice messages by removing most of the redundancy and silent periods in speech communications. Other users can share the same channel during the intervening times. Users share a physical channel in a TDMA system, where they are assigned time slots. All the users sharing the physical resource have their own assigned, repeating time slot within a group of time slots called a *frame*. If everything works well, the users cannot tell that they are using a TDMA system, and each of the users assigned a time slot within a frame will insist they have their own channel. A time slot assignment, therefore, is often called a channel. GSM sorts users onto a physical channel in accordance with simple FDMA techniques. Then the channel's use is divided up in time into frames, during which eight different users share the channel. A GSM time slot is only 577 μs, and each user gets to use the channel for 577 μs every 4.615 ms (577 $\mu s \cdot 8 = 4.615$ ms).

5.1.3 Space Division Multiple Access

Space-division multiple access (SDMA) is used in all cellular systems, both analog and digital. Indeed, cellular systems are distinguished from other trunked radio systems solely because they employ SDMA. Cellular systems allow multiple access to a common RF channel (or set of channels) on a per cell basis (i.e., according to a mobile's location on the landscape). A cellular radio system is a good illustration of SDMA. The limiting factor for this type of SDMA is the frequency reuse factor of the system. Frequency reuse is the chief concept in cellular radio, because cellular radio is an example of a frequency reuse system, where users simultaneously share the same frequency, with due regard for cochannel interference. Users must be separated by a distance sufficient to minimize the effects of cochannel interference. A set of frequencies used in a cell can be repeated in another cell elsewhere in the system, provided there is sufficient distance between the "identical" cells to prevent cochannel interference.

There are many clever and exciting innovations in the uses of SDMA in today's cellular systems. These include such schemes as (1) microcells, (2) umbrella cells, (3) sectored sites, and (4) smart antennas. These are all methods for dividing the space in which mobiles operate with greater resolution, thus shortening the distance between users without violating cochannel interference concerns.

1. Microcells are very-low-power base stations (about -5 dBm or 0.3 mW) placed in busy spots in a system.
2. Umbrella cells are very large cells designed to handle the considerable signaling traffic in a field of microcells.
3. When a cell site's coverage pattern is sectored into 120- or 60-deg sectors with directional antennas, greater capacity can be realized in the system due to the increased freedom for radio system engineers to alleviate cochannel interference.
4. Smart antennas are a relatively new innovation in analog cellular systems. Some are so sophisticated in their operation that they generate thin pencil beams between a base site and specific mobiles in a cell.

SDMA techniques hold the greatest possibilities for substantial capacity increases in current analog cellular systems, and will continue to delay the acceptance of digital cellular outside of Europe.

5.1.4 Code-Division Multiple Access

CDMA is new to cellular technology and is not used in GSM. A CDMA system is one of the proposals for the next generation of digital cellular systems in the United States. This system represents a dramatically different approach to providing cellular services, and is described in some detail in Chapter 13. Learning about such a system may seem irrelevant to the student of GSM, but the CDMA systems are, after all, competitors of GSM. Moreover, learning about CDMA techniques adds depth to our understanding of any cellular system.

CDMA puts all users desiring access to a trunked resource on the same RF channel at the same time. Just as in TDMA, cellular CDMA systems overlay their CDMA schemes on an FDMA plan. Users of a common frequency channel are separated from each other by superimposing a user-specific high-speed code on the modulation of each user. Since the separating code has the effect of spreading the occupied bandwidth of each user's transmissions, the system is called a *spread spectrum* system. Figure 5.1 shows how two CDMA "channels" may be generated, and Figures 5.1 and 5.2 illustrate how a special receiver, called a *correlator*, can distinguish one channel, or signal, from many others created in the manner illustrated in Figure 5.1.

We start in Figure 5.1 with two different sets of binary information: "A Data" and "B Data." We wish to transmit both data streams together over a channel and separate them at the receiver, thus giving each signal its own virtual channel. To do this, we give each data stream, A and B, its own character by modulo-2 adding (XOR) each with its own key: the "A Key" and the "B Key." In their respective cases, we get the "A Signal" and the "B Signal." Now we look in the receiver (correlator) to see what happens to the two signals.

The composite of the two signals generated in Figure 5.1 appears to a receiver as the waveform labeled "Composite A+B Signal" in the center of Figure 5.1. The waveform is simply the algebraic sum, bit for bit, of both the "A Signal" and the "B Signal." We recover the "A Data" stream from the composite waveform in two steps. First, we multiply the receiver's composite input with a copy of the corresponding key of the signal we want to recover: in this case, the "A Key." The waveform labeled "(A+B) * A Key" in Figure 5.1 is the result of this multiplication. Second, we integrate the resultant waveform, bit for bit, into the "Integrator Output" waveform. The sign of the integrator's output is tested at the end of every six key bit times against a threshold. The resulting integrator's sign output is the inverse of the original binary data, as illustrated in the last waveform of Figure 5.1. What happened to the data in the other signal? To answer this question, we look at Figure 5.2 where the composite waveform we left at the receiver's input in Figure 5.1 is multiplied with the "B Key." The procedure for recovering the "B Data" is exactly the same as was used to recover the "A Data," except that a different reference key is used.

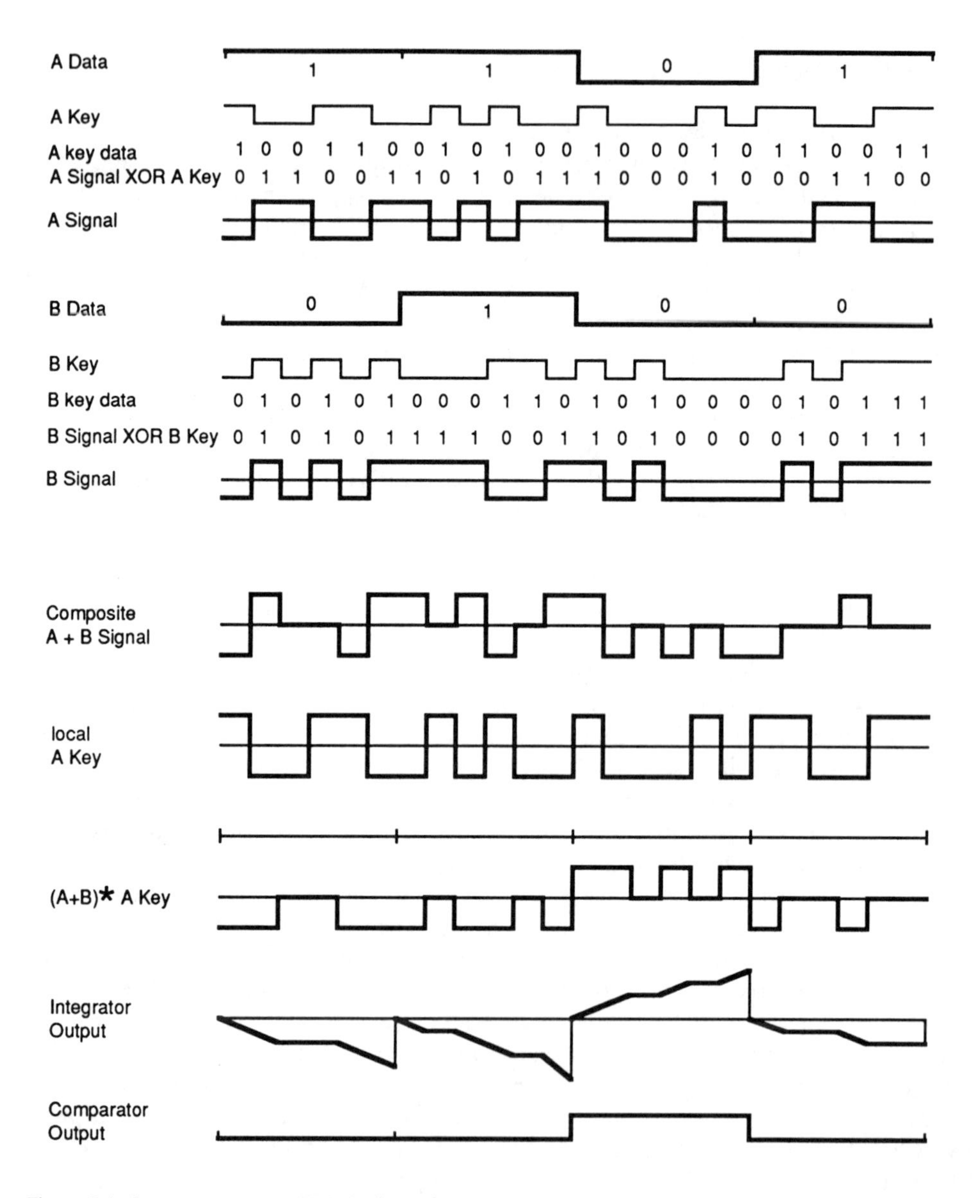

Figure 5.1 One among many CDMA channels.

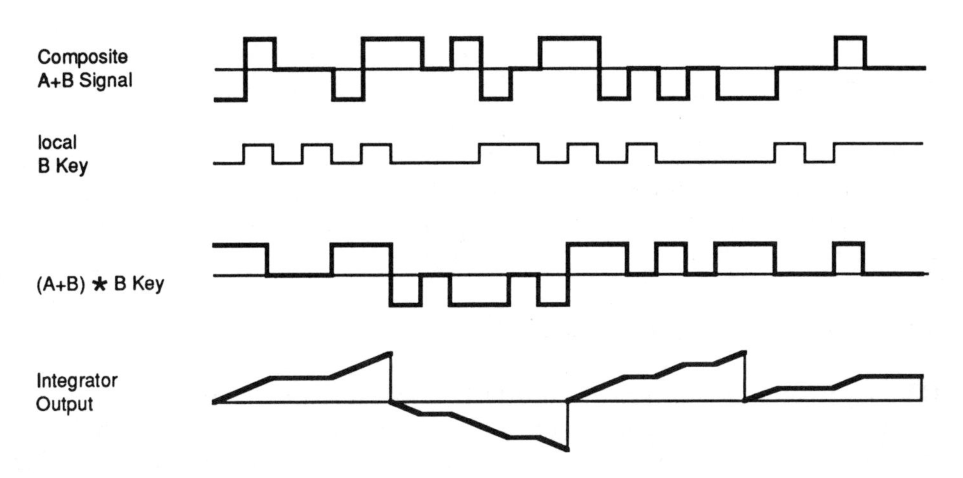

Figure 5.2 Recovering a different CDMA channel.

What happens when we try to recover data with the wrong key? This is not as silly a question as it seems, for the answer illustrates how CDMA receivers regard other channels and interference; they see them as the same (i.e., noise).

Figure 5.3 illustrates what happens if the wrong key is used: no valid data are recovered. This time we multiply the "Wrong Key" with the composite of the A and B signals, which appears at the receiver's input. Note that the integrator's output tends toward an average of zero volts. If the "Wrong Key" were the key to another valid data stream, then that other valid data would appear, inverted in this example, at the integrator's comparator output. Figures 5.1 through 5.3 are only illustrations of a fictitious CDMA system, and serve only to illustrate how multiple access by code could work. It should be clear to the reader that careful timing and synchronization are critical in such a system.

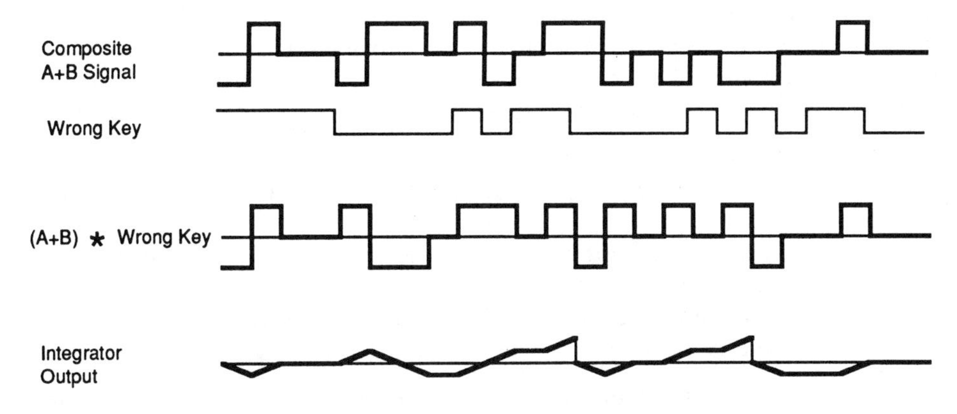

Figure 5.3 Using the wrong key.

5.1.5 Duplex Operations

Except in special situations, traffic moves between radios in a duplex mode, which means that for every transmission in one direction, a response is expected and then received in the other direction. There are two chief ways to set up duplex communications channels.

5.1.5.1 Frequency-Division Duplex

One way is frequency-division duplex (FDD). Since it is difficult and very expensive to build a radio that can transmit and receive signals at the same time on the same frequency, it is common to define a frequency channel with two separate operating frequencies, one for the transmitter and one for the receiver. All one needs to do to take advantage of a duplex offset in the channel definition is to add filters to the transmitter and receiver paths that attenuate the complementary function's operating frequency, thus keeping the transmitter's energy out of the receiver's input. A single common antenna could be used with such a filter system. The filter systems are called *duplexers* and allow us to use the channel (frequency pair) in *full-duplex* mode; that is, the user can talk and listen at the same time.

5.1.5.2 Time-Division Duplex

Many dispatched mobile radio systems, such as public safety systems, do not require their users to enjoy full-duplex operation. Radios in such systems transmit and receive on the same frequency, but never at the same time. This kind of duplex operation is called *half-duplex* operation, and all that is required for a half-duplex system to work is that a user give an indication that he or she has finished speaking, and is ready to receive a response from another user. The word *over* is often uttered at the end of a transmission, a button is released (which turns off the transmitter), and the receiver is connected to the antenna so that it can recover the expected response. Why are such awkward systems still used? There are many reasons:

1. Since only one frequency is used, only half the spectrum space is required.
2. Duplexers are heavy and expensive. Half-duplex operation eliminates the need for the duplexer.
3. Since the transmitter is off when the user is listening, battery life in hand-held radios is conserved and smaller batteries can be used.
4. Smaller and lighter batteries and the lack of a duplexer make portable half-duplex radios very rugged, low-cost, and popular.

It should be apparent that half-duplex operation is actually a type of time-division multiplexing. Indeed, another name for half-duplex operation is *time-divi-*

sion duplex (TDD). Radios in TDMA systems are usually TDD radios for all the same reasons and advantages that they bring to police radios: they are low-cost, lightweight, and rugged. This works because, as described in Section 5.1.2, voice traffic is digitized and then compressed (in time) in TDMA systems. As the received voice data are recovered in the receiver, is returned to an analog waveform, and grows to fill its original duration in time; the TDD process is hidden from the user. Cellular TDMA systems, however, retain the transmitter/receiver frequency pairs of their analog cellular cousins so that the receiver in a mobile phone can attend to other matters on other frequencies while the user is making a transmission. The TDD operations (e.g., saying "over") are handled by the radio itself.

5.2 THE RADIO CHANNEL

Section 5.1 explained some of the ways radio techniques can be used to lend order to the trunking process. Individual radios seek a radio channel on which to pass information with the help of trunked radio techniques. The remainder of this chapter explains how GSM orders communications on the physical layer. The collection of codes, techniques, schemes, and processes we use in order to make communications possible in the physical medium between the base station and the mobile station is called the *physical layer protocol*, or the *Layer 1 protocol*. The remainder of this chapter explains the myriad techniques and processes GSM uses to make sure data pass between the mobile terminal and the base station safely and clearly: Why are there so many aspects to this protocol? What is the nature of the radio channel? What aspects of the radio channel force us to use the various schemes and contrivances explained in this chapter?

5.2.1 The Problem

By the time information sent from a transmitter reaches a receiver, it is contaminated and distorted by five destructive influences: (1) the modulator, (2) the transmission medium, (3) noise sources, (4) fading phenomena, and (5) the demodulator. Everything that corrupts a signal originating at a transmitter on its way to a receiver is called the *channel*.

When we use an analog cellular phone at the edge of a cell's coverage area, we can hear the destructive influences of the channel. We can hear the hisses, crackles, pops, interference, and fading that sometimes annoy the users of analog cellular phones, but we quickly train ourselves to listen through all interference and distractions as we become accustomed to the convenience mobile communications affords. The same phenomena that cause the temporary annoyances for us in analog systems can completely destroy communications with digital radios. A single pop, for example, may go unnoticed as it obliterates a part of a syllable, but that same pop could invert a single data bit, which could hide the meaning of a whole frame of data if special measures are not taken to add redundancy to the data in the channel. Adding redundancy diminishes the importance of each individual bit in the channel. Engineers accustomed to wire-line tasks only feel comfortable with a 40-dB signal-to-noise ratio (SNR). If the SNR falls below 10 dB in a terrestrial wire-line system, there is cause for alarm and plans are made to replace the faulty facility. In the mobile environment, 20-dB fades occurring 100 times a second is the best that can be expected, and the typical case is much worse. The bit error rate (BER) on the mobile channel can be a million times higher than on a point-to-point digital microwave radio channel or a landline wire circuit.

5.2.2 Dealing With the Problem

Designing a communications system for the mobile environment is probably the most demanding task in communications. The literature is full of precise statistical terms and computer models that attempt to define the mobile channel, but all of these end up as simplifications that have to be tested with actual field trials at great expense. Attending engineers have to keep an attentive eye and steady nerves as they (1) make some assumptions, (2) try to understand the specific problems with statistical models, and (3) supervise a long and expensive field trial to temper their understanding and confirm the validity of their assumptions. When all the preliminary work is done, they design a host of tricks and contrivances to combat problems in the hostile mobile channel. Engineers new to mobile digital radio are usually surprised by the number and complexity of schemes used to protect data destined for the air interface.

5.2.3 Characteristics of the Radio Channel

The good news is that the 900-MHz mobile radio channel is linear; every other property of the channel is bad or even worse. One aspect of a radio channel is the occupied bandwidth caused by alterations in the amplitude, frequency, or phase of a carrier. Any of these three parameters of a carrier can be altered, and these alterations can carry information we measure in *bits* or *symbols* per second. The radio spectrum is a fixed and precious resource with almost incalculable value. System designers direct much of their work toward designing a system that gets all the information to be transmitted into a very narrow segment of the spectrum allocated by some regulatory body. The channel becomes a cheap hotel room in a city with no vacancies: someone checks into a poorly managed hotel after a long search in the night and finds that housekeeping has not cleaned up the assigned room after the former guests, and noisy neighbors can be heard clearly through the thin walls. There are two sources of trouble in the channel: noise and interference.

Transmitters on adjacent channels (the noisy neighbors) cause *adjacent channel interference*, and distant transmitters sharing the channel cause *cochannel interference*. We quantize the effects of interference with the carrier-to-interference ratio (C/I). Random noise in the channel (the dirty room) comes from natural and man-made sources, such as ignition systems, power distribution systems, and machines. We measure the effects of noise with the SNR.

Being a mobile channel, our spectrum hotel is also in a bad neighborhood, which is subject to frequent fires, earthquakes, and numerous incidences of break-ins and assorted criminal activities. We sort these additional problems into two groups: (1) those appearing when nothing is moving, and then add (2) some very destructive influences appearing when the mobile station moves.

5.2.4 Static Conditions

First, we consider the special case where neither the mobile radio is moving, nor is anything else moving nearby. The channel in this unusual case is an additive Gaussian white-noise channel and is governed by atmospheric propagations with a path loss possibility worse than that of an inverse fifth or sixth law. Whole bursts of data suffer errors, since the channel is subject to flat fades up to 40 dB. Because the channel is a frequency-selective one and there are many paths from the base station to the mobile; all the data are subject to multipath propagation, shadowing, and delay spread, where the delay spread can easily be several microseconds. Channel equalization is used to counter the resulting intersymbol interference at higher bit rates. Finally, the local receiver generates its own noise.

5.2.5 Dynamic Conditions

Next, if we allow the mobile station to move about the landscape among other mobile radios, we add the effects of terrestrial propagation, which is dominated by the most destructive influence of all: Rayleigh fading. Since different radio waves can follow a variety of paths to the mobile receiver, phase changes, which are frequency-dependent, may occur that are often subtractive at the receiver. Multipath fading is the phenomenon that causes a TV picture to flutter as an airplane flies overhead, offering a moving mirror for the RF path between the broadcast transmitter and the TV's receiving antenna. Rayleigh fading occurs with a statistical distribution called a *Rayleigh distribution*. Multipath causes the channel to suffer such deep and rapid fades that the channel's impulse response has to be estimated frequently. Other distant mobiles and base stations on the same channel can fill in these deep fades with cochannel interference.

5.2.6 Quantitative Analysis of the Radio Channel

The remarkable Shannon-Hartley channel capacity formula gives us guidance into how we can pass information through a channel:

$$C = B \log_2\left(1 + \frac{S}{N}\right)$$

where C is the capacity of the channel in bits per second or the maximum number of bits that can be sent per second. The channel's bandwidth is B Hz wide, and S/N is the ratio of the received signal power to the noise power. The wider the channel B, the more information we can pass through the channel, but this will decrease S/N. If we make S/N arbitrarily high, then the noise will be as small as we wish. Because the noise is very small, we can amplify the output of the channel as much as we need and clear the distortion with an equalizer. Primarily because of Rayleigh fading and interference, the typical mobile channel capacity C will never be as good as the formula suggests. We can still use this valuable formula to understand what we can do to get a radio system to work and why system designers employ the methods they do to increase the efficiency of their systems. If we use an alphabet of waveforms, each of which represents m bits, we call these waveforms *symbols*. If one of the M symbols (waveforms) in the alphabet is transmitted during a symbol period T_{symbol}, then

$$R = \frac{m}{T_{symbol}} \text{ [bps]}$$

where R is the data rate and m is the number of bits represented by each symbol. The larger we make M, the larger the variety of waveforms we use, and the greater the number of bits we can send during a symbol period T_{symbol}. Since the time to send one bit, T_{bit}, is the reciprocal of the data rate R,

$$T_{bit} = \frac{1}{R} \text{ [s/bit]}$$

The more bits we can represent by a symbol, m bits per symbol, the shorter the bit period will be.

$$T_{bit} = \frac{1}{mR_{symbol}} \text{ [s/bit]}$$

The greater R becomes, which could be R_{bit} or R_{symbol}, the greater the channel's bandwidth B must be in order to carry the information. Since spectrum is the limiting resource, system engineers are concerned with the bandwidth efficiency *R/B* or the information rate per hertz:

$$\frac{R}{B} = \frac{1}{BT_{\text{bit}}} \text{ [bps/Hz]}$$

The smaller we make the BT_{bit} product, the greater the information rate. We can decrease the bandwidth B or decrease the bit period T_{bit} in order to increase the bandwidth efficiency. GSM specifies Gaussian minimum-shift keying (GMSK) modulation with a BT_{bit} product of 0.3, where each symbol represents one bit. This product is often called simply the *BT product.* The bandwidth B of the Gaussian filter in the GMSK modulator is 81.3 kHz. The bit time T_{bit} (or simply T) is the reciprocal of the bit frequency f_{bit}, which is 270,833 bps.

$$T = 3.692 \text{ [}\mu\text{s]}$$

$$TB = (81.3E3)(3.692E - 6)$$

$$TB = 0.3$$

$$\frac{R}{B} = \frac{1}{BT}$$

$$\frac{1}{BT} = 3.33\text{[bps/Hz]}$$

The number of bits per second we can send over a channel of bandwidth B is called the *spectral efficiency*. Our best tool for increasing spectral efficiency is the *information density*, which we measure in bits per hertz. We can think of this as the *modulation agility* or *modulation muscle* in our system. GSM's GMSK modulation is an example of a two-level modulation scheme, where each symbol represents only one bit.

$$M = 2^m$$

$$2 = 2^1$$

There are two different waveforms for each value of the bit, 0 or 1; a positive voltage or a negative voltage are two examples. There is a trivial case where M is 1 and m is 0. In this strange case, the transmitter is sending a message already known by the receiver; there is only one possible message in the system. Going toward the other extreme, there are many modulation schemes in which each symbol represents more than one bit. Chapter 12 describes a digital cellular system where $M = 4$ and $m = 2$.

$$4 = 2^2$$

Though they seem attractive, these kinds of multilevel modulation schemes require expensive linear amplifiers in the transmitter and strain the receiver's demodulator. As m, the number of bits per symbol, increases, the variety of waveforms M increases exponentially. This requires a substantial increase in the precision of the demodulator.

5.2.7 Limiting Factors and Some Solutions

There are three limiting factors in a mobile communications system: (1) the available power, (2) the noise and interference, and (3) the overwhelming need to limit the bandwidth. Chapter 13 describes an interesting digital cellular system that is power-limited rather than bandwidth-limited. Though it still has to confine itself to a segment or band in the radio spectrum, a power-limited system makes the bandwidth B artificially large, at the expense of power, to lower the typical BER. GSM is a bandwidth-limited system in a very hostile mobile environment. Unless extraordinary measures are taken, the BER will be far too high to carry voice information. The solution is to remove as much of the redundancy from the voice data as possible, and then add all those bits back in an orderly fashion we call *channel coding*. The receiver uses these redundancy bits to reconstruct the damaged data it receives from the transmitter. There is a great deal of redundancy in voice data, and GSM removes almost all of it. The tradeoff in this process is that each of the voice bits becomes very important, and some are much more important than others. GSM protects the important bits with greater care than it protects the bits that are not so important.

As the mobile station moves among the cells, all the radios in the system are engaged in an orderly conversation among themselves, called *signaling*. These signaling messages contain, among other things, signal strength and BER reports, power adjustment and time alignment commands, handoff instructions, and channel sounding tests for equalizer adjustments. The signaling exercises supplement the modulation and channel-coding schemes to bring order to the destruction in the radio channel. We can compare the whole communications process in GSM to sending a cheap and ugly porcelain cup (voice data) in the mail to a friend. Even after we pack the cup in a huge box full of soft packing material (redundancy bits), seal the box tightly, and address it accurately (signaling messages), it arrives at its destination smashed into pieces. We therefore include a picture of our porcelain gift (redundancy bits) in the package, so that our friend on the receiving end can glue the mess back together again.

Another analogy illustrating the way communication takes place on the radio channel in a digital cellular system is playing a new musical creation to a friend over a very noisy phone line. We decide to play our creation on the guitar. The phone line is so noisy that our friend can scarcely hear our playing. To work around the situation, we yell out chord and timing instructions on the phone, and we confirm our friend's understanding of the instructions when he repeats them back to us. Eventually, after considerable yelling and checking, our friend will have complete instructions for playing our musical creation himself on his own guitar.

In the first analogy, if our friend can glue everything back together in some recognizable form that holds water, we are happy with the mail service. In the second analogy, if our friend can finally play our creation himself on his own guitar, we are satisfied with the phone service. Such is the state of communications on the mobile channel.

5.3 THE FREQUENCIES

Now it is time to look for a job with the transit system, to investigate the fine details of the physical layer in GSM. We start with the *FDMA* part of GSM and look at *channel numbering.*

5.3.1 Channel Numbering

The GSM system was the first digital cellular standard to be created and commercially implemented. As of the end of 1994, the system had been in commercial use in Europe for almost two years, and there were already further developments to migrate GSM into different and higher frequencies. These newer developments are mentioned in this chapter together with the frequency domain details of the more common 900-MHz GSM system. Except for the frequencies used, all the features of the various GSM implementations (DCS-1800, PCS-1900) are similar to the more common 900-MHz GSM system, and no further distinction is offered for them in this book. Some people argue that since the NADC standard was developed at the same time as GSM, we should not claim that GSM is the first digital cellular system. The authors of this book stand by their claim in favor of GSM, because the NADC system uses the original analog control channel in its underlying host analog system. The NADC system is a dual-mode system, not a purely digital system, as GSM is.

5.3.2 Primary GSM

The *primary GSM* system refers to the first generation of those systems currently installed in each European country, even though not all of the systems are fully operational everywhere. It uses two 25-MHz frequency bands in the 900-MHz range. The mobile station transmits in the 890- to 915-MHz frequency range, and the base station transmits in the 935- to 960-MHz range. The end points within the *physical layer* are the mobile station and the BTS. Depending on where we look within the hierarchy of the network, the MS-to-BTS direction is referred to as the *uplink* (ul), and the BTS-to-MS direction as the *downlink* (dl). This is the case within the air interface, and will be how we refer to directions in GSM. Just as is found in the popular analog systems, a duplex spacing of 45 MHz is used; the base station always transmits on the high side of the duplex frequency pair. A handy way to remember this is to recall that base stations are often found on top of buildings, which means that a BTS, therefore, transmits on a higher frequency than mobiles, which are found far down below the BTS on the ground. The frequency bands are divided into 125 channels with widths of 200 kHz each. These channels are numbered from 0 to 124. Within the system, only the *absolute radio frequency channel number* (ARFCN) is used, from number 1 to 124. Channel number 0 is used as a guard band between GSM and other services on lower frequencies. The absolute frequency, referred to in megahertz, is not used at all. With these parameters in mind, there are some easy formulas to describe the actual frequency of an ARFCN:

$$Ful(n) = 890.0\text{ MHz} + (0.2\text{ MHz}) \cdot n$$

and

$$Fdl(n) = Ful(n) + 45\text{ MHz}$$

where

$$n = \text{ARFCN},\ 1 \le n \le 124$$

Any frequency may be assigned to a mobile station by the base station from a selection of between 1 and approximately 16 frequencies. The number of channels a base station may have at its disposal depends on network planning considerations and the traffic density expected in the base station's coverage area. The assignments are made to these frequency channels with normal FDMA techniques (see Section 5.1). Within countries using any of the NMT-900, AMPS, or ETACS analog cellular systems, contention in the frequency ranges occurs, since those of the analog systems are nearly the same as those for GSM. The usual agreement between the operators is that they split the frequency assignments in half as long as the analog systems remain in use.

5.3.3 E-GSM

With the further development of the GSM standard, an additional range of frequencies has been made available to the system. For each of the two duplex frequency ranges, one for the forward direction and one for the reverse direction, an additional 10 MHz has been added to the bottom end of the bands, extending the frequency range another 50 channels. The numbering for these additional channels is from 974 to 1,023. This is done to avoid assigning one channel number twice within the same standard. Channel number 0 is returned to use in the *extended GSM* (E-GSM) systems, since it does not make sense to reserve it as a guard band within the E-GSM bands. Now, the lowest channel, number 974 (880.0 MHz), serves as the guard band. The formulas for converting an ARFCN in an E-GSM system into its corresponding frequencies are

$$Ful(n) = 890 \text{ MHz} + (0.2 \text{ MHz}) \cdot n, \text{ where } n = \text{ARFCN}, 0 \leq n \leq 124$$

and

$$Ful(n) = 890 \text{ MHz} + (0.2 \text{ MHz}) \cdot (n - 1024), \text{ where } n = \text{ARFCN}, 975 \leq n \leq 1023$$

and

$$Fdl(n) = Ful(n) + 45 \text{ MHz}$$

5.3.4 PCN or DCS-1800

From the GSM standard came the personal communications network, as it was called in its early days. When the ETSI finally completed the specification on this system, its official name became *DCS-1800*. In some countries, particularly the United Kingdom, it is still referred to as *PCN*. This system uses the same signaling or message exchange techniques as GSM, but the assigned frequencies are much higher, and there are nearly twice the number of channels available. Licenses have already been issued in the United Kingdom and Germany for DCS-1800 services, and trial systems have been tested (e.g., in Singapore, a trial system has been used to confirm the proper operation on the higher assigned frequency bands). DCS-1800 uses the frequency ranges of 1,710 to 1,785 MHz in the uplink direction, and 1,805 to 1,880 MHz in the downlink direction. From these figures it follows that the duplex spacing is 95 MHz with 374 channels of 200 kHz each. The channels are numbered from 512 to 885 to distinguish them from the channels in the primary and extended GSM frequency bands. The applicable frequency formulas for the DCS-1800 system are

$$Ful(n) = 1{,}710 \text{ MHz} + (0.2 \text{ MHz}) \cdot (n - 511)$$

and

$$Fdl(n) = Ful(n) + 95 \text{ MHz}$$

where

$$n = \text{ARFCN}, 512 \leq n \leq 885$$

5.3.5 PCS-1900

Sometime during the first half of 1995, a PCS standard will start to take form in the United States. One of the proposed standards for North American PCS is the DCS-1800 system shifted up 100 MHz into the 1,900-MHz range, hence its name *PCS-1900*. Except for the difference in frequency bands, this system is identical to the DCS-1800 standard. The frequency shift is required in the United States because of the presence of some point-to-point radio links on the 1,800-MHz band.

5.4 RADIO FREQUENCY POWER LEVELS

Radio equipment in GSM and DCS-1800 are distinguished from each other by, among other things, the different transmitter *power levels* they can produce. The equipment is thus classified by the power classes shown in Tables 5.1 through 5.3. Each of the power levels in the various classes are separated from each other by something greater than 2 dB. For GSM, the minimum mobile station power level is 20 mW (13 dBm), and for DCS-1800 it is 2.5 mW (4 dBm). The relationship between dBm and watts is reviewed in Chapter 9, Section 9.2.1.

In the next-generation GSM, which is called *Phase II* or *Phase 2*, additional diminished power levels are introduced. There are also some new microcell applica-

Table 5.1
Power Levels in the GSM System

Power Class	*Maximum Power of a Mobile Station/(dBm)*	*Maximum Power of a Base Station/(dBm)*
1	20W (43)	320W (55)
2	8W (39)	160W (52)
3	5W (37)	80W (49)
4	2W (33)	40W (46)
5	0.8W (29)	20W (43)
6		10W (40)
7		5W (37)
8		2.5W (34)

Table 5.2
Power Levels in the DCS-1800 System

Power Class	*Maximum Power of a Mobile Station/(dBm)*	*Maximum Power of a Base Station/(dBm)*
1	1W (30)	20W (43)
2	0.25W (24)	10W (40)
3		5W (37)
4		2.5W (34)

Table 5.3
Power Levels for the Micro-BTS in the GSM and DCS-1800 Phase II Systems

Power Class	*Maximum Power of a GSM Micro-BTS/(dBm)*	*Maximum Power of a DCS-1800 Micro-BTS/(dBm)*
M1	0.25W (24)	1.6W (32)
M2	0.08W (19)	0.5W (27)
M3	0.03W (14)	0.16W (22)

tions that call for some micro-BTS power levels. The reduced BTS power levels are shown in Table 5.3.

5.5 TRANSMISSION ON THE RADIO CHANNELS

GSM allows radios to share its channel resources in both the frequency and time domains. The details of all the available frequency channels and bands were shown in Section 5.3. The methods used to make additional allocations in the time domain are called *TDMA techniques*, and these add considerable complexity to the system when compared to analog cellular systems. The reward, however, is better performance and additional features without squandering additional bandwidth for each user.

5.5.1 TDMA Frame

Considering that the GSM channel spacing is 200 kHz, it would be rather wasteful for a system not to subdivide this resource any further, since regulatory bodies and operators continue to strive for increased efficiency in the use of spectrum. To achieve this, the GSM system makes use of TDMA techniques, with which each frequency channel is further subdivided into eight different time slots numbered from 0 to 7.

Each of the eight time slots is assigned to an individual user. A set of eight time slots is referred to as a *TDMA frame* (Figure 5.4), and all the users of a single frequency share a common frame. The frames will become more important in later discussions, as the complete system counts these frames and constantly refers to them. If a mobile, for example, is assigned time slot number 1, it transmits only in this time slot and stays idle for the remaining seven time slots with its transmitter off. The mobile's regular and periodic switching (on and off) of its transmitter is called *bursting*. The length of a time slot, which is equivalent to a *burst* from a mobile, is 577 μs, and the length of a TDMA frame is 4.615 ms ($8 \cdot 577\ \mu s = 4.615$ ms).

Compare the GSM spectrum allocation of 200 kHz per radio channel with the typical analog cellular system allocations of between 12.5 and 30 kHz. The use of TDMA techniques effectively reduces the allocation for each traffic channel to 200 kHz/8 = 25 kHz, which is equivalent to the typical analog system. This fact demonstrates that the GSM specification was selected for its additional quality and features rather than for increased capacity. The enhanced capacity TDMA affords, however, is an important feature of GSM, because to add the same GSM features and GSM quality to an analog system would require a significant increase in bandwidth for each user and render the "improved" analog system too inefficient to continue to operate.

Although the TDMA structure gives the system more capacity, there is a price to pay. If a mobile station transmits a burst every 4.615 ms, then the underlying frequency is 216.6 Hz (=1/4.615 ms), which is within the audible range. If, for example, a GSM mobile is operated close to a home stereo system, this frequency can be heard in the speakers. More serious is the impact on electronic devices such as hearing aids, cardiac pacemakers, or automobile electronics. Due to the relatively high power transmitted in the GSM mobile's bursts, they can have a significant

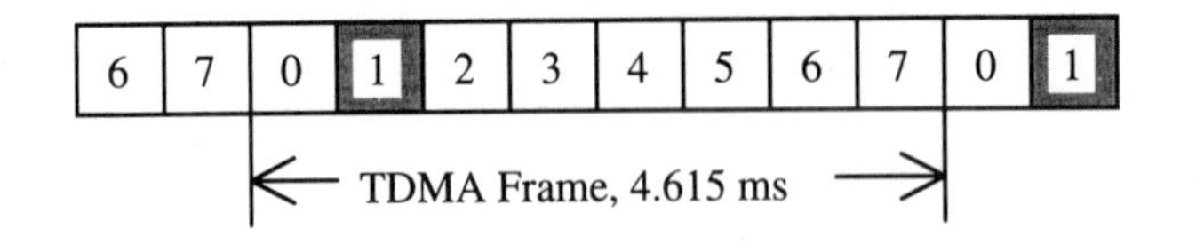

Figure 5.4 TDMA frame.

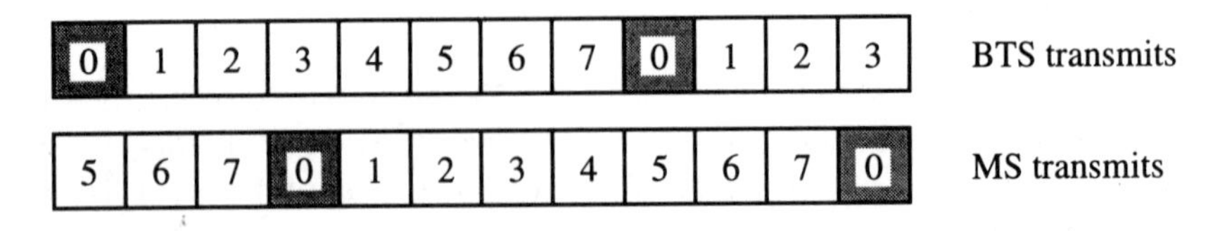

Figure 5.5 Time-division duplex in the GSM system.

influence. Some car manufacturers already suggest that a GSM mobile should only be used with an external antenna, because if used inside, the mobile phone may block or trigger the air bag or other important systems. Sorting out and clearing the risks and nuisances caused by GSM phones adds to the cost of designing and manufacturing them.

5.5.2 Time-Division Duplex

When using the TDMA technique, it is not necessary to transmit and receive signals at the same time in full-duplex mode. It is for this reason that TDD is introduced and used to great advantage in GSM (see Figure 5.5). The advantages for a mobile or a hand-held portable are numerous:

1. No need for a dedicated duplex stage (duplexer); the only requirements are to have a fast-switching synthesizer, RF filter paths, and fast antenna switches available;
2. Increased battery life or reduced battery weight;
3. Better quality (more rugged) and lower cost phones.

TDD tends to confuse those who have to repair digital phones. To make this feature less confusing for technicians and engineers and to increase some of the switching time specifications and lower costs, the GSM specifications stipulate that even though the time difference between transmit and receive functions is three time slots, the time slot numbering for both the BTS and the mobile station stays exactly the same as if both were using the same time slot at the same time. They are, of course, not doing so; the mobile station transmits three time slots later than the BTS.

5.6 PULSED TRANSMISSIONS

The requirement on the mobile station to transmit in only one single time slot and stay idle during the remaining seven time slots results in very tight demands on the switching on and off of the RF power. If a mobile station does not perform according to the specifications, it will disturb other mobile stations in adjacent time slots and on adjacent frequencies. The tendency for a pulsed radio to disturb neighboring frequency channels is called *AM splash*. The GSM specifications make sure a mobile's emissions remain in its assigned channel by specifying a power-versus-time template. An example of the power-versus-time template from the specifications is given in Figure 5.6. From this figure it can be seen that a mobile station has to switch on or off within only 28 μs and, during this short time, achieve a dynamic range of up to 70 dB. At the smaller power levels of 34 dBm and below, the dynamic range is a little bit more relaxed, since the absolute power level that must be achieved at

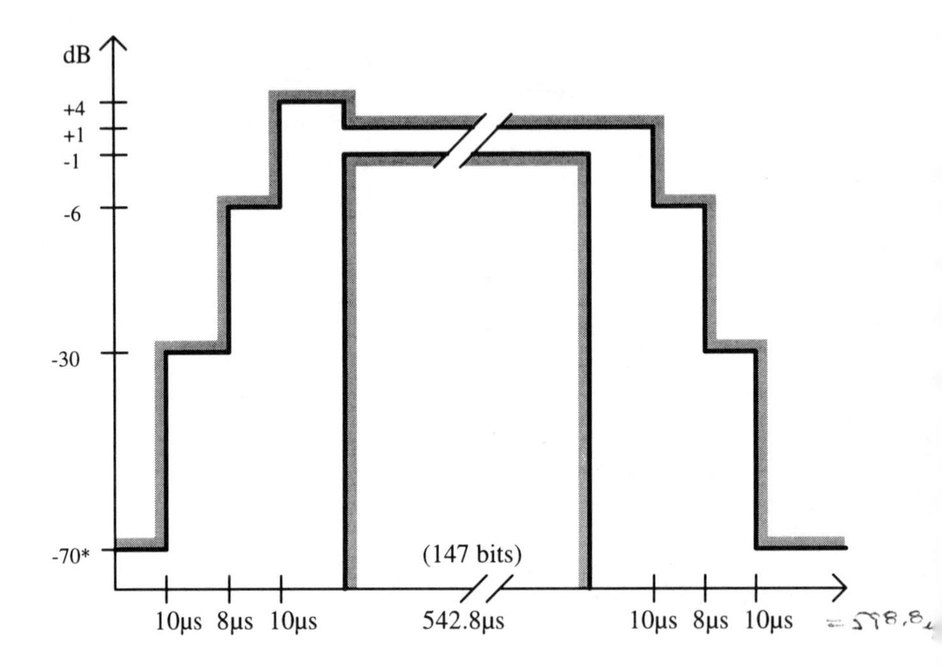

Figure 5.6 Power-versus-time template.

power-off is −36 dBm. After ramping up its RF power, a mobile station has 542.8 μs to transmit information. The information contained in a burst is described in further detail in Section 5.10.

This *pulsed transmission* is not only defined for the mobile station, but is also generally found in base stations. If only certain time slots of a physical channel are used, the base station also performs power-up and power-down ramping of its output. The advantage of this is that only those time slots and resulting RF carriers that are assigned to an active mobile station are transmitting; the others remain idle. There are some operational exceptions where the power ramping of base stations would disturb the proper operation of the GSM system. These exceptions will be pointed out, where appropriate, later in this chapter.

5.7 TIMING ADVANCE AND POWER CONTROL

Within a single cell, mobile stations can be found at different distances from the base station. Depending on the distance to the base station, the *delay time* and the attenuation of an individual mobile's signal is likely to be different from the delay and attenuation of any of the other mobile stations. It should be apparent by now that TDMA techniques rely heavily on the proper timing of the transmissions of bursts, as well as the correct reception of the bursts at the base station's receiver. To avoid the collision or interleaving of the signals from mobiles assigned to adjacent time slots, the base station performs measurements on the timing delay of each mobile station, and commands those mobiles with bursts arriving too late at the base site to advance their burst transmissions in time; late mobiles (mobiles farthest from the base station) are given a head start. This feature is called *timing advance.*

To compensate for the attenuation over different distances within the cell, the base station, at the same time it is making timing adjustments on mobiles, commands the mobiles to use different power levels in such a way that the power arriving at the base station's receiver is approximately the same for each time slot. The *power control* is performed in steps of 2 dB. This means that a mobile station more distant from the base site has to transmit at a higher power level than those close to the base site.

As mentioned in Chapter 3, Section 3.7, a mobile station has to perform measurements on the quality and power level of the link between the BTS and the mobile station. If the BTS discovers that a mobile station does not receive its signal at a sufficient power level for reliable downlink communications, it may also apply power control on its own RF output and transmit at different power levels in each time slot. BTS power control is an option for base stations and is not yet implemented in all brands and systems.

Figure 5.7 illustrates the problem of timing advance and power control on the uplink. Mobile A is farther from the BTS than mobile B is. Mobile A is assigned to time slot (TS) number n (TS_n), and mobile B is assigned to the time slot following the one assigned to A (TS_{n+1}). At the BTS's receiver, A's burst arrives with a lower level than B's burst, and A's burst is partially obliterated by B's burst. The details of how GSM compensates for these problems are given later in this chapter.

5.8 BURST STRUCTURES

The physical aspects of transmissions over the air within a GSM system were explored in Sections 5.3 through 5.7. These aspects were the RF channel schemes, access techniques, power control, and some timing considerations. We saw that information is moved between radios as data (ones and zeros) that are confined to time slots, and that there are eight time slots in a frame. The frame is something like a freight train with eight freight cars, and the time slots are the individual cars in the train. Now, as with freight trains, where some kinds of freight require special freight cars (e.g., tankers for fluids, hoppers for loose solids, and boxcars for general freight), different burst structures can appear in a time slot when a special kind of data (freight) has to appear on the GSM channel.

In Section 5.6 we saw one of the freight cars in the GSM freight train as the power-versus-time template and learned that there are only 542.8 μs allotted in each

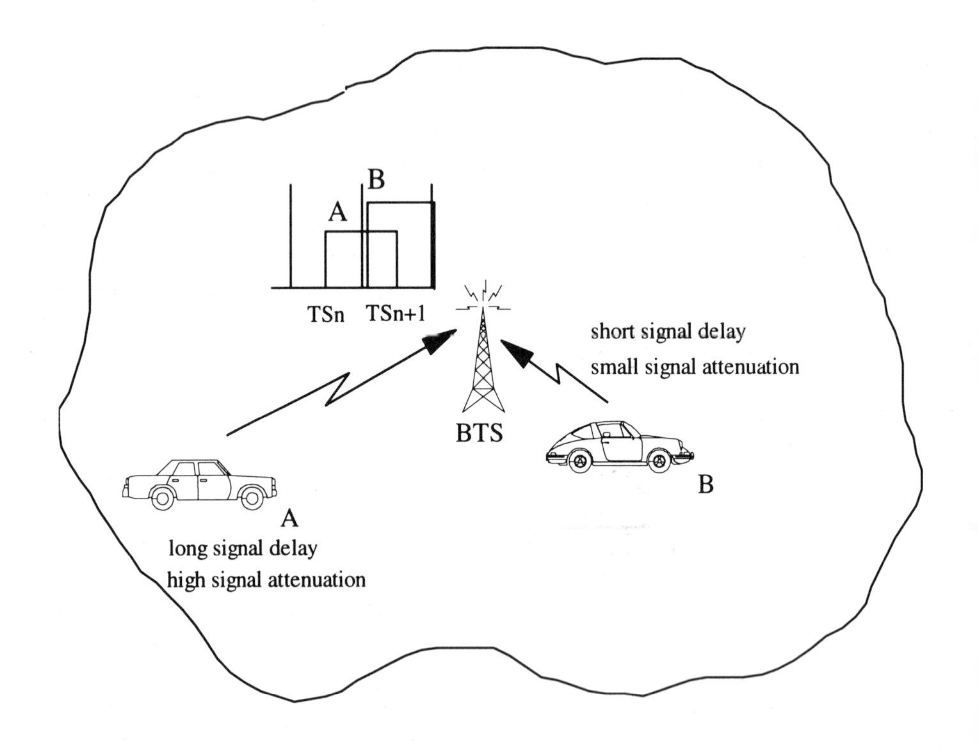

Figure 5.7 Timing advance and power control.

time slot for data transmissions. This short period accommodates 147 data bits; there are 147 bits in a GSM freight car. Actually, there are 148 bits in each time slot, but the time reserved for the first and last half bits is, as we will soon see, reserved for the on-off RF switching time. Even at this lowest logical level at the time slot or burst level, there are many structures, just as there are many kinds of freight cars. All of these burst structures will be thoroughly explained.

5.8.1 Normal Burst

Figure 5.8 shows the structure, in the time domain, of a *normal burst*. A normal burst is something like a boxcar in a freight train; it carries almost anything except special freight (acids, automobiles, sand, and explosives). A normal burst is the most common burst in the GSM system and is transmitted in one time slot either from the base station or from the mobile station. There are eight time slots in a TDMA frame. The actual user data (coded data) occupy only a portion of the time slot, and the remainder of the bits are reserved for a host of control functions and some demodulating aids.

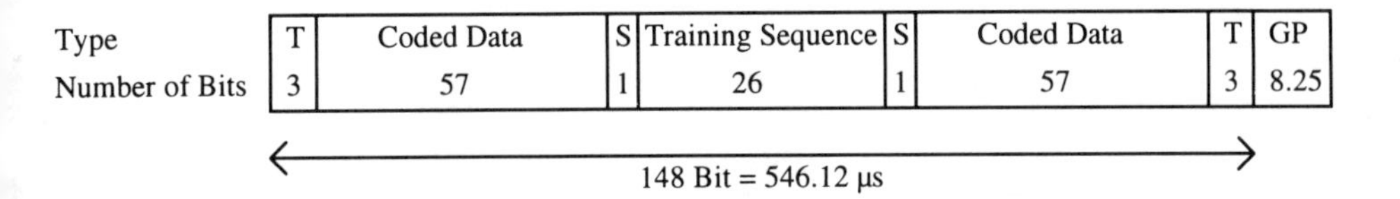

Figure 5.8 Structure of a normal burst.

Tail Bits (T). This small group consists of three bits at the beginning and the end of each burst and is used as *guard time.* The tail bit time covers the periods of uncertainty during the ramping up and down of the power bursts from the mobile in accordance with the power-versus-time template shown in Figure 5.6. The tail bits are always set to zero. Coincidentally, the demodulation process requires some initial zero bit values.

Coded Data. These two times, of 57 bits each, contain the actual transmitted signaling data, or user data. Included and mixed in with the user's payload data are channel-coding bits, which are used in the receiver to help recover the original data. For now, you should think of the coding bits as packing material that protects the freight. Large parts of the remainder of this chapter explain channel coding, or data protection.

Stealing Flag (S). These two bits are an indication to the decoder (in the receiver) of whether the incoming burst is carrying signaling data, which are usually messages the radios use to maintain the link between themselves, or whether the burst is carrying user data. The indicating flag is needed because signaling data are very important and go to different places than user data go. Another word for user data is *traffic.* For example, during a call, important signaling messages have to be exchanged to complete a handover. When it is time for a handover, the user data are substituted by signaling data. The exact coding and other characteristics of the signaling data will be explained in Section 5.14.

Training Sequence. This is a fixed bit sequence known to both the mobile and the base station, which lets radios synchronize their receivers with the bursts. Synchronization lets receivers interpret the recovered data correctly. It might not be fully clear why this bit pattern is required, since all the timing seems to be well defined so far. The reason we include the training sequence in each normal burst is to compensate for the effects of multipath fading. We learned a little about multipath and its destructive influence on TDMA systems in Section 5.2.

Recall that *multipath propagation* results from reflections of the transmitted signal from houses, mountains, and other obstacles. The result is that the same signal takes different paths to the receiver, where each path yields valid signals with different relative time delays and different attenuations. Figure 5.9 shows this effect in principle. The main signal from the mobile station arrives directly at the receiver (the BTS)

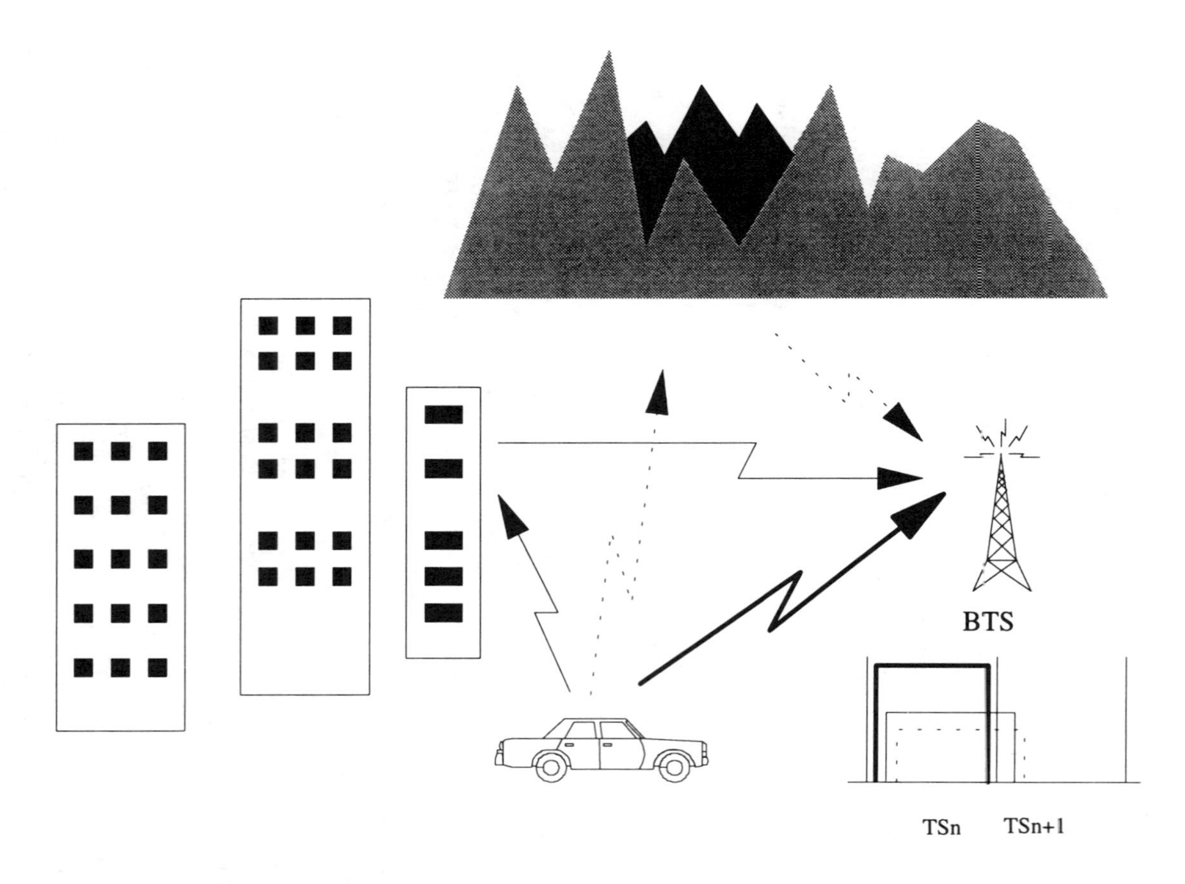

Figure 5.9 Multipath propagation in the GSM system.

with no delay and exactly within its assigned time slot, as depicted with the dark arrow. A second signal, depicted with the thin arrows, is reflected from buildings, and arrives somewhat later with a certain attenuation. The third signal, depicted with dashed arrows, is reflected from the mountains and has an even greater delay and attenuation. Looking at the figure, it does not seem difficult to distinguish the signals, but the signals typically arrive at the receiver overlapping each other, with nearly the same timing and power levels. The result is a *smearing* or *delay spread* of the recovered data in the receiver. To assist the receiver in separating the different signals, to sharpen and make the recovered data clear, the training sequence is used. There are eight different sequences defined in GSM. All the radios in a particular cell share the same training sequence. Great care was taken to make sure these bit sequences cannot be repeated within the normal coded data part of the burst, and to make the training sequence a unique part of the burst.

The part of the receiver that clears up the distorted and "fuzzy" data and which needs the training sequence to do so is called an *equalizer*. An equalizer is a filter that melds the different signals together into a single nonambiguous signal. The equalizer does this by first looking at the distorted training sequence in each time slot it sees and then adjusting its own filter characteristics to get the original, clean training sequence back again. The equalizer knows what all the training sequences are supposed to look like, and the network tells the equalizer which training sequence to look for in a particular cell. If the training sequence is cleaned up, then all the other bits in the time slot must be clean too. Think of an equalizer as a set of spectacles with corrective lenses for the receiver. The longer the possible delays, the fancier an equalizer has to be. The equalizer in GSM radios can compensate for timing delays of up to 16 μs.

In other words: The training sequence helps the equalizer, which is part of the receiver section of a digital radio, to demodulate the bit content of the data section in the burst. Being a bit pattern that is known to the receiver, the equalizer detects the filter function (so-called *impulse response*) imposed on the modulated signal. This filter function is due to the transmission over the radio interface, including its multipath and Doppler effects. By applying the *inverse* filter function (the equalizer does this by calculating appropriate digital filter coefficients) to the signal part containing the data (two times 57 bits), the equalizer then regenerates the actual modulated symbols in the signal [1].

Guard Period (GP). It probably seems odd to specify *fractions* of a bit, so instead it should be considered as a defined time (measured in bits), rather than as actual data bits. No data are transmitted during the guard period, which is reserved for the ramping time. Taking the bit length defined in the system as 3.69 μs/bit, the guard period can be calculated as 8.25 bits $\cdot$ 3.69 μs/bit = 30.4 μs, which is approximately the time used during power ramping (Figure 5.6). During this time, two consecutive bursts from two mobiles may overlap (i.e., the previous burst ramps down and the current burst ramps up). No data are transmitted during the ramp time (GP), and communication is not disturbed while radios are ramping their RF power outputs.

This explanation of the parts of a normal burst shows that the introduction of TDMA techniques requires additional control functions on the radio path. These additional functions represent an overhead in the system and are part of the price we pay for improved performance.

5.8.2 Random Access Burst

Within a cell, strict timing has to be maintained in order for the bursts from mobile stations to arrive at the base station within their assigned time slots. The assumption is that the link has already been established. What happens *before* the link is established? How can you adjust and synchronize something that does not exist yet? There must be an initial situation during which the base station can make a preliminary rough estimate on timing advance settings for a mobile, a way for the base station to measure the time delay a mobile's burst is experiencing. The delay is proportional to the distance between the base and the mobile, and the distance can change.

If this measurement were taken on a normal burst, there would be a great likelihood that such bursts from many mobiles would overlap each other at the base station, particularly when mobiles are transmitting from the edge of a large cell. To avoid this useless situation, the mobiles use a shorter burst during initial access, which takes the maximum cell radius into account. Even if a mobile station is at the border of a large cell, its shortened burst would still not overlap onto any adjacent normal bursts. The burst type used for this purpose is called the *random access burst*. It gets this curious name from the fact that mobile stations transmit this type of burst at random times and only when the mobile is trying to gain initial access to the system.

Figure 5.10 shows the content of a random access burst. The significance of the bits within the burst are the same as that explained for the normal burst. The *synchronization sequence* has the same significance as the training sequence. The obvious difference is that the synchronization sequence is much longer because the

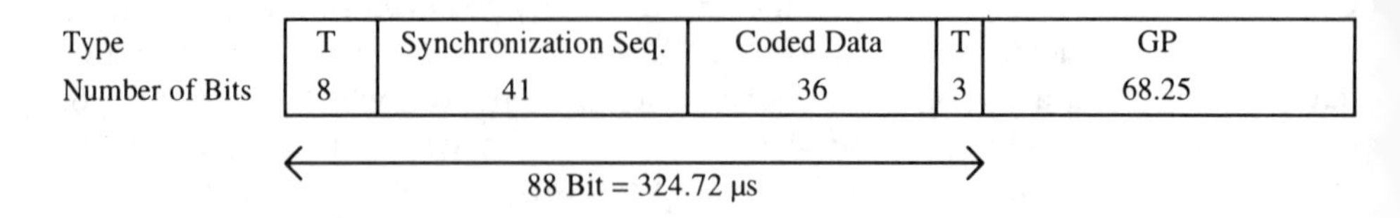

Type	T	Synchronization Seq.	Coded Data	T	GP
Number of Bits	8	41	36	3	68.25

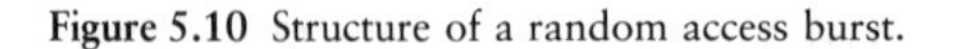

Figure 5.10 Structure of a random access burst.

equalizer needs more information; it needs to take a longer look to synchronize properly with a new signal.

Look at the random access burst's long *guard period.* Take the guard period's actual time as 68.25 bits · 3.69 μs/bit = 252 μs. Perform a rough calculation for the maximum distance (to the BTS) that the random access burst can survive. Ignore propagation conditions and assume the speed of the radio waves is similar to the speed of light. The result is a maximum distance of 252 μs · $(3 \times 10^8$ m/s$)$ = 75.5 km. Now radio waves have to travel twice the distance between stations in order to have an effect on the link: one way from the BTS to the mobile station, where the mobile station synchronizes with the system's timing, and then the other way from the mobile station back to the BTS. To make sure the random access burst will not collide with a normal burst in the same cell, the maximum allowed distance between mobile station and BTS is half the maximum delay, which is only 37.75 km.

5.8.3 Frequency-Correction Burst

Since timing is a critical need in the system, the base station has to provide the means for a mobile station to synchronize with the master frequency of the system. To achieve this, the base station transmits, during certain known intervals, a pure sine wave signal for the period of exactly one time slot. What a comfort this must be for the mobile. Due to the nature of the type of modulation used in GSM, this can be accomplished by simply sending a fixed sequence of zeros (000. . .) in the time slot. See Section 5.15 for a thorough discussion of this technique. The mobile station has precise knowledge of when to expect a *frequency-correction burst.* (See Figure 5.11 for its structure.) Depending on the quality of the clock reference inside the mobile station, the mobile station's design engineer can determine how often it is necessary to resynchronize the mobile station (how often to look for and acquire the frequency-correction burst).

5.8.4 Synchronization Burst

When a mobile station starts to synchronize with the network, it first looks for and detects only the frequency where the base channel is located. The mobile does not

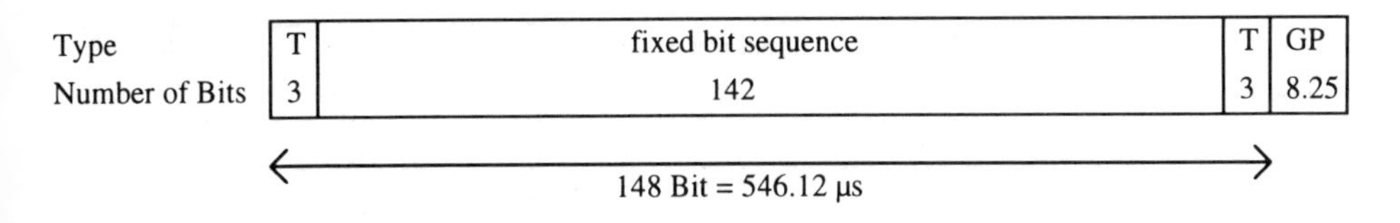

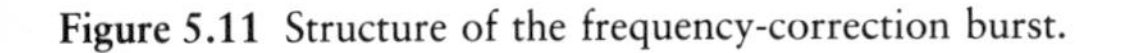

Figure 5.11 Structure of the frequency-correction burst.

yet have a key with which to demodulate and decode the information provided in the forward base channel, which is information that contains some valuable system parameters. As was explained previously, the key is one of the eight defined training sequences. The base tells the mobile which key to use with the *synchronization burst.* Figure 5.12 shows the content of this burst type, which is similar to the normal burst. The difference is the longer synchronization sequence and the presence of some diminished coded data. The coded data contain the base station information code (BSIC) indicating the current training sequence (base station color code (BCC)) and the national color code (NCC), and another figure indicating the so-called shortened TDMA frame number. The significance of the TDMA frame number will be explained in Section 5.11.

5.9 CELL SIZE

Several parameters in GSM confine the maximum cell size to approximately 35 km. The maximum timing advance is 63 bits (0 to 63). The duration of a single bit is 3.69 μs. Since the path to be equalized is a two-way one, the maximum physical distance between a BTS and a mobile is half the maximum delay, or 70 km/2 = 35 km. Recall that the random access burst can accommodate a maximum delay over a distance of 75.5 km, so that these bursts can appear at the BTS receiver with a good possibility that they will not be covered by another mobile's normal burst. This happy situation cannot be guaranteed in larger cells.

5.10 LOGICAL CHANNELS

Let us say you are a freight clerk in a large company that makes dc muff wuffers as distinguished from your competitor, your next door neighbor, who also makes muff wuffers, but of the ac variety. Your employer, you see, makes the somewhat superior dc muff wuffers. Business is good; everyone wants dc muff wuffers, and you have lots of dc muff wuffers to ship. The GSM freight train (eight freight cars per train) stops right outside your loading dock precisely on the hour, every single hour of the day. The GSM freight train has a good service reputation and competitive rates. You decide to use the GSM shipping services for your dc muff wuffers, but

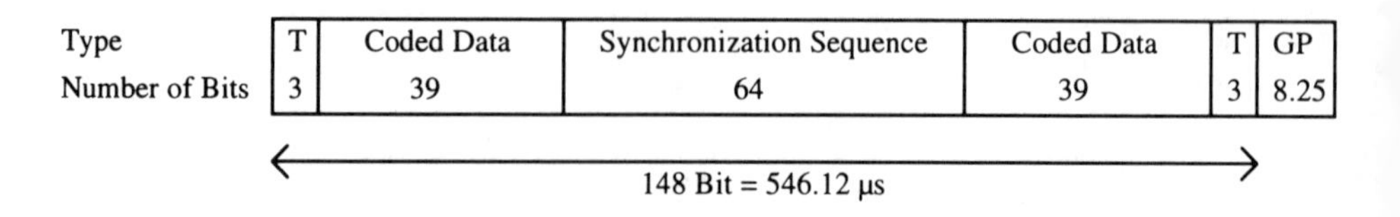

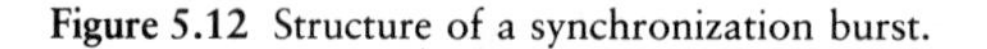
Figure 5.12 Structure of a synchronization burst.

learn that, since you have so many dc muff wuffers to ship, and the GSM train is so popular, that many trains will be needed to move your shipment. Only one freight car in each train (the last car, car number 7) is reserved for your shipment each hour; the other cars are filled with goods from other satisfied customers of the GSM freight train. Your neighbor has been filling car number 4 in each and every GSM freight train with ac muff wuffers for months. To get your dc muff wuffers out to their destination on time, you set up a logical sequence of dc muff wuffers on your loading dock, all packed securely on pallets, labeled, and ready to be loaded onto car number 7 each hour.

The loaded, packed, and labeled pallets full of dc muff wuffers are arranged on your dock as a single *logical channel.* The contents of this channel (pallets of dc muff wuffers) are loaded onto a *physical channel* to their destination. The physical channel, therefore, becomes the eighth car on each and every GSM freight train, which stops at your dock every single hour. Each of those eighth cars in each GSM train (frame) is your physical channel until your dock is empty of dc muff wuffers. Note that the eighth car is car number 7; the first car is car number 0.

With the concept of logical channels, we are getting farther away from the physics of the signals in GSM and closer to the information carried. The way we move information depends on the type of information we move. A shipping clerk has no difficulty with this concept. DC muff wuffers are packed onto pallets, cleaning solvent is sealed in drums, and circus animals go into cages. Different types of information can exist in the system on different types of logical channels. The contents of the different logical channels can appear in any physical channel (frequency and time slot), but once a physical channel is assigned the task of carrying the contents of a logical channel, the assignment should not change. Some special kinds of freight must go into special cars (special time slots, e.g., frequency-correction burst). The two expressions *physical channel* and *logical channel* should not be confused.

A logical channel carries signaling data, or a user's data. The data, of whatever kind, are mapped onto a physical channel. The manner in which the data are mapped onto the physical resource depends on the data's content. One should be more careful with important data than with more trivial data. Important data have a higher priority than routine data. The mapping schemes yield a few logical channel structures, which we regard as combinations of structures. There are seven combinations of logical channels that can be mapped onto their allowed physical channels.

The GSM system distinguishes between *traffic channels,* which are reserved for user data, and *control channels,* which are used for network management messages and some channel maintenance tasks. Paying passengers on an airplane are *traffic data,* while flight attendants and pilots are *control data.* When the expression *channel* is used in the following discussion, the reference is to logical channels (rather than physical channels) unless specified otherwise.

The traffic channels allow a user to transmit either speech or data (e.g., computer data). Depending on the type, speech or data, different channels are called for.

- The *traffic channel/full-rate speech* (TCH/FS) is the channel currently in use for the transmission of speech. The net speech rate is 13 kbps.
- The *traffic channel/half-rate speech* (TCH/HS) is intended for future use. The idea is to double the capacity of the system by compressing the data by a factor of two (half the current data rate). The doubling of speech capacity will be achieved with no degradation in speech quality.
- TCH/F9.6/4.8/2.4 is used for the transmission of data at rates of 9.6/4.8/2.4 kbps. Whether the specified data here are fax data or computer data from a file depends on the equipment used on both sides of the link. The data rate used in the system depends on the capability of the mobile station. For each data rate, a different coding scheme is used, which requires, of course, appropriate software capabilities within the mobile stations. Currently, there are developments for the implementation of a GSM transceiver inside a laptop computer for data or fax transmission. Such a transceiver may plug into a laptop's PCMCIA port, and the transceivers would use this type of traffic channel.
- TCH/H4.8/2.4 transmits data on a half-rate traffic channel. The implementation of this service depends on the availability of the half-rate traffic channel.

Control channels do not move users' voice data, fax data, or even computer data. Control channels move the data that the network and the radios need in order to make sure all the traffic moves about in the system reliably and efficiently. Depending on their tasks, there are four different classes of control channels: (1) *broadcast channels*, (2) *common control channels*, (3) *dedicated control channels*, and (4) *associated control channels*.

The *broadcast channels* (BCH) are transmitted only by the base station and are intended to provide sufficient information to the mobile station for it to synchronize with the network. Mobiles never transmit a BCH. There are three types of BCH:

- The *broadcast control channel* (BCCH) informs the mobile station about specific system parameters it needs to identify the network or to gain access to the network. These parameters are, among others, the location area code (LAC), the MNC (to identify the operator), the information on which frequencies the neighboring cells may be found, different cell options, and access parameters.
- The *frequency-correction channel* (FCCH) provides the mobile station with the frequency reference of the system. This logical channel is mapped onto the frequency-correction burst. The FCCH only appears on the frequency-correction burst, and the frequency-correction burst only contains the FCCH. Here is an example of where the physical channel, the logical channel, and the burst type may become blurred, and where readers should exercise caution in their study.
- The *synchronization channel* (SCH) supplies the mobile station with the key (training sequence) it needs to be able to demodulate the information coming from the base station. The SCH is mapped onto the synchronization burst.

The *common control channels* (CCCH) support the establishment of a dedicated link between a mobile and a base station. These channels provide the tools to perform the channel (or call) setups and can originate from the network or mobile. There are three types of CCCH:

- The *random access channel* (RACH) is used by the mobile station to request a dedicated channel from the network. The base station never uses the RACH. The RACH is mapped onto the random access burst and contains the first message sent to the base station. A measurement on the delay of the mobile station still has to be performed before the link can be permanently established.
- The base station calls individual mobile stations within its cell on the *paging channel* (PCH).
- The mobile station gets information from the base station on which dedicated channel it should use for its immediate needs from the *access grant channel* (AGCH). Along with this comes information about the timing advance that should be set. The message in the AGCH is a response from the base station to a mobile's RACH message. The principle procedures will be explained in the practical examples in Chapter 5.12.

The *dedicated control channels* (DCCH) are used for the message transfers between the network and the mobile station, not for traffic. They are also used for low-level signaling messages between the radios themselves. The network messages are needed for the registration procedure or for a call setup. The lower level signaling messages are used for channel maintenance. The network sometimes has to be involved with channel maintenance tasks.

- The *standalone dedicated control channel* (SDCCH) is intended for the transfer of signaling information between a mobile and a base station.
- The *slow associated control channel* (SACCH) is always used in association with either a traffic channel or SDCCH. If a base station assigns a traffic channel, there will always be an SACCH assigned with the traffic channel. The same holds true for an SDCCH. The purpose of the SACCH is channel maintenance. The SACCH carries control and measurement parameters or routine data needed to maintain a link between the mobile and the base station. The radios carry on their own conversations with each other on the SACCH.

In the *downlink* direction, a base station transmits a reduced set of system information parameters to keep the mobile up to date on the latest changes in the system. These are similar to the data transmitted on the BCCH, with some additional control parameters to command the mobile station to use a specific timing advance value or another power level.

In the *uplink* path, the mobile station reports the results of the measurements performed on the neighboring cells. These are put into a measurement report and sent to the network, and they assist the network in the handover decisions. The mobile station also reminds the network which timing advance and power level settings it is using.

- The *fast associated control channel* (FACCH) can carry the same information as the SDCCH. The difference is that the SDCCH exists on its own, whereas the FACCH replaces all or part of a traffic channel. If during a call there is need for some heavy-duty signaling (e.g., the mobile station passes from one cell to another cell), then the FACCH appears in the place of the traffic channel so that a handover can be accomplished. The FACCH is used to transmit these more lengthy signaling messages. To do this, the FACCH substitutes itself for the traffic channel, it steals bursts from it and indicates its presence with the stealing flags.

There are other common abbreviations for the channels mentioned here. In accordance with the ISDN, a TCH/FS, including the data channels, is also referred to as a *bearer-mobile channel* (Bm). A TCH/HS, again including the data channel, is called a *low-mobile channel* (Lm). A signaling channel is called a *data-mobile channel* (Dm). The suffix *m* is confined to GSM and usually refers to specific characteristics implemented for the mobile or *modified* environment, and which differ from the ISDN implementation for these channels. The *m* is also found in the abbreviation for the air-interface (Um).

Table 5.4 gives a summary of the different logical channels and the transmitting directions they use in the GSM system.

5.11 FRAME STRUCTURES

The GSM freight train is popular, has lots of different kinds of freight to move, and runs an extraordinarily tight schedule. Trains must arrive at and then depart from

Table 5.4
Summary of the Different Logical Channels and Their Directions

Logical Channel	*Transmitting Direction*
TCH	MS ↔ BTS
FACCH	MS ↔ BTS
BCCH	MS ← BTS
FCCH	MS ← BTS
SCH	MS ← BTS
RACH	MS → BTS
PCH	MS ← BTS
AGCH	MS ← BTS
SDCCH	MS ↔ BTS
SACCH	MS ↔ BTS

the customer's loading docks precisely on time, or the whole system will collapse. To make sure everything is on time and that no freight is missed or delivered to the wrong place, everything is ordered into *frames*. This additional organization is a scheduling aid.

Information, both traffic and signaling, is ordered into frames before it is mapped onto time slots. Frames are carefully organized into structures that appear as certain channel combinations, one after another, so that receivers can recognize what kind of data should be present at any time with minimum delay and error.

5.11.1 Channel Combinations

As mentioned in Section 5.10, channel combinations are always used in certain combinations on physical channels. These combinations are described below. To make the description easier to understand, RACH, PCH, and AGCH have been grouped together as CCCH. This is valid, since they all deal with the assignment of a channel to the mobile station. The abbreviation for these combinations use roman numerals according to the specifications:

I	TCH/FS + FACCH/FS + SACCH/FS;
II	TCH/HS(0,1) + FACCH/HS(0,1) + SACCH/HS(0,1);
III	TCH/HS(0) + FACCH/HS(0) + SACCH/HS(0) + TCH/HS(1) + FACCH/HS(1) + SACCH/HS(1);
IV	FCCH + SCH + CCCH + BCCH;
V	FCCH + SCH + CCCH + BCCH + SDCCH/4 + SACCH/4;
VI	CCCH + BCCH;
VII	SDCCH/8 + SACCH/8.

Unfortunately, even with the knowledge of these *channel combinations*, it is not clear how these channels are mapped onto a physical channel. Each channel combination requires one single physical channel. The TDMA technique creates up to eight physical channels on one carrier. It is possible, therefore, to put different channel combinations onto one carrier; one combination for each assigned—repeating—time slot. How this is actually done will be shown later.

In a manner similar to the TDMA frame structure that allows time slots to be ordered on a carrier, there are also some *multiframe* structures made of a fixed number of TDMA frames that allow logical channels to be ordered into time slots. There is a big difference between the logical channels that carry speech data and those that carry signaling data. A *26-multiframe* structure is used for the traffic channel combinations, and a *51-multiframe* structure is used for the signaling combinations. If, in the following, the expression *frame* is mentioned, it only refers to a TDMA frame with no regard for which time slot it may occupy.

5.11.2 Traffic Channel Frame Structure (26-Multiframe)

TCH/FS (Combination I)

The first 12 frames (Figure 5.13) are used to transmit traffic data. They may also be used for data transmissions at 9.6, 4.8, or 2.4 kbps. One frame for the SACCH follows, and then another 12 frames for more traffic data. The last frame stays idle, and nothing is transmitted in it. This idle frame gives the mobile station time to perform other tasks such as measuring the signal strength of neighboring cells or its own cell. The total length of a 26-frame structure is 26 · 4.615 ms = 120 ms.

TCH/HS (Combination II or III)

For the transmission of *half-rate speech* channels, it is possible to pack two of them onto one 26-multiframe structure (Figure 5.14). Within the single frames, the two channels are transmitted in an alternating manner, one logical channel using every other TDMA frame. Frame number 25, in this case, is reserved for the SACCH of the second half-rate channel; each half-rate channel has its own SACCH. This structure is used if two half-rate channels are assigned at the same time, as would be the case for combination III. If only one half-rate channel is needed, then combination II applies. In combination II, every other frame is an idle frame, during which times the mobile station enjoys the possibility of performing other tasks.

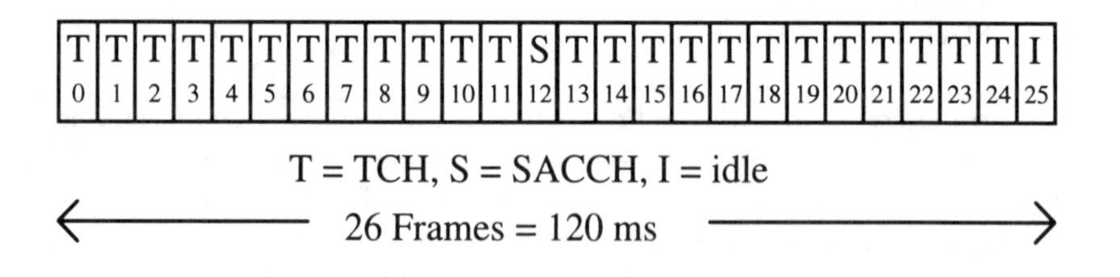

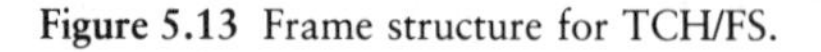
Figure 5.13 Frame structure for TCH/FS.

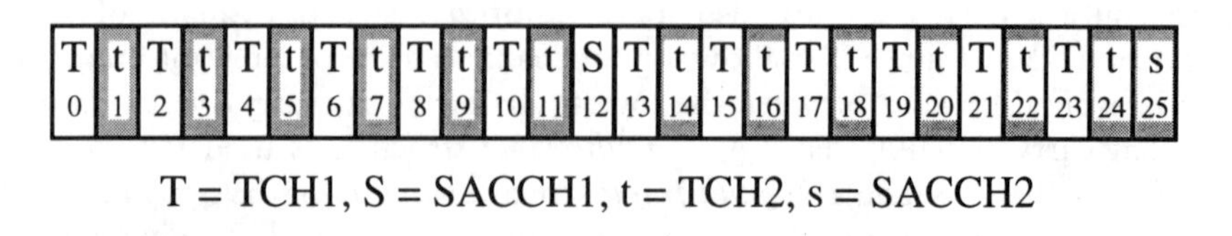

Figure 5.14 Frame structure for TCH/HS.

The FACCH was mentioned earlier as a signaling channel that holds a higher priority over routine traffic data. When the network needs to, it will substitute the FACCH for all or part of the traffic channel. This means that the FACCH takes the position of a traffic channel for as long as it is necessary, which will be for as long as the FACCH is required to transport important signaling data. The more signaling data, the longer the FACCH exists. The decoder in the target receiver sees that an FACCH substitution has been made by simply looking at the stealing flag.

5.11.3 Signaling Frame Structure (51-Multiframe)

Signaling frames do not carry user data. The *51-multiframe* structure is a little bit more complex than the 26-multiframe variety, since the former incorporates four different channel combinations, each of which requires a different structure.

FCCH + SCH + CCCH + BCCH (Combination IV)

This channel combination (Figure 5.15) is very similar in function to the *forward control channel* in analog networks. All channel types used in this combination occur in the direction BTS to mobile station, or mobile station to BTS. There are different structures for both cases, one for the *downlink* and another for the *uplink*. In the downlink direction, combination IV offers lots of space for the CCCH, which may be either a PCH (to call a mobile station) or an AGCH (to assign a channel to a mobile station). It does not matter which channel occupies which position in the 51-frame structure, it only depends on the signaling needs of the cell. The FCCH and the SCH are, however, always in consecutive frames. This rather simple construction makes it easier for a mobile station to synchronize—in frequency—with the base channel before attempting to synchronize with time and data. The uplink path is only used by mobile stations to transmit the RACH on random access bursts, which request other signaling channels from the base station.

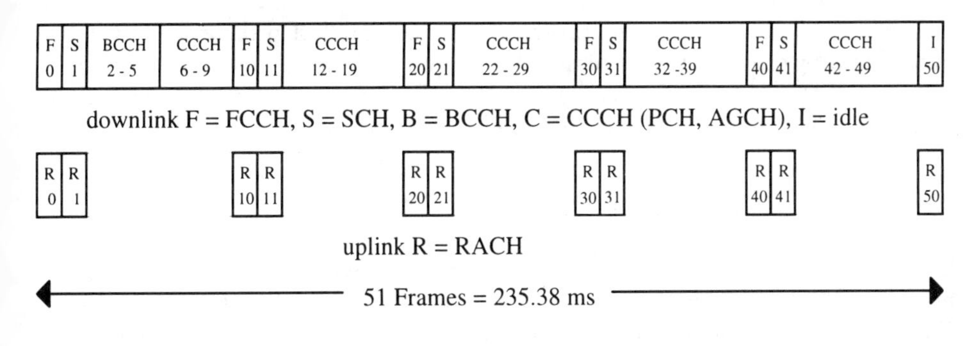

Figure 5.15 Frame structure for channel combination IV.

This channel combination is normally used for cells with several carriers and with a large amount of expected traffic on the CCCHs (paging, channel requests, and channel assignments). It may be assigned to a cell only once, since the FCCH and the SCH are peculiar to the base channel. It can be transmitted on any frequency available to the cell on time slot number 0. The frequency on which this combination is transmitted is used as a reference in the neighboring cells to mark it as an adjacent cell; that is, mobiles in neighboring cells perform their periodic measurements on this frequency during time slot 0.

FCCH + SCH + CCCH + BCCH + SDCCH/4 + SACCH/4 (Combination V)

This is the minimum combination for smaller cells with only one or two transceivers. The use of this combination follows the same rules as the use of combination IV (i.e., only once in a cell and always on time slot number 0). Combinations IV and V exclude each other. The expression SDCCH/4 + SACCH/4 means that it is possible to assign up to four DCCHs with their required associated channels. These are referred to as *subchannels;* that is, SDCCH(2) reads as "SDCCH subchannel two." Figure 5.16 shows the frame structure for this combination.

There are two things that catch one's eyes in this figure. One is that the positions for the SDCCH for corresponding subchannels are at a certain distance from each other: for the uplink direction it is 15 frames and for the downlink it is 36 frames. The intention behind this is to reduce the command response cycle to one multiframe. If, for example, the base station commands the mobile station to authenticate itself, the response can be sent only 15 frames later. The same applies for the other direction, with the only difference being that the network has more time allocated for a response. This need should be obvious, since the distances within the network for the signals are far greater. The same applies for channel combination VII, with eight SDCCHs.

The other fact we notice is that there are two multiframe structures drawn together in Figure 5.16. The reason for this is that an SACCH is only transmitted on every other version of the multiframe. One might think that there is a possibility of transmitting only two frames of the SACCH every multiframe, but the information is spread over four frames, and they logically belong to each other. There is half as much SACCH time as there is corresponding SDCCH time.

CCCH + BCCH (Combination VI)

If a base station manages a huge number of transceivers, it is probable that the number of CCCHs provided by combination IV is not enough to handle the work. To service a very large number of base stations, it is possible to assign additional control channels in combination VI. The assignment of combination VI only makes

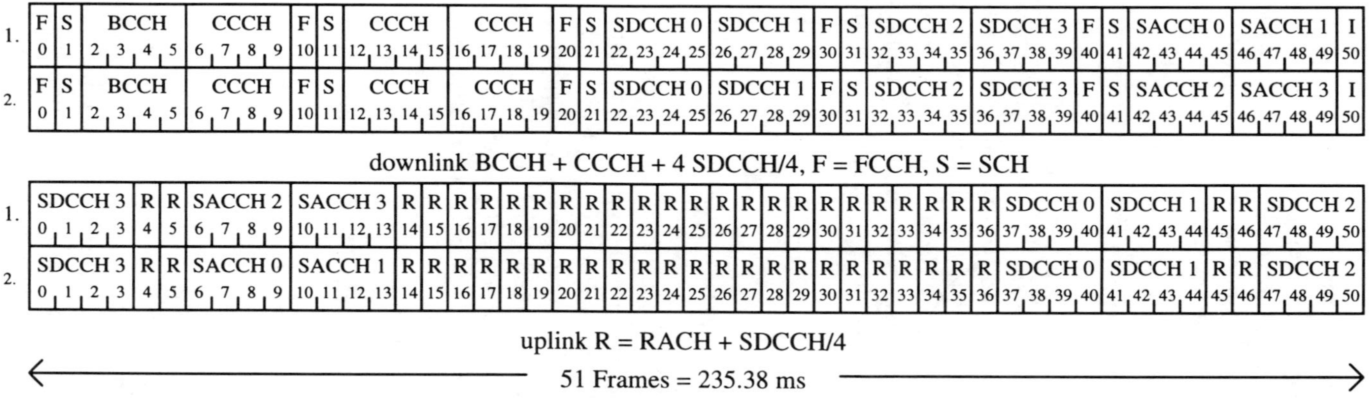

	0	1	2 3 4 5	6 7 8 9	10	11	12 13 14 15	16 17 18 19	20	21	22 23 24 25	26 27 28 29	30	31	32 33 34 35	36 37 38 39	40	41	42 43 44 45	46 47 48 49	50
1.	F	S	BCCH	CCCH	F	S	CCCH	CCCH	F	S	SDCCH 0	SDCCH 1	F	S	SDCCH 2	SDCCH 3	F	S	SACCH 0	SACCH 1	I
2.	F	S	BCCH	CCCH	F	S	CCCH	CCCH	F	S	SDCCH 0	SDCCH 1	F	S	SDCCH 2	SDCCH 3	F	S	SACCH 2	SACCH 3	I

downlink BCCH + CCCH + 4 SDCCH/4, F = FCCH, S = SCH

	0 1 2 3	4	5	6 7 8 9	10 11 12 13	14	15	16	17	18	19	20	21	22	23	24	25	26	27	28	29	30	31	32	33	34	35	36	37 38 39 40	41 42 43 44	45	46	47 48 49 50
1.	SDCCH 3	R	R	SACCH 2	SACCH 3	R	R	R	R	R	R	R	R	R	R	R	R	R	R	R	R	R	R	R	R	R	R	R	SDCCH 0	SDCCH 1	R	R	SDCCH 2
2.	SDCCH 3	R	R	SACCH 0	SACCH 1	R	R	R	R	R	R	R	R	R	R	R	R	R	R	R	R	R	R	R	R	R	R	R	SDCCH 0	SDCCH 1	R	R	SDCCH 2

uplink R = RACH + SDCCH/4

← 51 Frames = 235.38 ms →

Figure 5.16 Frame structure for channel combination V.

sense when it is combined with combination IV, since combination VI merely adds additional control capacity to the existing combination IV resources. While combination IV always occupies time slot 0, combination VI is assigned to time slot 2, 4, or 6. The combination VI multiframe structure is similar to combination IV. The only difference is that there are no FCCHs and SCHs in combination VI.

SDCCH/8 + SACCH/8 (Combination VII)

If a cell uses the combination IV signaling assignment (e.g., along with combination VI), it does not yet provide any signaling channel on which the mobiles perform such basic tasks as a call setup and registration. Combination VII is intended to provide routine signaling capability to the cell. The expression SDCCH/8 + SACCH/8 indicates that eight different DCCHs may be used with eight SACCH resources in this combination, and thus may serve for eight parallel signaling links on one physical channel. The frame structure for combination VII is shown in Figure 5.17.

5.11.4 Cell Broadcast Channel

The *cell broadcast channel* (CBCH) is not really a channel combination, but is rather an additional feature of the GSM system. The CBCH supports a part of the group of *short-message services* (SMS), sometimes also referred to as *point-to-omnipoint*. Through the CBCH, an operator has the ability to transmit messages to its subscribers. The broadcasts may be traffic information, bad weather alerts, or other items of general interest. When a cell provides this kind of service, the BCCH must supply instructions necessary to receive the CBCH. The CBCH is always mapped on the second subslot of the SDCCH independently of whether it is transmitted on channel combination V or VII. The CBCH is only transmitted in the downlink direction, and the mobile does not acknowledge any of the messages.

5.11.5 Combinations of the 26- and 51-Multiframes

It takes 26 TDMA frames to transmit all 26 parts of a 26-multiframe one time slot at a time. Similarly, it takes 51 TDMA frames to send all 51 parts of a 51-multiframe one time slot at a time.

Up to now, the traffic channel multiframe (26-multiframe structure) and the signaling data multiframe (51-multiframe structure) have been explored as structures by themselves, but it was not explained how both types are combined together on a single repeating TDMA frame (eight time slots per frame). How can a 51-multiframe be fed into time slot 0, and a 26-multiframe be fed into, say, time slot 5 at the same time? How can both of these structures coexist without some empty time slots

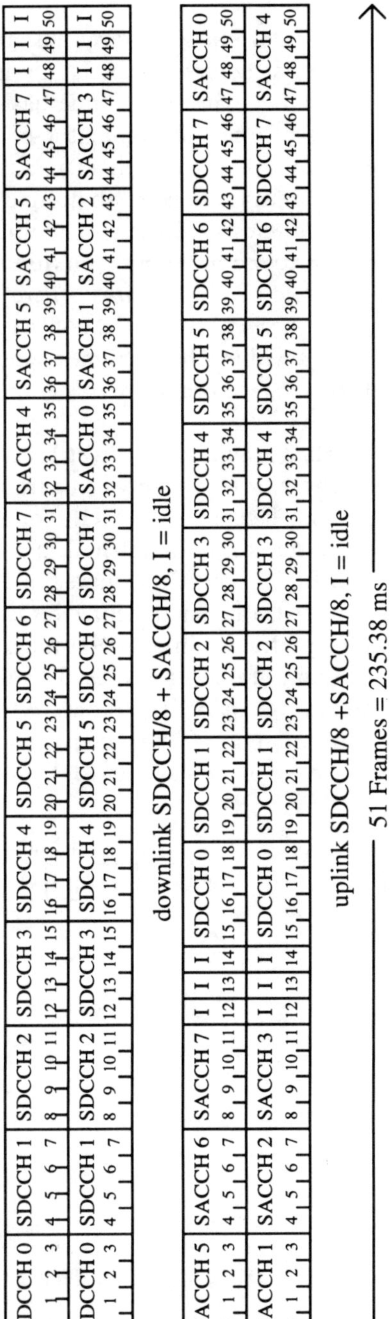

Downlink

	0–3	4–7	8–11	12–15	16–19	20–23	24–27	28–31	32–35	36–39	40–43	44–47	48	49	50
1.	SDCCH 0	SDCCH 1	SDCCH 2	SDCCH 3	SDCCH 4	SDCCH 5	SDCCH 6	SDCCH 7	SACCH 4	SACCH 5	SACCH 5	SACCH 7	I	I	I
2.	SDCCH 0	SDCCH 1	SDCCH 2	SDCCH 3	SDCCH 4	SDCCH 5	SDCCH 6	SDCCH 7	SACCH 0	SACCH 1	SACCH 2	SACCH 3	I	I	I

downlink SDCCH/8 + SACCH/8, I = idle

Uplink

	0–3	4–7	8–11	12	13	14	15–18	19–22	23–26	27–30	31–34	35–38	39–42	43–46	47–50
1.	SACCH 5	SACCH 6	SACCH 7	I	I	I	SDCCH 0	SDCCH 1	SDCCH 2	SDCCH 3	SDCCH 4	SDCCH 5	SDCCH 6	SDCCH 7	SACCH 0
2.	SACCH 1	SACCH 2	SACCH 3	I	I	I	SDCCH 0	SDCCH 1	SDCCH 2	SDCCH 3	SDCCH 4	SDCCH 5	SDCCH 6	SDCCH 7	SACCH 4

uplink SDCCH/8 +SACCH/8, I = idle

← 51 Frames = 235.38 ms →

Figure 5.17 SDCCH/8 + SACCH/8.

appearing somewhere? The task does not seem easy, because 26 does not divide evenly into 51. To combine both structures onto the air interface, a new frame format is introduced: the *superframe*. The superframe has a length of 51 · 26 = 1,326 frames, which is the least common denominator of both, and is the least number of TDMA frames (eight time slots each) that can absorb the contents of all available 26- and 51-multiframes and still finish with no spare time slots remaining. One superframe can accommodate either 26 each 51-multiframes or 51 each 26-multiframes. Look at Figure 5.18 to see how these multiframes fit together.

To complete the figure we add the *hyperframe*. There is nothing particularly special about the hyperframe; it just consists of 2,048 superframes. The system sometimes refers to frame numbers within a hyperframe context, and the hyperframe represents the most comprehensive structure in the system and lasts for nearly 3.5 hours before it is repeated.

When describing the signaling frame structure, it becomes apparent that it is important to know exactly which frame is currently being transmitted, since it would make a difference whether a mobile listens to SDCCH(0) or SDCCH(2). To remove the possibility of ambiguity, the frames are numbered in a special way: there are three counters, and we call them T1, T2, and T3. Counter T1 counts the superframes. Whenever a superframe is completed, T1 is incremented by 1. T1 has values between 0 and 2,047; there are 2,048 superframes in a hyperframe. T2 counts the speech frames, which only occur in 26-multiframe structures. T2's value, therefore, ranges from 0 to 25. Finally, T3 counts the signaling frames, which are 51-multiframe structures. Similarly to the traffic counter, T3's contents can be anything from 0 to 50. At some starting time, all three counters are set to 0, and then the frames start

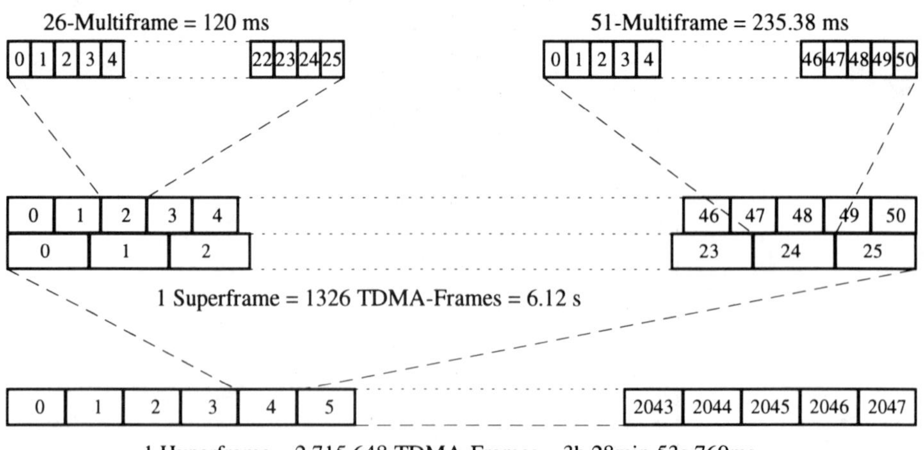

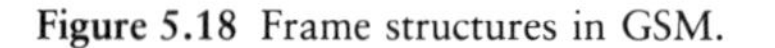
Figure 5.18 Frame structures in GSM.

to be transmitted. Whenever a speech or a signaling multiframe structure is finished, its respective counters (T2 and T3) are reset to 0 and start again. After 1,326 TDMA frames, both T2 and T3 are finally reset together and start counting again from 0 at that time. This marks the duration of one superframe. When the first superframe is completed, T1 increments by 1 count. T1 only resets after 2,047 counts, which takes well over 3 hours to do, and this is the duration of a hyperframe. If one knows the values in the T1, T2, and T3 counters, then one knows exactly what is in each and every time slot at that instant, providing one knows what kind of multiframe was assigned to each of the eight available time slots in the TDMA frame.

A shortened frame number consists of the current T2 and T3 values, and this shortened frame identification number is transmitted in the SCH to provide the mobile station with an initial indication of the prevailing frame structure at the time. With only these two numbers (T2 and T3), it is an easy task for the mobile to look for the BCCH and the system information. The mobile knows that the information is transmitted when timer T3 holds, for example, the values 2 through 5 (see Figure 5.15). It is very important for a mobile station to have this information about the exact timing on the frame structures. From this knowledge, it knows when to listen for something and when to make appropriate transmissions. Think of the T2 and T3 counters as a pocket-sized schedule for the GSM train system.

It is often difficult to understand how the transition from a signaling channel to a traffic channel works when they are located in different frame structures. How can a mobile, which has a fixed time slot assignment, recover its traffic data and still hear any signaling data addressed to it? One should remember that the 26-multiframe or 51-multiframe structures apply to a single time slot within one TDMA frame. If a mobile station is commanded to stop listening to the signaling structure in time slot 4 and, instead, to recover the traffic channel in time slot 2, it will do so, but then it also has to take the different structures in these two time slots into consideration. The timers are always running. Since they are, in fact, always running, it is easy to move from one frame structure to another one. Figure 5.19 illustrates an example of the mapping of the different structures onto single distinctive time slots. In this example, three time slots have been assigned for the mobile's review: 0 to the BCCH (combination IV), 2 to the traffic channel (combination I), and 5 to the SDCCH/8 (combination VII). Time slots 0 and 5 hold the 51-multiframe structure, whereas time slot 2 holds the 26-multiframe structure. These two structures are mapped together onto a superframe structure in which the different time slots still retain their own original identities, as shown in the enlargement.

5.12 EXAMPLES OF HOW A MOBILE STATION BEHAVES

A little theory is necessary before we can even hope to understand how a GSM mobile behaves on the air interface within a network, and the authors have labored

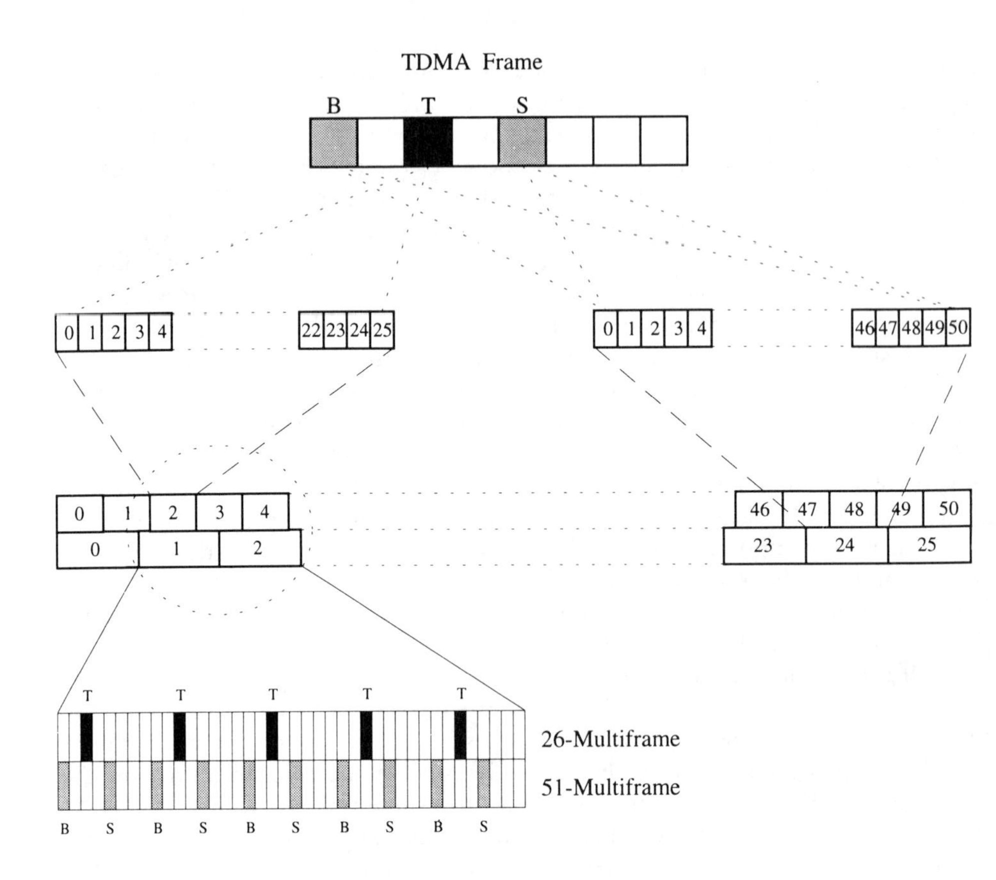

Figure 5.19 Time slot allocation with mapping onto different channel structures.

mightily in the first 11 sections of this chapter to make the details of the physical layer intuitive. To fix your understanding in your mind, we offer some practical examples of mobile behavior, as if observing a strange metalo-plastic animal in a galactic zoo. Of particular interest to radio watchers are three behaviors: (1) synchronizing with the network, (2) location updating, and (3) call establishment (MTC).

5.12.1 Synchronization With the Network

When a mobile station is turned on, it has to orient itself within the network. The mobile does this in three steps. First, it synchronizes itself in frequency, then in time. Finally, it reads the system and cell data from the base channel or, more specifically, from the BCCH. This procedure is purely passive; no messages are exchanged.

The first task is to find the frequency where the FCCH, SCH, and BCCH are being transmitted. In the GSM system, a base station must transmit something in each time slot of the base channel. Even if these time slots are not allocated to communication with any mobiles, the base station has to transmit predefined *dummy bursts*, especially defined for this purpose, in the idle time slots of the base channel. If the base station, taxed with the responsibility of broadcasting the base channel, fills all its time slots, then the power density for this frequency is higher than that for any of the other channels in the cell, which may have only a few time slots out of eight allocated. This peculiarity of the base channel makes it easy for a mobile to find its frequency. The mobile simply scans for the physical channels with the highest apparent power levels. After finding one of them, the mobile searches for the FCCH. The FCCH is easy to find once the base channel is located. If the reader observes a base channel's emissions on a spectrum analyzer, he or she would first see a "bump" of what seems to be noise, the base channel's data, filling approximately the full 270-kHz bandwidth of the channel. Closer inspection of the display would disclose the presence of a little "pip" of energy rising slightly above the noise of the base channel, where the pip is about 67 kHz above the center of the emissions filling the base channel's bandwidth. The pip is the FCCH (see Figure 5.20). If the FCCH

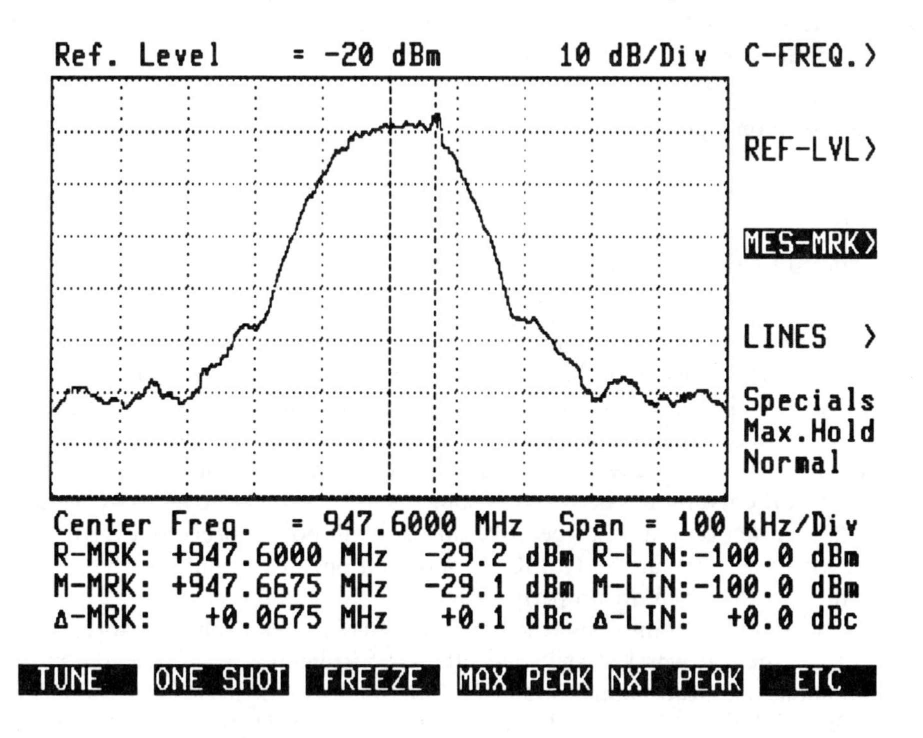

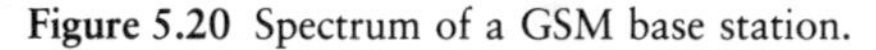

Figure 5.20 Spectrum of a GSM base station.

is not present, then a broadcast channel has not been found and the mobile tunes to and investigates the channel with the next highest power level. This is repeated until an FCCH is finally found.

After the mobile synchronizes with the system in the frequency domain, it proceeds to do the same in the time, or data, domain. The mobile uses the SCH for this second step, but it has already found the FCCH, so it already knows that the SCH will follow in the next TDMA frame (see Figure 5.15). From the SCH, the mobile gets information about the current frame number and the cell's training sequence.

With this information displayed on the SCH, the BCCH is an open book for the mobile station, and it reads about the location of the cell, any cell options of interest, and how to access this particular base station. Up until this instant, the mobile is listening busily but passively. All three of the synchronization steps take somewhere between 2 and 5 seconds to accomplish, and can (under certain circumstances) take up to 20 seconds. The acquisition time depends on the mobile station's design, its type, and whether the mobile station was last switched off within this cell (rather than somewhere else). When a mobile is switched off, it stores some information about the cell on its SIM card. This information includes the frequency of the base channel and the location of the cell. If the mobile is turned back on within the same cell, it already knows where to look for the base channel; the synchronization process is supposed to be much faster.

5.12.2 Location Updating

Two different conditions cause the mobile station to perform a *location updating procedure*:

- Forced by the network;
- Movement to a new location area.

A network can force a mobile station to perform a location update when it is switched on. This is accomplished with a flag set in the system information transmitted on the BCCH. If all mobile stations have to register themselves after being turned on, then the network has exact knowledge of which mobile stations are currently active, as well as in which cell they can be found. This is an important way to reduce signaling traffic within the network. If someone from the public network wants to call a mobile station that is turned off, this call will not be processed further, but is blocked right away at the GMSC. Mobile stations always inform the network when they are switched off properly.

If the mobile station is switched on in a different area from that stored on the SIM card (where it was last switched off), or if it enters a new area, the mobile station initiates a location updating procedure to inform the network about its new location, which the network needs if a call has to be routed from the public network to the mobile station.

The principle of location updating is illustrated in Figure 5.21 along with the logical channels that are used during the procedure. Before the location update messages can be exchanged, the mobile has to request a signaling channel on which to exchange the messages. The mobile starts its channel request with a RACH, which it places on a random access burst. After it sends the burst, the mobile listens to the AGCHs from the base. If there is no response within a certain period of time, the random access burst is repeated.

Upon receipt of the AGCH (in which there is a description of a dedicated channel to go to), the mobile moves onto the new channel, which is now a dedicated channel between the mobile and the base.

On the new channel—the SDCCH—the mobile station tells the network that it wishes to perform a location update. Before the network processes this humble request any further, it demands that the authentication procedure be performed. If the authentication is okay, the network assigns the new location area and makes note of the mobile's new location as it enters this information into the relevant registers, namely the VLR and the HLR. If necessary, the network assigns a temporary identity (TMSI) to the mobile, or it renews the old one. Now that the location update procedure is performed, the signaling channel is no longer needed, and the dedicated SDCCH is released for others to use.

5.12.3 Call Establishment (Mobile-Terminated Call)

The principle of a *mobile-originated call establishment* was shown in Chapter 3. For a little variety, the other direction is examined in Figure 5.22. If a mobile station

log. Channel	Mobile Station Base Station	
RACH	⟶	Channel request
AGCH	⟵	Channel assignment
SDCCH	⟶	Request for location updating This is already transmitted on the assigned channel
SDCCH	⟵	Authentication request from the network
SDCCH	⟶	Authentication response from the mobile station
SDCCH	⟵	Request to transmit in the ciphered mode
SDCCH	⟶	Acknowledgment of the ciphered mode
SDCCH	⟵	Confirmation of the location updating, including the the optional assignment of a temporary identity (TMSI)
SDCCH	⟶	Acknowledgment of the new location and the temporary identity
SDCCH	⟵	Channel release from the network

Figure 5.21 Principle of location updating.

log. Channel	Mobile Station Base Station	
PCH	←	Paging of the mobile station
RACH	→	Channel request
AGCH	←	Channel assignment
SDCCH	→	Answer to the paging from the network This is already transmitted on the assigned channel
SDCCH	←	Authentication request from the network
SDCCH	→	Authentication response from the mobile station
SDCCH	←	Request to transmit in the ciphered mode
SDCCH	→	Acknowledgment of the ciphered mode
SDCCH	←	Setup message for the incoming call
SDCCH	→	Confirmation
SDCCH	←	Assignment of a traffic channel
FACCH	→	Acknowledgment of the traffic channel
FACCH	→	Alerting (now the caller gets the ringing sound)
FACCH	→	Connect message when the mobile is off-hook
FACCH	←	Acceptance of the connect message
TCH	↔	Exchange of user data (speech)

Figure 5.22 Principle of call establishment—mobile-terminated.

is switched on and already updated, it is in a state we call *idle updated*. In this state, the mobile passively monitors the BCCH and the CCCH, which is the PCH. If the mobile is called from the public network, the base station will issue a paging message on the PCH to which a channel request from the mobile is the appropriate response. From now on, the MTC procedure follows nearly the same rules as already described for location updating. One of the differences is the first message on the assigned SDCCH signaling channel, which in the MTC case is the "answer to a page" message. Then some more messages follow on the SDCCH to set up the call until a traffic channel is finally assigned. From the instant the mobile and the base switch to the traffic channel, the remaining signaling messages are transmitted on an FACCH. Since the signaling is not finished yet and speech data are not yet transmitted, the FACCH is not yet displacing any traffic data, as the FACCH often has to do. When the call is finally connected, no further dedicated signaling messages need to be exchanged, and the traffic channel assumes the routine purpose for which it is intended.

With this description of the call establishment procedure, one advantage of the dedicated signaling channels (SDCCH) should be apparent. Since one traffic channel and its associated control channels (FACCH and SACCH) use up a whole time slot, it is possible to put four (combination V) or even eight (combination VII) different mobile stations onto one time slot to handle all their signaling needs. This means that for pure signaling traffic, the capacity of the system is even higher than it would be if mobiles were assigned dedicated time slots for everything they do. This is also the reason why the increased effort, complexity, and overhead associated with all the channel structures does not hurt the system, but helps make GSM even more efficient.

5.13 SPEECH CODING IN GSM

The most important service offered to the user of a cellular mobile network is speech transmission. The teleservice voice telephony is the main revenue generator for cellular operators and justifies the enormous efforts and investments that are necessary to install and maintain such networks.

The general technical requirement is simple: transmit voice signals at an acceptable level of quality. In conventional analog radio systems, the continuous low-frequency voice signal, also referred to as the *baseband signal*, modulates (modifies) a radio frequency carrier. The information (the voice signal) can be found in the variations in the (1) amplitude, (2) phase, or (3) frequency of the transmitted radio signal. The relevant techniques used to modify the carrier are called *amplitude modulation* (AM), *phase modulation* (PM), and *frequency modulation* (FM). See Section 5.15 of this chapter for a more rigorous examination of the particular types of modulation. In the receiver, the modulation process is reversed so that the original continuous baseband signal is retrieved. Disturbances in the radio channel add their own influence to the deliberate variations of one or more of the characteristics of the radio carrier. The additional modulation caused by the channel (see Section 5.2) yields audible effects, which can distract or annoy the user.

Now, what is the situation in a digital transmission system? In the earlier parts of this chapter, we learned that the GSM system uses a TDMA scheme and thus does not continuously transmit information on one selected channel. The information is transmitted within pulses, so that the content, the representation of the originally continuous audio signal, is compressed in the time domain when it is transmitted over the radio path. Inside the receiver, the information content is decompressed, or expanded, in order to regenerate the continuous audio signal.

The distinct logical states in a digital system, the ones and zeros (the bits), are transmitted with some special modulation techniques instead of a continuously changing aspect (modulation) of the carrier. We need a binary representation of the voice signal. Due to the restricted transmission capacity on the radio channel (Section 5.2), it is desirable to minimize the number of bits we need to transmit. The device that transforms the human voice into a digital stream of data suitable for transmission over the radio interface and regenerates an audible analog representation of the received data (voice) is called a *speech codec* (speech transcoder, speech coder/decoder). The speech codec is part of every mobile station designated for voice transmission. There are mobile stations that may not transmit speech, but instead transmit data services or facsimile information. As a counterpart to the mobile station in the network, the speech transcoding function is allocated to the BTS. Even though the speech transcoding function is the responsibility of the BTS, the speech codecs may be physically placed in the BSC. It is common for the various functions of the BTS and the BSC to be freely placed in one or the other depending on the manufacturer's decision.

The full-rate speech codec, used in GSM, uses the full complement of channel time allotted to a mobile, one time slot in each TDMA frame. It was developed after a call for proposals. Candidate codecs were submitted, and a selection was made from six different proposals. The winner was a codec named *RPE-LTP*, which stands for *regular pulse excitation* and *long-term prediction*. There will, eventually, be a standardized half-rate speech codec. The half-rate codec will cut in half the amount of data we need to adequately represent human speech sounds, and so allow twice as many users to share the same time slot in a TDMA frame. Two half-rate mobile stations can use the same resource a single full-rate mobile uses by simply taking alternate turns transmitting in the assigned time slot; each half-rate mobile uses every other TDMA frame (see Figure 5.14).

The remainder of this section offers a general view of the principles of speech processing in GSM. This is an extraordinarily rich and complex subject with a vast universe of techniques and possibilities. A full and detailed explanation of the subject is well beyond the scope of this book. The more ambitious reader should refer to the *GSM Technical Specifications* [2–5], which contain the applicable definitions and the most comprehensive description on the subject.

5.13.1 Requirements for Speech Transcoding in GSM

The simple and direct conversion of an analog signal into a digital representation as a sequence of bit symbols, and the reverse process, is an easy and routine task these days. All sorts of analog-to-digital converters (ADC) and digital-to-analog converters (DAC) are available at low cost, and their implementation, within normal technical ranges and applications, is a skill that no longer intimidates engineers. But these simple digitizing tasks are not the only ones employed to transform analog baseband signals into digital ones and back to analog again. As one of the key functions in a digital radio system, speech transcoding in GSM must meet the following additional requirements:

- The *redundancy* inherent in the baseband signal of the human voice shall be significantly reduced. There is a great deal of redundancy in the sounds of human language, and if most of this is removed, there is plenty of time remaining for others to use the channel. The remainder of the voice encoding process leaves us with the minimum information content that is necessary to reconstruct the signal in a receiver. The restricted capacity for carrying data (information) that can be allocated to a single channel in cellular radio makes this reduction essential. More data would mean more bandwidth or higher order modulation schemes which are not robust (Section 5.15).

- The quality of *speech transmission* under good radio conditions (i.e., those conditions during which there are no disturbances that would produce errors in the bit transmission and reception) shall be at least equal to the quality expected in conventional cellular radio systems under the same conditions.
- The pauses in the normal flow of conversations shall be detected in order to suspend (optionally) the radio transmissions during such periods. This feature will reduce air traffic, reduce interference among the cells, and extend the battery life in hand-portable phones. This function is called *discontinuous transmission* (DTX).

5.13.2 How Speech Transcoding Works in GSM

The block diagram in Figure 5.23 shows all the components required in the process of speech transcoding.

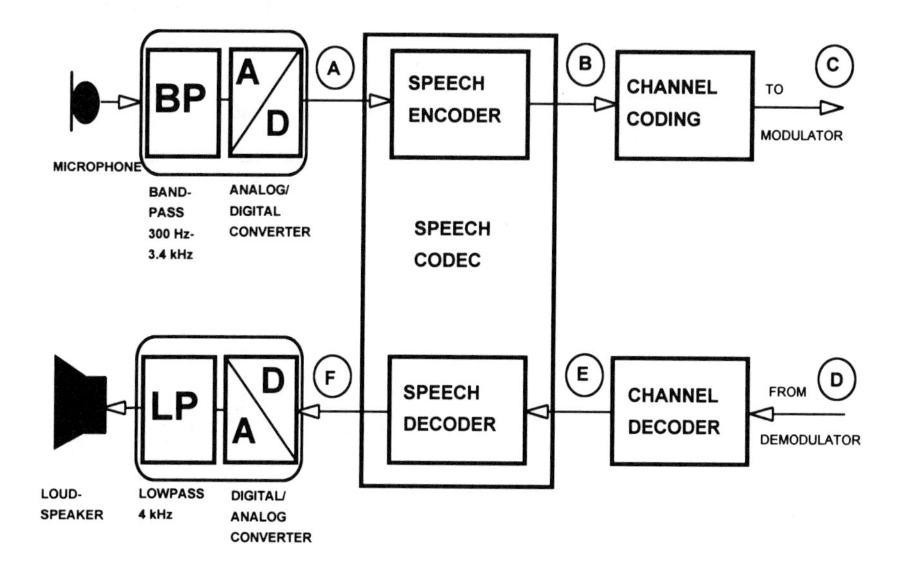

Figure 5.23 Audio signal processing.

Sound (human voice) is converted to an electrical signal by the microphone. To *digitize* this analog signal, it has to be sampled. If we convert the signal into data right away, we will force the ADC to do more work than is really necessary. To reduce the work, the signal is filtered, so that it only contains frequency components below 4 kHz. Baseband voice signals in telephony are restricted to the minimum bandwidth (300 Hz to 3.4 kHz) sufficient for the unambiguous and distinct recognition of a voice. We sample the signal after filtering. Every 125 μs, a value is sampled from the analog signal and quantized by a 13-bit word. The 125-μs sampling interval is derived from a sampling frequency of 8 kHz, which is 8,000 samples per second. The preliminary process of sampling and analog-to-digital conversion is shown in Figure 5.24, which shows a crude three-bit quantization of a sampled signal, which allows only $2^3 = 8$ different quantization levels.

The analog audio signal is limited to frequency components below 4 kHz. Why do we need a sampling signal of 8 kHz? Why not 4 kHz? The sampling theorem states that the sampling rate has to be at least twice the maximum signal frequency in order to reconstruct the original signal with minimal distortion. A standard sampling frequency is 8 kHz for audio signals in telecommunications.

A signal quantized by 13 bits is represented by $2^{13} = 8{,}192$ quantization levels. A sampling rate of 8,000 samples per second means that at the output of the ADC (position A in Figure 5.23), the converter delivers a data rate of $8{,}000 \cdot 13$ bps = 104 kbps. This interface in the speech-coding process is also called the *digital audio interface* (DAI). In the case of a speech transcoder application in the network side (speech transcoder in the BTS or BSC) it is possible, and even common practice

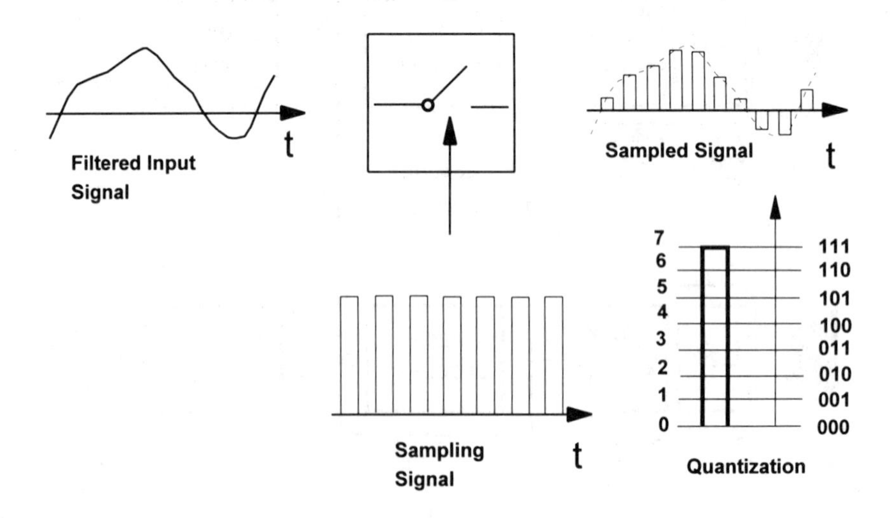

Figure 5.24 Analog-to-digital conversion by sampling and quantizing.

(ISDN), that the speech data are already represented by a digital bit stream, so that a dedicated analog-to-digital conversion is not necessary.

Now, 104 kbps is a data rate far too high to be economically transmitted over the radio interface. The speech coder will have to do something to significantly reduce this rate by extracting irrelevant components in the data stream at the DAI. The speech coder has to search for excess baggage we can safely remove from the bit stream scheduled for transport over the radio path. GSM uses two processes to strip redundant fat from the data representing voice traffic.

5.13.2.1 Linear Predictive Coding and Regular Pulse Excitation Analysis

Every 20 ms, 160 sampled values from the ADC are taken and stored in an intermediate memory. An analysis of a set of data samples produces eight filter coefficients and an excitation signal for a time-invariant digital filter. This filter can be regarded as a digital imitation of the human vocal tract, where the filter coefficients represent vocal modifiers (e.g., teeth, tongue, pharynx), and the excitation signal represents the sound (e.g., pitch, loudness) or the absence of sound that we pass through the vocal tract (filter). A correct setting of filter coefficients and an appropriate excitation signal yields a sound typical of the human voice.

This procedure, so far, has not performed any data reductions. The reductions come in further steps, which take advantage of certain attributes of the human ear and vocal tract (e.g., that sound levels are perceived on a logarithmic scale). The 160 samples, transformed into filter coefficients, are divided into four blocks of 40 samples each. Each block represents a 5-ms period of voice. These blocks are sorted into four sequences, where each sequence contains every fourth sample from the original 160 samples. Sequence number 1 contains samples 1, 5, 9, 13, . . ., 37, sequence number 2 contains samples 2, 6, 10, 14, . . ., 38, sequence number 3 contains samples 3, 7, 11, 15, . . ., 39, and sequence number 4 contains samples 4, 8, 12, 16, . . ., 40. The first reduction in data comes when the speech encoder selects the sequence with the most energy.

This linear predictive coding (LPC) and regular pulse excitation (RPE) analysis has a very short memory of approximately 1 ms. A more long-term consideration of neighboring (or adjacent) blocks in time is not performed here. There are numerous correlations in the human voice, especially in long vowels such as the *a* in *car* or the *oo* in *fool*, where the same sound recurs in succeeding 5-ms samples. Taking the similarity of sounds between adjacent samples (adjacent 5-ms blocks) into account can significantly reduce the amount of data required to describe the human voice. This second reduction task is performed by a LTP function.

5.13.2.2 Long-Term Prediction Analysis

The LTP function accepts a sequence selected by the LPC/RPE analysis. Upon accepting a sequence, it then looks among all the previous sequences passed to it (which still reside in another intermediate memory for 15 ms) for the earlier sequence that has the highest correlation to (bears the greatest resemblance to) the current sequence. We can also say the LTP function looks for the one sequence from among those already received that is most similar to the sequence just received from the LPC/RPE. Now it is only necessary to transmit a value representing the difference between the two sequences, along with a pointer to tell the receiver on the other end of the radio channel which sequence it should select among its recently received sequences for comparison. The receiver knows which differential values it has to apply to which sequences. The transmission of the whole sequence is not necessary, only the difference between sequences. This further reduces the data on the channel.

The speech coder issues a block of 260 bits (a *speech frame*) once every 20 ms (position B in Figure 5.23). This corresponds to a net data rate of 13 kbps, a data reduction of a factor of eight. Speech transcoding is a task that requires a large number of calculations at high speeds. It is, therefore, an ideal application for digital signal processing (DSP) techniques.

The data coming from the speech codec are *channel-coded* (see Section 5.14) before they are forwarded to the modulator in the transmitter. The channel coder, curiously, adds some redundancy back into the data stream, but does so in a very careful and orderly way so that the receiver on the other end of a noisy channel can correct bit errors caused by the channel. The receiver needs the extra bits the channel coder adds to perform this interesting function, which is explained in Section 5.14. Channel coding almost doubles the data rate to 22.8 kbps.

The bits in a speech frame contain enough information to enable the speech decoder in the receiver to reconstruct the speech signal from the data obtained from the channel decoder (position E in Figure 5.23).

Before channel coding, the 260 bits in each speech frame are sorted into different classes according to their function and importance. The different classes receive different attention from the channel coder so that the excess redundancy added back into the data stream is efficient and useful. There are three classes of speech bits. The most important bits are the class Ia bits, which are 50 bits out of the 260 in each speech frame. The Ia bits describe the filter coefficients, block amplitudes, and the LTP parameters. Since these parameters have to be regarded as the most important ones, the class Ia bits receive the greatest attention by the protection given to them in the channel-coding process. Next in importance are the class Ib bits (132 bits out of the 260 speech frame bits), which consist of RPE pointers, RPE pulses, and some other LTP parameters. Least important, but nice to have, are the class II bits (78 out of the 260 bits), which contain the RPE pulse and filter parameters. The class II bits do not receive any protection from the channel-coding process.

The 104-kbps bit stream, at position F in Figure 5.23, represents a reproduced digital speech signal. This point in the speech-coding process (position F) is part of the DAI. Finally, the DAC transforms the 104-kbps bit stream into an analog signal.

The lowpass filter just before the loudspeaker is necessary for the suppression of sampling effects.

Transmission errors may be compensated for to a certain degree in the speech decoder by interpolation. Interpolation, in this context, means that filter coefficients are kept in a memory and used when new ones are not available. Uncorrected bit errors found among class Ia bits cause the whole speech frame to be discarded. The discarded and missing frame is extrapolated from the last valid speech frame. Short-term disturbances can be bridged by the speech codec. The preceding channel decoding process (error correction mechanism) further enhances the system's resistance to reception problems, so that a degradation in quality will not be noticed by the user.

5.13.2.3 Discontinuous Transmission

As mentioned earlier, another requested feature of the speech transcoder is the detection of pauses in speech. When a pause is detected, we discontinue or suspend radio transmissions for the duration of the pause. The use of this feature is a network option. The DTX option tends to reduce interference in adjacent cells and to mobile stations close to the base site. Since transmit time is further reduced when DTX is used, the power consumption of hand-held terminals is reduced, which gives users the option of fitting their terminals with smaller batteries. Pauses in normal speech occur at a rate that makes speech appear to have about a 50% duty cycle. This means that a telephony channel is only used for speech transmission about half the time a speaker is using the phone.

The possibility of invoking DTX functions have extended the original speech codec specifications to include two additional features:

- *Voice activity detection* (VAD) to determine the presence or absence of speech at the microphone. This is not as easy to implement as it may sound, for it has to work well even when there is a high level of background noise, such as in a car.
- The total absence of sound in the ear piece would annoy the user at the receiving end of a radio channel; the handset appears to be dead, and users tend to speak too loudly when there is total silence in the ear piece. There needs to be a minimum of conventional background noise present during pauses, and this minimal background noise is sometimes descriptively called *presence*. This is accomplished by transmitting *silence descriptor* (SID) frames at a rather slow rate of once every 480 ms. Upon receiving the SID frame, the receiving speech decoder has to fake an existing wire-line connection by generating some background noise. The noise is called *comfort noise*, and comfort noise gives the system "presence."

As an aid in review, Figure 5.25 shows the logical blocks involved in the encoding, transmission, and the decoding of speech in GSM.

5.14 CHANNEL CODING

Even though there are many happy customers, and business is growing, all is not well with the GSM freight train company. The tracks are a mess, some bridges are washed out on certain routes, frequent landslides litter the tracks, children are seen playing on the tracks on all the routes, and cows lounge on the tracks at all hours. Timing, however, is king on the GSM freight train, and tardiness is unknown among its happy customers. How does the company succeed? Skill and raw courage, together with tolerant and forgiving customers is how. The train engineer is well known for the incredible speeds at which he sometimes takes corners, and he slams on the brakes and blasts the horn regularly. Customers, therefore, are obliged to pack their wares thoroughly and carefully; the more important the goods, the more liberal the packing material should be. Most customers pack much more freight than they need to. So bumpy and hazardous is the ride, that if half of the goods arrive at their destinations in one piece, the customers are delighted. These forgiving customers, obsessed with on-time performance as they are, gladly submit to the GSM freight train's occasional requests for complete replacements of several shipments destroyed in the frequent and spectacular train crashes.

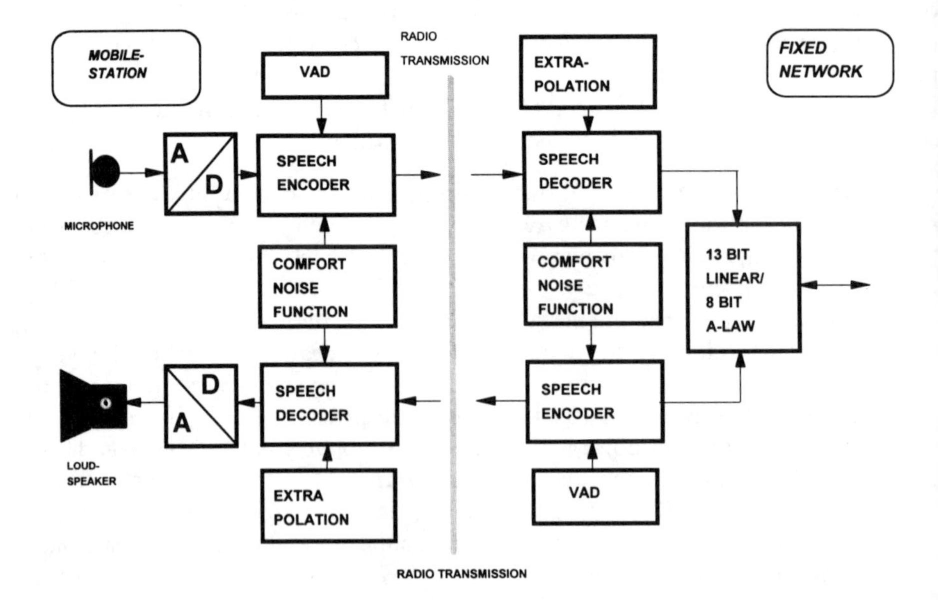

Figure 5.25 Speech processing functions used in GSM.

It is difficult to place too much emphasis on how destructive the RF channel is to data and how important the elaborate measures engineers take with protecting the transmitted data really are. The protection (packing materials and spare shipments) are called *channel coding.*

Channel coding is another vast field of endeavor that absorbs the careers of many talented scientists and engineers. A comprehensive treatment of this almost magical subject is, as was the case with speech coding, beyond the scope of this book. The ambitious reader will find a large selection of excellent books on the subject in an appropriate university library. A brief but worthy introduction to the subject is found in [6]. A major feature of digital data transmission is the myriad techniques used to protect data or speech through *coding.* Coding adds additional bits to the original payload to provide a means of protecting the original information. This gives the data more security, since it is possible to identify and even correct (within certain limits) data corrupted in the RF path. A very simple channel-coding scheme is to break the data stream to be transmitted into little blocks or words, and then add a single bit to each block, which tells the receiver if the word (block) is correct. Figure 5.26 is an illustration of this simple *parity* method. To each seven bits an eighth is added such that the sum of all the ones (1s), including the added parity bit, is even. If, on the receiving side of the channel the parity sum is not correct (the sum is odd), then the receiver knows that one of the last seven transmitted bits (excluding the parity bit) is wrong. This is an extremely simple way to code data that is only suitable for use in wire-line networks; it would never work in a radio link. It is only possible to detect if one bit is wrong. If two of the received bits are wrong, the condition cannot be detected. If one bit is wrong, there is no way to correct it, because we do not know which of the seven bits is inverted. For all bad words received, we have no choice but to ask for a retransmission of the word, or simply discard it. The channel-coding mechanisms used in GSM are far more elaborate than this simple parity scheme, and they are explained in this section.

5.14.1 Coding of Speech Data

It was shown in the chapter on speech coding that the speech quality depends to a considerable extent on an error-free transmission of the class Ia bits. Their importance

	information (seven bit)	even parity bit
1st Word	0 1 1 0 1 0 1	0
2nd Word	1 1 0 0 0 1 0	1
...	...	...
xnd Word	0 1 1 1 0 0 1	0

Figure 5.26 Even parity to detect errors.

means these bits have to be very thoroughly protected so that errors in their transmission or reception can be detected, and the specific errors corrected. The class Ib and class II bits are not as important and are not as carefully protected as the class Ia bits.

Speech data are channel-coded in two steps. First the class Ia bits are *block-coded.* This is a *cyclic code* and is used simply for error detection. It adds three parity bits to the speech data, which gives an indication to the decoder, on the other end of the radio channel, if errors have occurred that have not been detected and corrected. This is the first stage for the coding function, and the reverse is the last one for the decoder. The second stage of the decoder takes care of error detection; it audits the preceding error correction function in the decoder. If the block code detects an error in the class Ia bits, then the entire 260 bits in the block are abandoned rather than passed on to the speech codec. When a block is discarded in this way, the speech codec is informed that a speech block was discarded and that it should try to interpolate the subsequent speech data. This will give better speech quality than if the speech codec has to reproduce a sound from corrupted data.

The second function in channel coding is convolutional coding, which adds redundant bits in such a way that the decoder can, within limits, detect errors and correct them. This code is applied to both the class Ib and class Ia bits (including the parity bits generated from the class Ia bits). For a code to be able to correct errors, a certain number of additional bits have to be added to the payload bits. The added bits are called *redundancy bits.* The convolutional code employed here uses a rate of $r = 1/2$ and a delay of $K = 5$. This means that five consecutive bits are used for the calculation of the redundancy bits, and that for each data bit an additional redundant bit is added. Before the information bits are encoded, four bits are added. These bits are all set to zero, and are used to reset the convolutional code. Since we always use five bits to calculate the appropriate redundancy bits (in this GSM application), we need the trailing four zeros for the last data bit.

Figure 5.27 shows how the *convolutional code* is generated, and Figure 5.28 gives an example of the generator in Figure 5.27. The reader should consider Figure 5.27 as a little machine that outputs two bits for each single bit fed into it. When the machine starts, all the memory cells M are reset to zero. The speech bits come in the input port. Each time a bit enters the machine on the left, two bits come out through the two ports provided on the right of the machine. The chart in Figure 5.28 is a log of the state of the machine in Figure 5.27 after the machine cycles on the first five bits. In both these figures, it is assumed that immediately after this example, the next (subsequent) block of voice data has to be coded. In the absence of this block, some additional zeros are entered, which will, in normal circumstances, come from the subsequent block.

Figure 5.29 shows the complete channel-coding scheme for all the speech bits. It is interesting to note that the class II bits, which are the least important ones, are not protected at all. In this scheme, 189 bits enter the convolutional coder, and 2 ·

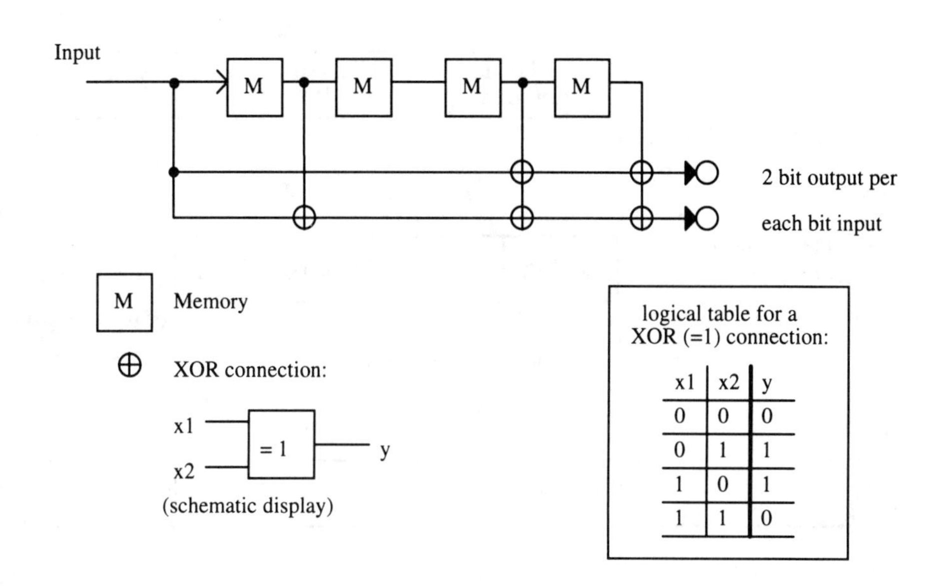

Figure 5.27 Convolutional coding.

Bit stream (input)	1 0 1 1 0 0 0 1 1 0 1 0 1
Adding of four 0 bits (M)	1 0 1 1 0 0 0 1 1 0 1 0 1 **0 0 0 0**
Delay of one bit (M2)	0 1 0 1 1 0 0 0 1 1 0 1 0 1 0 0 0 0
Delay of two bit (M3)	0 0 1 0 1 1 0 0 0 1 1 0 1 0 1 0 0 0 0
Delay of three bit (M4)	0 0 0 1 0 1 1 0 0 0 1 1 0 1 0 1 0 0 0 0
Delay of four bit (M5)	0 0 0 0 1 0 1 1 0 0 0 1 1 0 1 0 1 0 0 0 0
1st stage (M + M4 + M5)	1 0 1 0 1 1 0 0 1 0 0 0 0 1 0 0 0
2nd stage (M + M2 + M4 + M5)	1 1 1 1 0 1 0 0 0 1 0 1 0 0 0 1 1
Output of the convolutional code	1110111001110000011000100001001010

Figure 5.28 Example of a convolutional code.

189 = 378 bits come out, and 78 class II bits are added to the 378 bits to yield 456 bits. This is exactly 4 times 114, and 114 is the number of coded bits within one burst, or 8 times 57, which is the number of bits put onto eight subblocks. Subblocks are explained in the next section.

5.14.2 Reordering, Restructuring, and Interleaving

It was noted that 456 bits will fit perfectly into four time slots (bursts), but if these coded data were inserted into four consecutive bursts, then the whole speech block

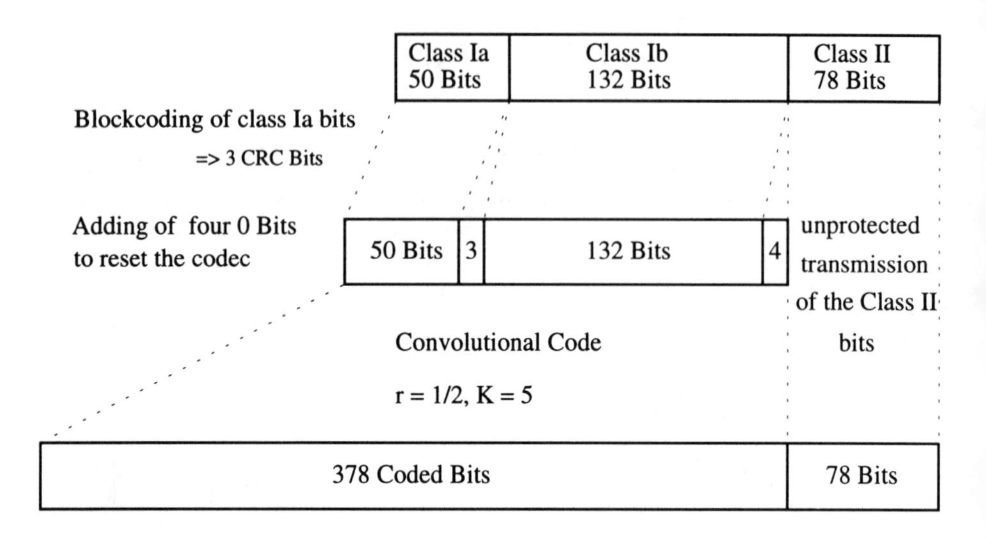

Figure 5.29 Block and convolutional coding of full-rate speech data

would be too susceptible to the loss of a burst. Whole bursts are lost regularly on radio channels; it could happen, for example, if the mobile passes through a tunnel or if just about any other type of interference occurs. To avoid the rather high risk of losing consecutive data bits, these data are spread out over more than four bursts. The block (456 bits) is spread onto eight bursts in subblocks of 57 bits each. A subblock is defined as either the odd- or even-numbered bits of the coded data within one burst.

The data are not put in an ordered row into these subblocks, but are reordered (shuffled) before they are mapped onto the time slots. The shuffling is orderly, so the bits can be put back in their proper order in the receiver. This further decreases the possibility of a whole group of consecutive bits being destroyed in the radio channel. Convolutional codes, you see, are much better at repairing individual bit errors than they are at repairing burst errors or errors in a sequence of neighboring bits. The 456 bits are subdivided onto the eight subblocks in the following way. Bit number 0 goes into subblock 1, bit number 1 goes into subblock 2, and so on until all eight subblocks are used up. Bit number 8 ends up in subblock number 1 again (see Table 5.5). The first four subblocks are put into the even-numbered bits of four consecutive bursts, and the second four subblocks are put into the odd-numbered bits of the next consecutive four bursts (see Figure 5.30). Upon receipt of the next speech block, a burst then holds contributions from two successive speech blocks. This whole procedure of putting the bits into subblocks is referred to as *reordering* and *restructuring*, and the mapping of the subblocks onto the eight bursts is called *diagonal interleaving*.

Table 5.5
Reordering Scheme for a Traffic Channel TCH

Bit Number of the Coded Bits	*Position Within the 26-Frame Structure*	
0 8 448	Even bits of burst N	(No. 0, 4, 8, 13, 17, 21)
1 9 449	Even bits of burst $N + 1$	(No. 1, 5, 9, 14, 18, 22)
2 10 450	Even bits of burst $N + 2$	(No. 2, 6, 10, 15, 19, 23)
3 11 451	Event bits of burst $N + 3$	(No. 3, 7, 11, 16, 20, 24)
4 12 452	Odd bits of burst $N + 4$	(No. 4, 8, 13, 17, 21, 0)
5 13 453	Odd bits of burst $N + 5$	(No. 5, 9, 14, 18, 22, 1)
6 14 454	Odd bits of burst $N + 6$	(No. 6, 10, 15, 19, 23, 2)
7 15 455	Odd bits of burst $N + 7$	(No. 7, 11, 16, 20, 24, 3)

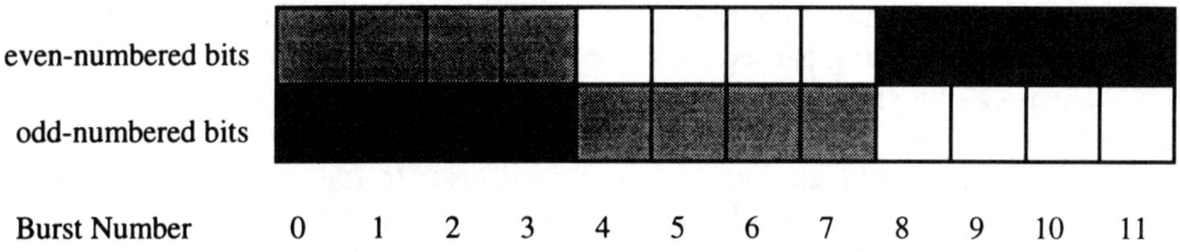

even-numbered bits
odd-numbered bits
Burst Number 0 1 2 3 4 5 6 7 8 9 10 11

Figure 5.30 Diagonal interleaving of speech data (blocks of the same color belong to each other).

If a traffic channel is sacrificed (stolen) by an FACCH, the theft is indicated by the *stealing flags* within the burst, where the first stealing flag, when set to 1, indicates that the even-numbered bits are occupied by the FACCH, and the second flag indicates that the odd-numbered bits of the burst are stolen by the FACCH. The coding scheme for the FACCH is the same as for the other signaling data, as described later on in Section 5.14.4. It should be apparent by now why it is necessary to tell the channel decoder when data are being stolen from a traffic channel. It is because the coding scheme for both types of information—traffic or signaling—is different.

5.14.3 Channel Coding of Data Channels

The coding scheme for the user *data channels* is very complex. The reason for this is that these user bits must be protected even better than the traffic channel data representing speech data. A figurative example: In the former case, a single error could represent a decimal point change, which could change, for example, 1 pound to 1,000 pounds. A single bit error in speech data will, under the worst circumstances, only result in the loss of a speech frame. The speech transcoder merely repeats the sound represented by the previous speech frame.

Each data rate has a different interleaving scheme and different parameters for the accompanying convolutional code. From the five different data channels (TCH/F9.6, TCH/F4.8, TCH/F2.4, TCH/H4.8, and TCH/H2.4), only two will be explained as examples: the TCH/F9.6 because it is the one with the highest data rate, and TCH/F2.4, because there are several reasons that this might become the most popular one.

5.14.3.1 Channel Rate of 9.6 kbps

Even though the rate 9.6 kbps refers to the user's transmission rate, the actual rate is brought up to 12 kbps through channel coding in the terminal equipment; 12 kbps is the rate delivered to the mobile station. Since it is independent of GSM, the coding within the terminal equipment will not be explained in this book, except to mention that this coding is used for error detection in a wire-line environment.

The user's bit stream is divided into four blocks of 60 bits each, for a total of 240 bits, which are coded together in a convolutional code. In contrast to the coding of the speech data, a block code is not applied prior to the convolutional coding, because the error detection is already performed within the terminal equipment. As one must always do with convolutional codes, four zero bits are added to the 240 data bits to reset the decoder. The parameters for the convolutional code are the same as those used for speech coding ($r = 1/2$, $K = 5$). The convolutional code accepts, then, 244 bits and issues 488 coded bits. Unfortunately, 488 bits do not fit easily in the 456 bits per block scheme we saw with coded speech data; there is an excess of $488 - 456 = 32$ bits. We *puncture* the 488 bits, according to a certain rule, to reduce their number by 32; the 32 punctured bits are not transmitted at all.

Each individual bit in the user's data files holds more importance than it does in the user's speech data. The fading errors in radio channels, therefore, tend to be more troublesome in data applications than in speech applications. The interleaving scheme, then, for data applications is deeper and more complex than the one used for coded speech blocks. The blocks are spread over 22 bursts, which is almost a complete traffic channel frame (Figure 5.13) with its SACCH and idle frame. Even though the interleaving covers 22 frames, it is referred to as a *19-burst interleaving plan*. Where does the number 19 come from? It comes from the early days of GSM specs, when a subblock was an entity of 114 bits (a time slot), which was spread over 19 bursts in a regular way with 6 bits per burst ($6 \cdot 19 = 114$). Since 456 bits cannot be easily accommodated in 22 bursts, a rather complex mapping scheme is performed on the 456 bits. The scheme breaks up the 456-bit block into 16 parts of 24 bits each ($16 \cdot 24 = 384$), 2 parts of 18 bits each ($2 \cdot 18 = 36$), 2 parts of 12 bits each ($2 \cdot 12 = 24$), and 2 parts of 6 bits each ($2 \cdot 6 = 12$). The parts add up to $384 + 36 + 24 + 12 = 456$, which is the user's coded data block. A burst (time slot) contains information from either 5 or 6 consecutive data blocks; that is, 4 parts of

24 bits each (96 bits) and either 1 part of 18 bits (96 + 18 = 114), or 1 part of 12 bits each and 1 part of 6 bits (96 + 12 + 6 = 114).

The first and the twenty-second bursts contain 6 bits each (2 · 6 = 12). The second and twenty-first bursts contain 12 bits each (2 · 12 = 24). The third and twentieth bursts carry 18 bits each (2 · 18 = 36). Now 6 bursts are used up, and 16 bursts remain of the 22 we started with. We put, then, 24 bits in each of the fourth through nineteenth bursts (16 · 24 = 384). All 456 bits are accommodated in the 22 traffic channel frames. The structure repeats itself starting at every fourth burst so that the bits are distributed over 22 time slots (bursts) diagonally, similar to the plan used for the speech frames, as depicted in Figure 5.31. There are k bursts depicted, each with n contributions of user data. Each burst carries five or six different data contributions, depending on which frame in the traffic channel multiframe is referred to. Then 22 frames can carry 5.5 information blocks. Far to the right of the bursts, not shown in Figure 5.31, the SACCH (TDMA frame number 12) and the idle frame (TDMA frame number 25) retain their traditional functions and do not carry any user data.

If, during a data transmission session, some signaling transfers are required, this is handled with the FACCH, which steals information from the traffic (data) channel. FACCH information is very important, and it cannot wait for the slow and complicated deep interleaving (depth of 22) of the data it is displacing. The FACCH is interleaved with a depth of only 8 (similar to a normal TCH). The user's data are simply replaced by the FACCH data, and the user's data are lost. Given the depth of *interleaving* and the rather short duration of the FACCH message, it is likely that the *convolutional code* can restore the lost data bits in the receiver.

As an aid in review, the entire coding scheme is summarized in Figure 5.32. We see that the user's data starts at 9.6 kbps and is expanded to 12 kbps by the customer's equipment. Convolutional coding requires 4 extra bits and doubles the number of bits to 488. Counting on the power of the convolutional code, we remove 32 of the coded bits before we spread the remaining 456 bits among 22 bursts according to a complicated interleaving plan.

5.14.3.2 Channel Rate of 2.4 kbps

A change in the *interleaving* scheme requires a completely different Layer 1 implementation. This makes it more difficult and expensive to provide different data services with a single mobile station. An easier approach, in this respect, is the 2.4-kbps data rate, which uses the same interleaving scheme as the common traffic channel; it requires no changes in the Layer 1 procedures. There is, of course, a change in the convolutional code parameters: it uses $r = 1/6$ and $K = 5$. The data rate coming from the terminal equipment is at a rate of 3.6 kbps, which already includes the terminal equipment's error correction coding. The customer's coded data are divided into

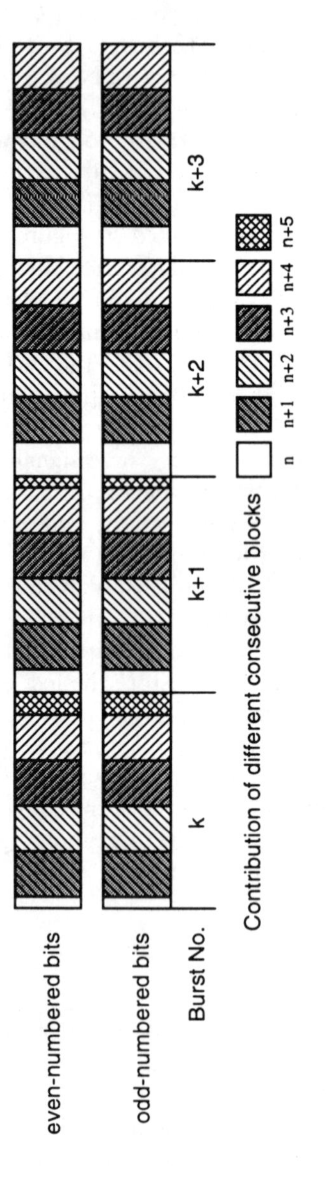

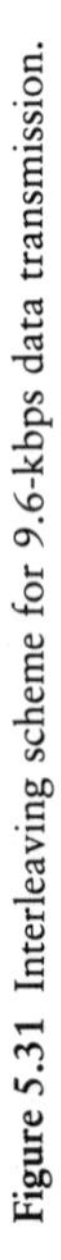

Figure 5.31 Interleaving scheme for 9.6-kbps data transmission.

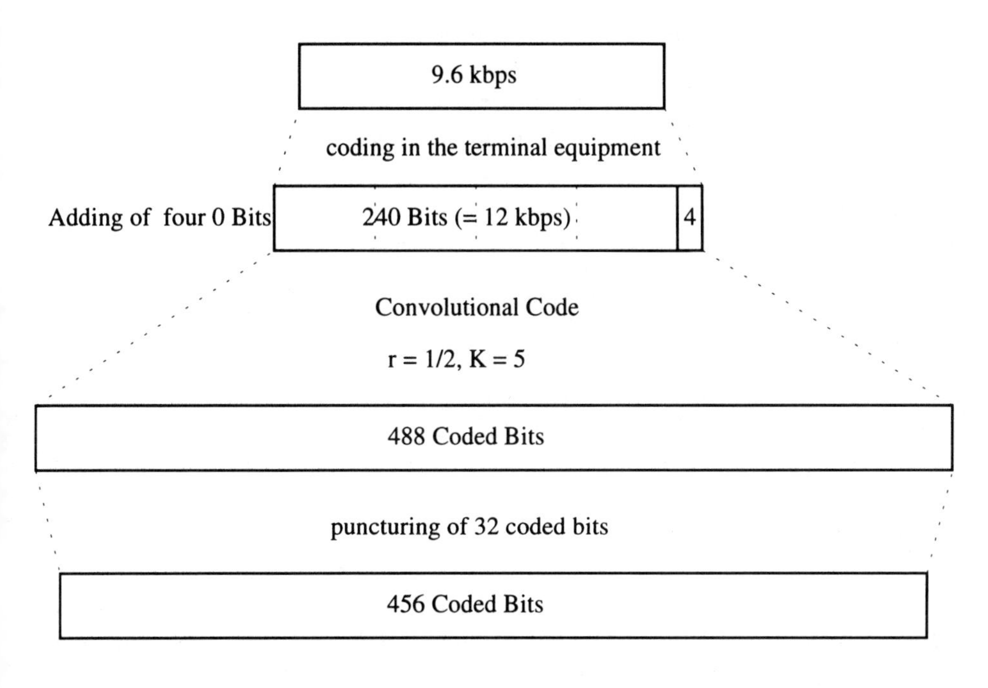

Figure 5.32 Coding scheme for 9.6-kbps data transmission.

72-bit blocks to which, again, four zero bits are added, since $K = 5$. The convolutional code transforms the $72 + 4 = 76$ input bits into $76 \cdot 6 = 456$ coded output bits, which are mapped onto eight subblocks in the same way they are in the speech frame case. A key advantage is that this is an ideal application for mobile data applications, since the complete Layer 1 is available as a single-chip solution these days. (See Figure 5.33 for a display of the convolutional coding of the 2.4 kbps data channel.)

5.14.4 Channel Coding of Signaling Channels

Signaling data are more important than any user data that may appear on the network. The passengers (user data) in an airplane may have all paid full fare, the plane may be safe, and the accommodations exceptional, but nobody is going anywhere without the a pilot and crew (signaling data).

Signaling information contains a maximum of 184 bits, which have to be encoded. It does not make a difference whether the type of signaling information to be transmitted is mapped onto a BCCH, PCH, SDCCH, or SACCH. The format always stays the same. Special formats are reserved for the SCH and RACH, and the FCCH requires no coding at all.

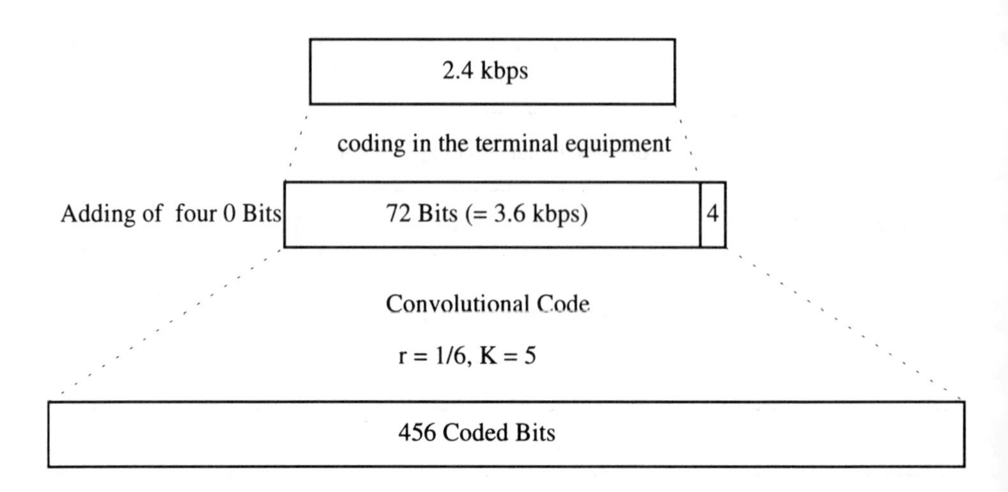

Figure 5.33 Convolutional coding for the 2.4-kbps data channel.

The significance of each of the 184 bits is the same, and no distinction is made between them as was the case with the three classes of speech bits. The coding scheme, again, is divided into two steps (see Figure 5.34). The first step is a *block code*, which is dedicated to detecting and correcting *bursty errors*. These are errors that occur when large parts of a burst, or even a complete burst, are lost or corrupted

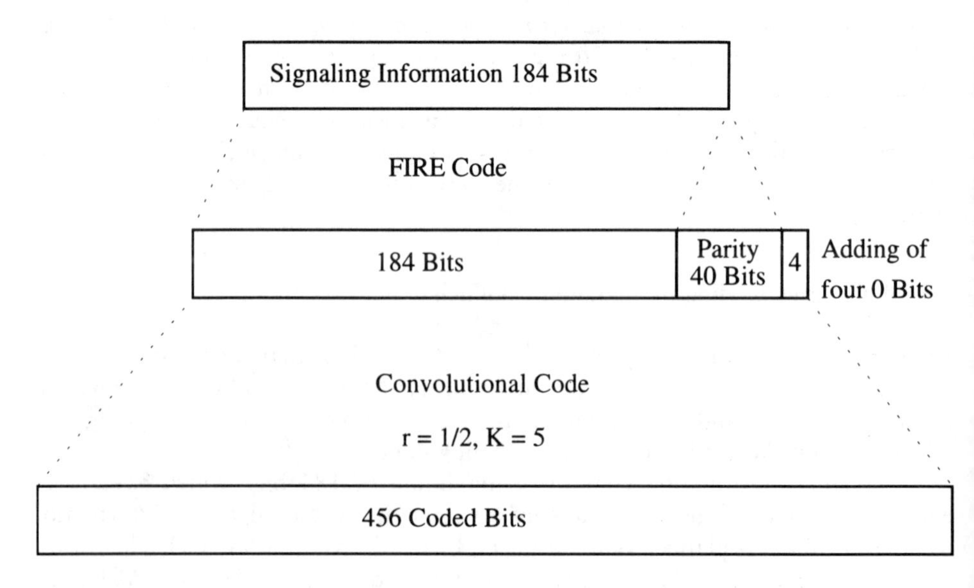

Figure 5.34 Coding scheme for signaling channels.

in the radio channel. The block code belongs to the family of *fire* codes. It adds 40 parity bits to the 184 data bits, resulting in 224 fire-coded bits.

The second step is the convolutional code which uses the same parameters as those used for the traffic channel ($r = 1/2$, $K = 5$), and 4 zero bits are added to the 224 bits to yield 228 bits. The convolutional code doubles the number of bits to be transmitted to 456 bits. We have seen this number before, and it fits well into eight subslots, with 57 bits in each. The coded data are interleaved over four bursts. The first four subblocks are packed onto the even-numbered bits of four consecutive bursts, and the second four subblocks are mapped onto the odd-numbered bits of the same four consecutive bursts (see Figure 5.35 and Table 5.6).

One peculiarity, which was stated without further explanation in the earlier section on frame structures (Section 5.11), becomes clear now. Recall the 51-multiframe structure where there have always been four consecutive TDMA frames filled with either BCCH, CCCH, SDCCH, or SACCH. The reason behind this is that it takes four frames to carry a complete block of signaling information on one signaling channel. Using four frames means we do not have to find a place to store intermediate data. Now recall the 26-multiframe structure (Figures 5.14 and 5.15) and note that the SACCH occupies only one TDMA frame in 26 frames. This means

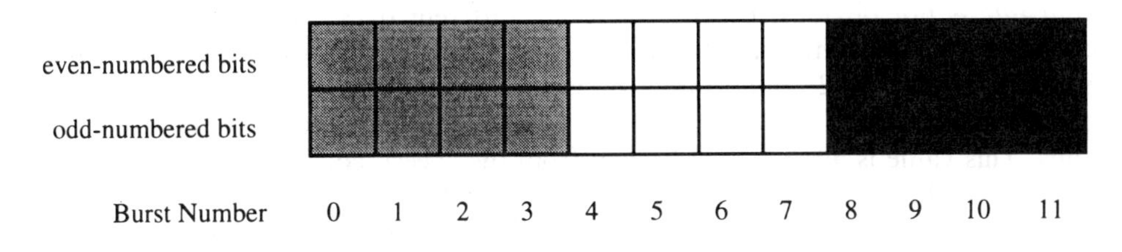

Figure 5.35 Blockwise interleaving scheme for signaling data (blocks of the same color belong to each other).

Table 5.6
Reordering Scheme for a Signaling Channel

Bit Number of the Coded Bits	*Position Within the Frame Structure*
0 8 448	Even bits of burst N
1 9 449	Even bits of burst $N + 1$
2 10 450	Even bits of burst $N + 2$
3 11 451	Even bits of burst $N + 3$
4 12 452	Odd bits of burst N
5 13 453	Odd bits of burst $N + 1$
6 14 454	Odd bits of burst $N + 2$
7 15 455	Odd bits of burst $N + 3$

Table 5.7
Data Rates, Repetition Rates, and Delay Through Interleaving of the Bearer Channels

Channel Type	*Net Data Rate (bps)*	*Repetition Rate (ms)*	*Delay Due to Interleaving (ms)*	*Gross Data Rate (kbps)*
TCH/FS	13,000	20	38	22.8
FS/data, 9.6 kbps	12,000	5	93	22.8
FS/data, 4.8 kbps	6,000	10	185	22.8
HS/data, 4.8 kbps	6,000	10	93	11.4
FS/data, 2.4 kbps	3,600	20	185	22.8
HS/data, 2.4 kbps	3,600	20	38	11.4

Table 5.8
Data Rates, Repetition Rates, and Delay Through Interleaving of the Signaling Channels

Channel Type	*Net Data Rate (bps)*	*Repetition Rate (ms)*	*Delay Due to Interleaving (ms)*	*Gross Data Rate (kbps)*
FACCH/FS	9,200	20	38	22.8
FACCH/HS	4,600	40	74	11.4
SDCCH	782	236	14	1.932
SACCH/TCH	382	480	360	0.950
SACCH/SDCCH	391	471	14	0.968
BCCH	782	236	14	1.932
AGCH	782	236	14	1.932
PCH	782	236	14	1.932
RACH	34	236		

5.14.6 Differences in the Speech Quality in Analog and Digital Systems

We have enough knowledge, now, to understand why digital systems are generally superior in their performance compared to analog systems. The quality of the channel is mostly a function of the difference between the signal and the noise floor. This is often expressed as the SNR (see Section 5.2).

In analog systems, the speech is directly and linearly frequency-modulated onto the carrier. Harmful influences on the radio path, such as *reflections, multipath propagation, noise impulses*, and simple *interruptions*, tend to modulate the amplitude of the received signal, and thus have an audible influence on speech quality. The lower the signal's amplitude when measured against the noise at the receiver, the less tolerance the link has for amplitude variations in the signal. There is, then, a nearly linear relationship between decreasing speech quality and the distance between the mobile station and its base station. Users of analog systems routinely deal with the diminished quality of the radio link by using the inherent redundancy in human speech. Users merely listen through the noise and interference caused by the channel. Eventually, the interference and noise can become so great that the concentration required for users to recover the spoken information becomes excessive, and they hang up, thus freeing the channel for other users.

In digital systems, these influences are not as linear in their degradation of voice quality. The coding and error protection applied, before the information is imparted to the transmitter's carrier, tends to make the speech quality constant until the SNR, which is an inverse function of the distance between radios, falls to a certain low value, at which point the channel simply fails to be of any use at all. This is to say that the *speech quality* in a digital system remains at a relatively high level for a much longer period of time as the distance between radios grows; there is little degradation in speech quality just before the radio link fails. Even as the carrier and its modulation from a digital radio can suffer the same destructive influences as that from an analog transmitter, the channel coding and other error protection measures keep speech quality rather high. At a certain distance from the base station, or at a certain interference level, the decoding and error correction functions can no longer keep up with the destruction of the data, and the quality of the connection degrades very quickly.

Figure 5.37 shows this relationship of the signal quality to the distance from the base station for both analog and digital signals.

5.14.7 Ciphering

Ciphering is not channel coding. Ciphering is performed after the encoding of the different logical channels, and it is done independently of whether a channel is a signaling channel or a traffic channel. The reasons for and the uses of ciphering were mentioned in Section 3.8.2. Now we have a look at how it works by taking a look at the *ciphering algorithm*. The details of ciphering are shrouded in secrecy, so we show only its principles of operation. The ciphering algorithm is a pure *XORing* of the coded Layer 1 data with a key. In the receiver, we XOR the ciphered data with the same key to get the clear Layer 1 data back again. A special key (ciphering sequence) is used together with the burst numbers, which the radio has full knowledge of thanks to the counters T1, T2, and T3. The data appear on the Um interface as a nearly random bit sequence. It is impossible for anybody to decode the data without lots of information, most of which never appears on the radio channel. Table 5.9 shows the principle of ciphering as it applies the XOR function to the data on either end of the radio channel.

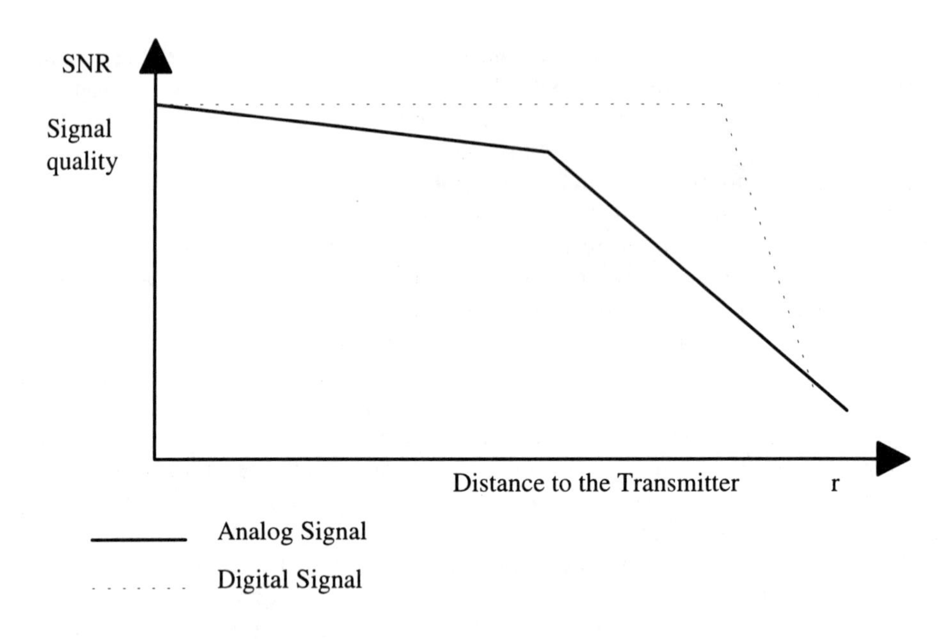

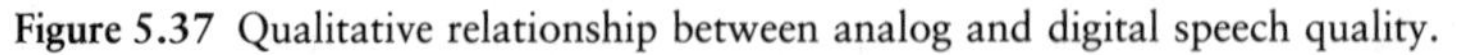

Figure 5.37 Qualitative relationship between analog and digital speech quality.

Table 5.9
Principle of Ciphering

Plain data	0 1 1 1 0 0 1 0 1 0 0 0 1 1 1 0 0 1 1 0 1 ....
Ciphering sequence	0 0 0 1 1 0 1 0 1 0 1 0 0 0 1 1 0 1 1 1 0 ....
XORed	
Ciphered data (transmitted)	0 1 1 0 1 0 0 0 0 0 1 0 1 1 0 1 0 0 0 1 1 ....
Ciphering sequence	0 0 0 1 1 0 1 0 1 0 1 0 0 0 1 1 0 1 1 1 0 ....
XORed	
Recovered data	0 1 1 1 0 0 1 0 1 0 0 0 1 1 1 0 0 1 1 0 1 ....

5.15 DIGITAL MODULATION

In this section on *modulation*, we explore the iron, steel, wheels, axles, and grease of the GSM freight train. We arrive, finally, at the lowest level of the physical interface, Layer 1, between radios in the GSM system. We digitized and encoded our voice, removed some useless redundancy from the resulting speech data, and added some more useful redundancy to replace it. We have organized and protected all of our signaling messages. We have ordered all of our payload data, both user and signaling data, into multiframes so that they can be mapped onto time slots and bursts. We have even included the option of scrambling or ciphering the data just before we paint them onto the physical TDMA frame structure. All that remains is to modulate a radio carrier with the myriad bits representing the Layer 1 data.

When we superimpose a binary waveform on an RF carrier, we call the process *digital modulation.* We can impart digital information to a carrier by altering any of the carrier's characteristics, amplitude (AM), frequency (FM), or phase (PM), synchronously with the information. The characteristics of the channel over which the carrier must transport the binary waveform and the application intended for the waveform determine which of the modulation methods are used. Because the mobile channel is primarily characterized as a frequency-selective *fading channel* (see Section 5.2), FM or PM is used instead of AM, since angle modulation is less sensitive to the amplitude changes that occur in a fading channel.

To reflect the binary on-off nature of digital modulation, the various modulation methods are often called *keying* methods: *amplitude-shift keying* (ASK), *frequency-shift keying* (FSK), and *phase-shift keying* (PSK). The *keying* term recalls the earliest days in radio when the amplitude of a transmitter was keyed on and off with a *key*, which was a special type of hand-operated switch, which, owing to its peculiar shape and construction, adapted itself well to creating the characteristic dot and dash codes radio operators used in those days. It is interesting to note that with digital cellular, we are returning to more traditional modulation techniques after a sojourn into analog modulation lasting many decades.

Whatever digital modulation scheme we use, we need an appropriate modulator in the transmitter and a corresponding demodulator in the receiver. This modulator-demodulator pair is often called a *modem* when the modulating information is binary. The world is full of radio transmitters that modulate their carriers linearly in amplitude (AM), frequency (FM), or phase (PM); they are, however, seldom referred to as modems, since the modulating information is not digital. Since this book is a system-level discussion of GSM digital cellular radio, we will confine our examination of digital modulation to the system level, referring to specific modulation and demodulation techniques only for clarification.

5.15.1 ASK, FSK, and PSK

If we shift the amplitude of a carrier with binary information, we are employing ASK. If we switch the amplitude of our carrier between their full-on and full-off conditions, we are shifting between two carriers: a carrier that is always present and one that is never present. This kind of modulation is still used today on the HF bands (3 to 30 MHz) among amateur radio operators, ships at sea, and other applications requiring low cost and long-range communications.

If we shift the carrier's frequency with our binary information, this is equivalent to shifting between two or more carriers (or transmitters) of different frequencies. This is FSK and is widely used in analog cellular systems for signaling functions.

There is no limit to the number of carrier frequencies we can shift between, but the use of two frequencies, quite close together, is the universal implementation of FSK.

As with FSK, we can shift between various carriers differing from each other only in their relative phase (PSK). There are many varieties of PSK, and each is broadly distinguished from the others by the number of allowed places on the unit circle (within cycles of the carrier) on which we can change the phase.

5.15.2 I/Q Representation

Instead of the traditional time domain diagrams we see on oscilloscopes, or the often more useful depictions in the frequency domain on spectrum analyzers, we can, if we choose, represent a carrier as a counterclockwise rotating vector in the *I/Q plane*, where *I* represents the inphase part of the vector at any time and *Q* the quadrature contribution. The length of the vector represents the output level on some scale of the generator creating the rotating vector. If we rotate a carrier's vector on the I/Q plane, we will resolve a cosine function in the *I* direction and a sine function in the *Q*-direction. Though of no immediate value for us, we can observe an unmodulated carrier on some kind of machine capable of resolving the *I* and *Q* components, as shown in Figure 5.38. If the carrier has "slow" FSK superimposed on it, we will see no change in the display; the rate of rotation of the vector changes as the frequency

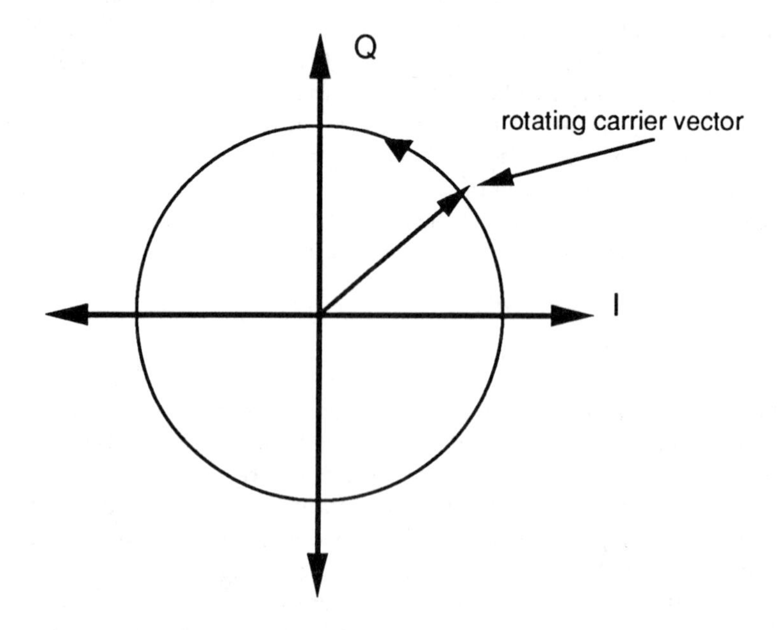

Figure 5.38 Constant wave or frequency modulation.

shifts. We see only a circle traced out by the rotating vector. If we were to increase the power from the transmitter generating the I/Q display, we would see the diameter of the circle increase. If the carrier is not modulated at all, then it is a continuous wave (CW) signal, and it would appear exactly as shown in Figure 5.38.

If we use the same I/Q display to observe ASK, we may see what is depicted in Figure 5.39. The figure can only show us the depth of amplitude shifting, and Figure 5.39 illustrates an AM depth of something much less than 100%.

Whenever we quickly change any characteristic of a carrier, amplitude, frequency, or phase, we will always get corresponding additional amplitude changes, together with occupied bandwidth spreading artifacts. The relative amount of these artifacts, which we call AM splash, is proportional to the suddenness with which we make the changes in the carrier.

5.15.3 Binary Phase-Shift Keying

If we flash an imaginary strobe light one or more times onto the vector as it rotates in the I/Q plane, and if we make the flashing strobe precisely synchronous with the vector's rotation, then we have a very useful display for various forms of PSK. As an example, we consider binary phase-shift keying (BPSK), depicted in Figure 5.40, which is PSK with only two carriers out of phase with each other to select from.

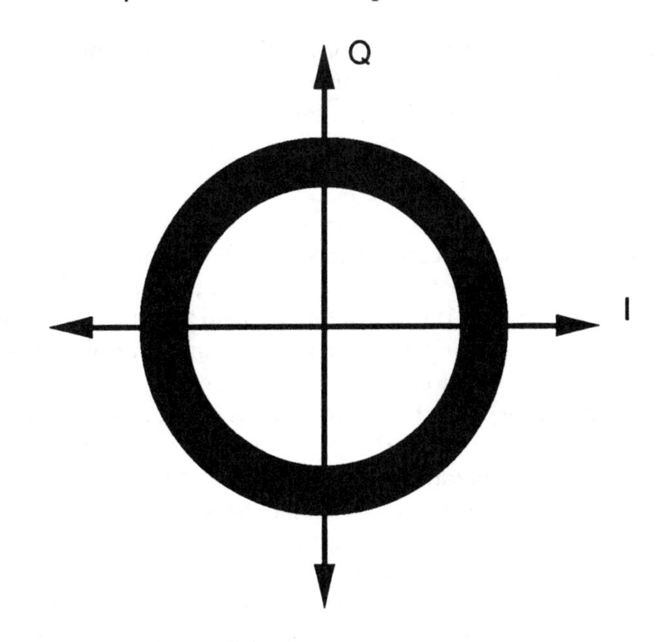

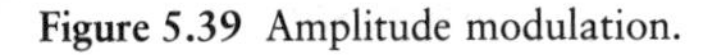
Figure 5.39 Amplitude modulation.

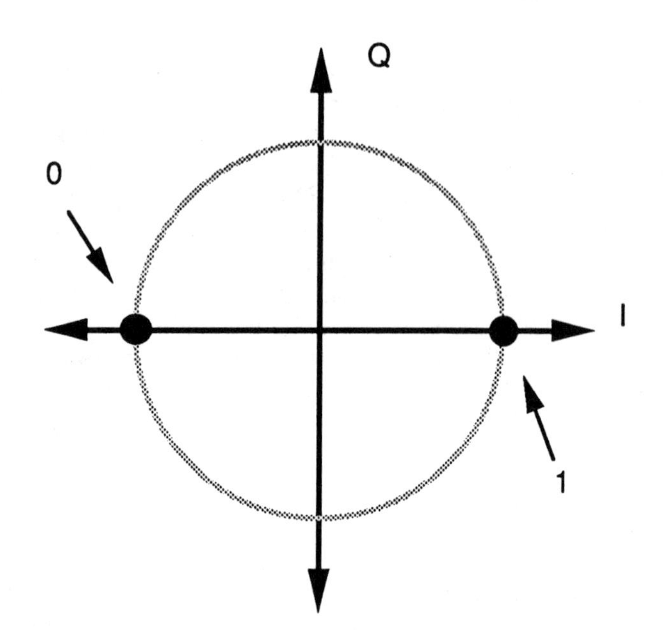

Figure 5.40 Binary phase-shift keying.

The shift between these two carriers occurs at the data rate, and the carriers differ from each other in phase by 180 deg, or π rad. Note how easy it is to represent this type of modulation with an I/Q diagram.

5.15.4 Quadrature Phase-Shift Keying

We are not limited to only two carrier phases when we endeavor to generate PSK. If we give ourselves four carrier phases to select from, and if we decide to separate the allowed phase changes into the four different quadrants of a circle, then we have quadrature phase-shift keying (QPSK). We gain maximum resolution in our ability to demodulate this kind of waveform on the receiving end if we make all four carriers exactly 90 deg ($\pi/2$ radians) out of phase with each other. An I/Q display for QPSK is shown in Figure 5.41, which clearly shows that there are two BPSK generators in quadrature, one BPSK generator in each of the *I* and *Q* branches. Figure 5.41 is drawn in a way that discloses the data that may have created the phase changes in the carrier. Each of the four allowed phase change points is marked on the figure with its corresponding data representation, but the circle the carrier's vector traces is missing. The data, which are pairs of bits, are assigned to the phase points in an arbitrary way. All possible changes, from one pair of bits to another, are plotted on the figure with an arrow, and the little circular ones in the corners show the four

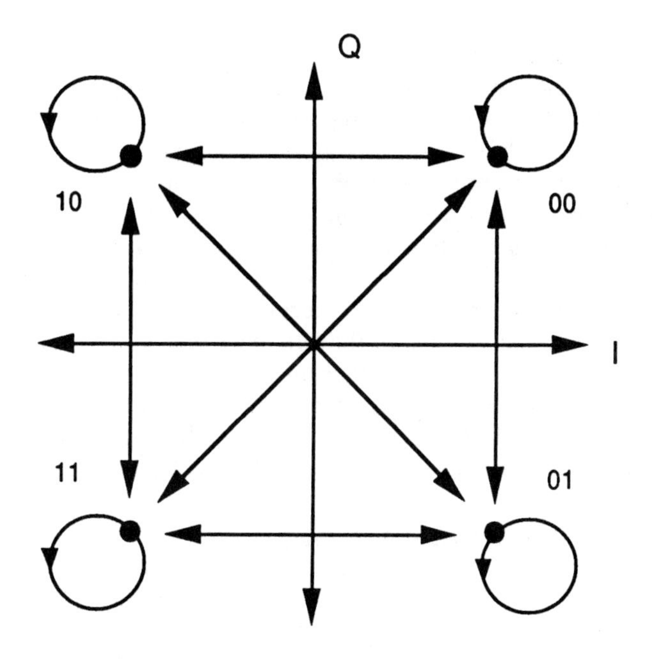

Figure 5.41 Quadrature phase-shift keying.

possible no-change conditions. The longer I and Q references are not data change indicators. The reader should take careful note that two bits are represented by a single phase change in Figure 5.41. Each phase change is called a *symbol*, and each symbol, in the case of Figure 5.41, represents two bits. It is interesting to note that these kinds of higher level modulation schemes (schemes that have more than two possible source versions of the carrier) are often generated in quadrature. This means that we can create QPSK, for example, by generating appropriate BPSK waveforms shifted 90 deg from each other, and then add them together to yield QPSK.

5.15.5 Bits and Symbols

Another reason we refer to higher level modulation schemes as such is that we can multiply the number of bits we can transport on a carrier by simply adding more versions of the carrier to shift between. Each carrier version, then, can represent whatever we like. If there are four possible carrier phases, as is the case in Figure 5.41, each phase can represent two bits: 00, 01, 10, and 11. Each of the four possible carrier phases represents two bits, and we group each of these two bits into symbols. At the receiver, each QPSK phase represents a symbol, which the receiver is instructed to resolve into two bits depending on which of the carrier phases is observed. In BPSK (Figure 5.40), each of the two phases of the carrier represents only one bit.

The number of bits that can be represented by a symbol can be very high. Take the case of 16-symbol quadrature amplitude modulation (16-QAM), which brings together phase and amplitude shifts to yield 16 distinct phase and amplitude settings in the I/Q plane for each symbol. Each symbol represents four bits. There is a practical limit to the number of bits we can map onto each symbol. As the number of bits in each symbol increases, so does the signal's susceptibility to noise and the increased requirements for a more accurate receiver. The greater the number of bits in a symbol, the smaller must be the difference the carrier can assume between phases. Interference in the radio channel creates amplitude changes on a carrier, and small changes in the carrier's phase are easily masked by these amplitude changes. The modulation specified in GSM employs only one bit per symbol, but Chapters 12 and 13 explore some digital cellular systems that pack two bits into each symbol.

5.15.6 Offset Quadrature Phase-Shift Keying

It is easy to believe that PSK leaves the transmitter's RF output level constant, that there is nothing to add AM to the carrier. This is not true. Any sudden change in the phase (or any aspect) of the carrier will add high-frequency components to the carrier's output. If we are obliged to confine our emissions to a relatively small bandwidth, then the amplitude of a carrier must change to compensate for the wider distribution of power during the phase change. Another way of looking at this is to regard the two different phases of the carrier, before and after the phase change, as two separate signals of different phase. The two signals will tend to cancel each other to the extent they differ in phase, thus altering the amplitude of the combined waveform. Any amplifier called upon to handle these kinds of digitally modulated signals must, therefore, be linear enough to handle the amplitude variations as well. This problem can be partially alleviated by invoking various schemes to reduce the tendency to produce amplitude changes in PSK. One partial solution is to avoid the case where there will be total cancellation of the combined output from the transmitter, such as would occur with a 180-deg phase change from one symbol to the next symbol. This will occur, in the QPSK case, if both bits in a symbol are allowed to change at the same time (Figure 5.41). We can avoid this, as shown in Figure 5.42, by offsetting the data going to the I and Q BPSK generators by one bit (1/2 symbol), thus preventing the possibility of two bits changing together. The origin, where there is no power emitted from the transmitter, is avoided.

5.15.7 Noise and Errors

As a signal moves from the transmitter to the receiver, noise will be added to the signal. In the case of digital modulation, the noise will appear to have been added

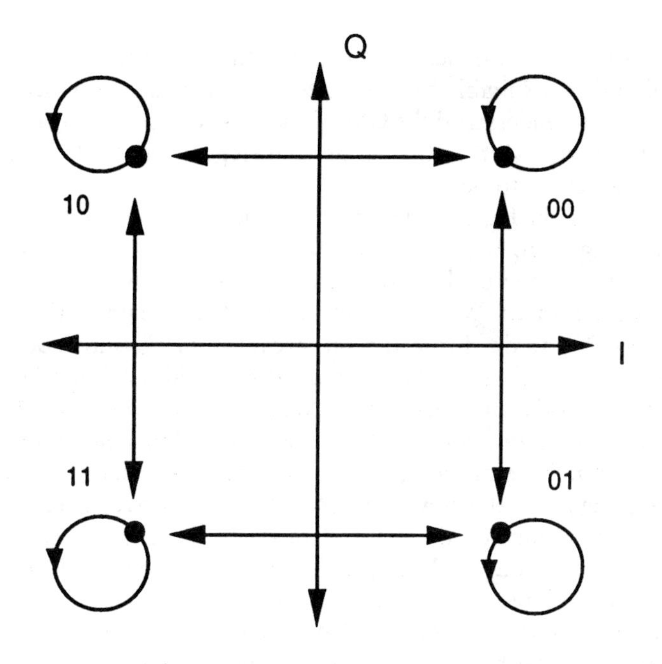

Figure 5.42 Offset quadrature phase-shift keying (OQPSK).

to either the *I* or *Q* component of the signal, or, more likely, to both the *I* and *Q* components, in such a way as to change the amplitude of the components. If the added noise is of sufficient amplitude, the resulting position of the vector at the phase change time may appear in another place in the I/Q plane reserved for a different symbol. The more bits mapped into each symbol, the shorter the distance between each point in the I/Q diagram reserved for the symbols, and the less noise that can be tolerated. As the ratio of the energy (E_b) in each bit to the noise power per unit bandwidth (N_0), E_b/N_0, increases, the less of a problem errors will be. It is a characteristic of bandwidth-limited systems to always strive to increase the E_b/N_0 ratio as a countermeasure against errors caused by noise.

5.15.8 Gaussian Minimum-Shift Keying

The modulation specified for GSM is GMSK with $BT = 0.3$ and rate 270 5/6 kbauds. We will see what all of this means shortly. GMSK is a type of constant-envelope FSK, where the frequency modulation is a result of a carefully contrived phase modulation. The most important feature of GMSK is that it is a constant-envelope variety of modulation. This means there is a distinct lack of AM in the carrier with a consequent limiting of the occupied bandwidth. The constant amplitude of the GMSK signal makes it suitable for use with high-efficiency amplifiers.

An easy way to understand the GMSK signal is to first investigate its precursor, *minimum-shift keying* (MSK). We will use the steps in Figure 5.43 to see how an MSK signal could come to be.

The waveforms in Figure 5.43 are all aligned together in phase. Little scales are placed in the figure to help make the phase relationships between the waveforms clearer. We start with the payload data stream ("data"), which will modulate the carrier in an MSK fashion. We see 10 bits of the data stream, which we declare to be 1101011000. Next, we divide the bit stream ("data") into odd and even bit streams: ("odd bits" and "even bits"). In creating "odd bits" and "even bits," we hold each alternate odd and even bit in "data" for two bit times. If we were to apply these two waveforms, "odd bits" and "even bits," to a *quadrature modulator*, we would have OQPSK. Staggering "odd bits" and "even bits" already helps us on our way to creating a waveform with minimal AM. For convenience, we make "odd bits" and "even bits" take on the values 1 or -1. In the GSM case, if the data rate (in waveform "data") is 270.833 kbps, then the staggered "odd bits" and "even bits" waveforms will have half the rate, 135.4 kbps.

The fourth and fifth waveforms in Figure 5.43 are the high-frequency and low-frequency versions, respectively, of the carrier. Since MSK is a form of FSK, we need two versions of our final carrier differing only in frequency. We shift between these

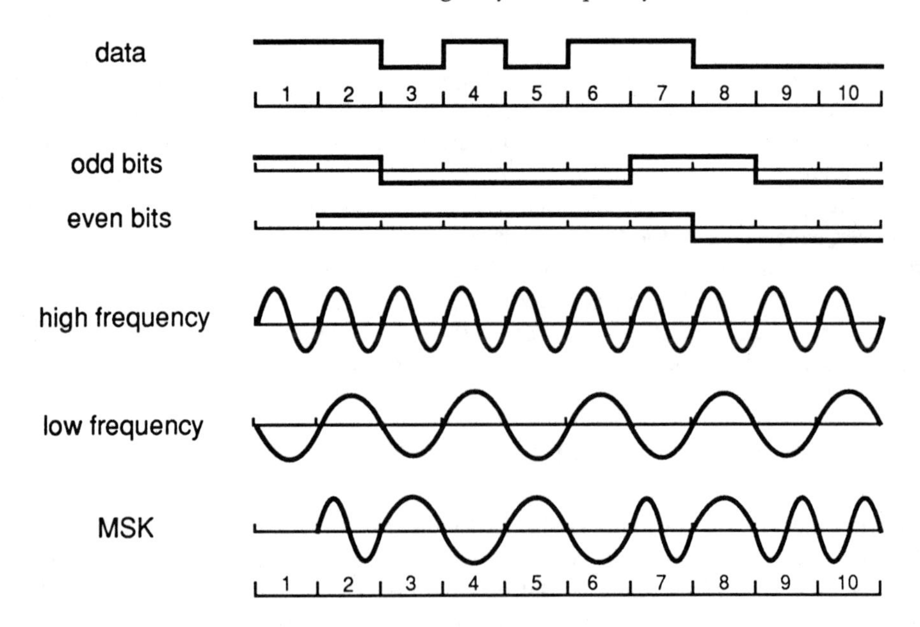

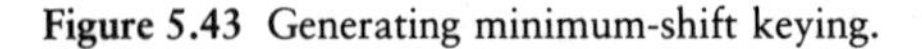
Figure 5.43 Generating minimum-shift keying.

two carrier frequencies to create our MSK signal. We create the MSK signal, starting with bit number 2, with the help of Table 5.10. At any instant, we regard the "odd bits" and the "even bits" values, look them up in Table 5.10, and follow the rules for the MSK output at that instant. Pick either the high- or low-frequency versions of the carrier, and—at the place in Figure 5.43 corresponding to the instant under consideration—turn that version of the carrier upside down, or not, according to the "sense" (+ or –) instructions in Table 5.10.

The resulting MSK waveform appears in waveform "MSK," which is the fifth waveform in Figure 5.43. Note the relatively smooth phase transitions from one frequency (and phase) to the next.

There is one last step we must consider when we build an MSK waveform. The two carrier frequencies, "high frequency" (H) and "low frequency" (L), we shift between (in the MSK case we slide between them) should be as close together in the frequency domain as possible and still remain orthogonal over the bit time T_{bit}. Two waveforms, $\sin \omega_H(t)$ and $\sin \omega_L(t)$, are orthogonal if

$$\int_0^{T_{bit}} (\sin \omega_H t)(\sin \omega_L t)\, dt = 0$$

The rather curious term, *orthogonal*, for these kinds of functions comes from the fact that any vector can be described by its three orthogonal components in the *X*, *Y*, and *Z* directions. In radio work, orthogonal can mean: "easily distinguished one from another," or "not easily confused with each other." Readers of Edwin Abbott's delightful nineteenth century work, *Flatland*, will have no trouble with this concept [7]. In the GSM case, the two frequencies, "high frequency" and "low frequency," differ from each other by the data rate of the modulating waveforms. This is one-half the 270.833 bit rate, or a maximum of 135.4 kHz. The two carrier versions, then, appear equidistant on either side of the assigned carrier for the channel, or 67.7 kHz above or below the assigned channel frequency. Thus, the modulation index (η) becomes 67.7/135.4 kHz = 0.50.

Table 5.10
MSK Truth Table

Digital Input		*MSK Output*	
Bit Value		*Frequency*	*Sense*
Odd Bit	*Even Bit*	*High or Low*	*+ or –*
1	1	High	+
–1	1	Low	–
1	–1	Low	+
–1	–1	High	–

To make a GMSK signal from an MSK signal, we need only filter the stretched data waveforms (each is 135.4 kbps) with a *Gaussian filter* of an appropriate bandwidth defined by the BT product (see Figure 5.44). In the GSM case, BT is 0.3, which makes $B = 81.3$ kHz when T is 3.7 μs ($T = 1/270{,}833$).

The reader is invited to go through the exercise in Figure 5.43 and Table 5.10 again, but this time with a null data stream (0000000000), and then with a 0101010101 data stream. In either case, we will get either the "high frequency" or the "low frequency" waveforms back in "MSK." This is exactly what we would expect if we wanted to generate a pure unmodulated carrier such as we need during the FCCH time. These two examples demonstrate the FM nature of MSK, as well as the need for a few leading zeros when the data arrives at the receiver (Section 5.8). GMSK, however, so smoothes the phase transitions from one bit time to the next that the sharp transitions are almost completely gone. GMSK, then, can be called *smooth MSK*, or *ultrasmooth FSK*. This makes sense, since the frequency shifting in GMSK comes from carefully steering the phase of the carrier in quadrature in such a way as to yield an even more phase-disciplined waveform than we see in the "MSK" waveform in Figure 5.43.

There are almost as many ways to generate a GMSK signal as there are engineers to build the modulators. The example in Figure 5.45 illustrates a method employing a quadrature process.

There is, however, only one data input in Figure 5.45. How can we get the OQPSK effect we saw in Figure 5.43 with only one data path? Figure 5.46 shows a function called *differential encoding*, which serves a function somewhat similar to the MSK truth table (Table 5.10). The process of differential encoding of the modulating data, as depicted in Figure 5.46, is used for two reasons. First, it avoids the necessity of coherent demodulation. Differential encoding removes the need for the receiver to have its own reference phase, as the data are not hidden in the phase

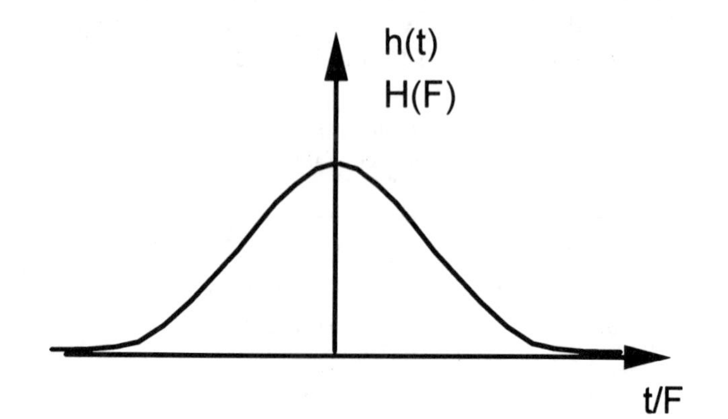

Figure 5.44 Characteristics of a Gaussian filter.

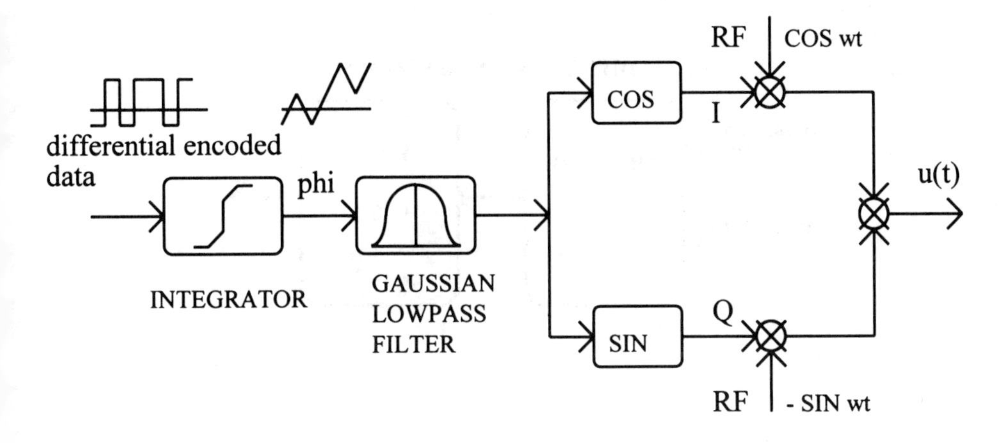

Figure 5.45 GMSK modulator.

itself, but in the phase changes. Second, it assigns a polarity (e.g., voltage) to the symbols. In this case the 0 and 1 symbols become +1 and –1 symbols.

MSK is measured with its frequency trajectory or its corresponding phase trajectory. Figure 5.47 plots the frequency trajectory we would expect from an MSK signal; there are only two versions of the carrier, differing only in their frequency, and they change when the payload data changes.

Figure 5.48, however, plots the corresponding phase changes in MSK. Though there are some sharp corners in the plot, there are no sudden shifts in the phase, say from +90 to 0 deg. There is not much AM splash with these kinds of phase changes.

Now, all we have to do is filter our data with a Gaussian filter to reduce even further the AM splash and occupied bandwidth. Figure 5.49 represents the MSK and GMSK cases for both the difference-encoded payload data and the resulting phase trajectories. Note the additional smoothing of the phase changes brought about by rounding the data's corners in the Gaussian filter.

In summary, any modulated carrier, even our GMSK carrier G(t) can be expressed with relation to time as

$$G(t) = a(t)\cos(\omega_0 t + \theta(t))$$

where $\omega_0(t)$ is the carrier frequency, $\theta(t)$ is the phase, and $a(t)$ is the amplitude. GSM dictates that $\omega_0(t)$ is decided by the network at handoff time and is the FDMA aspect of GSM. The power $a(t)$ is controlled to prevent cochannel and adjacent-channel interference. In the base station, $a(t)$ is constant until adjusted by the network. In the mobile, $a(t)$ varies so that the mobile's RF output ramps up to a network-decided level during each time slot assigned to the mobile. All that remains for the modulation

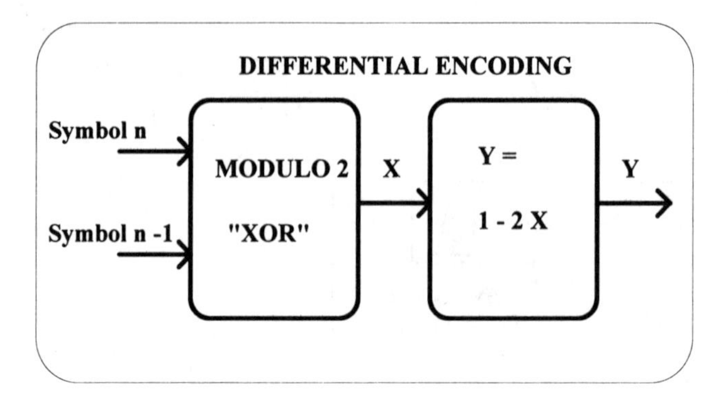

n	n-1	X	Y
0	0	0	+1
0	1	1	-1
1	0	1	-1
1	1	0	+1

LOGICAL TABLE FOR DIFFERENTIAL ENCODING

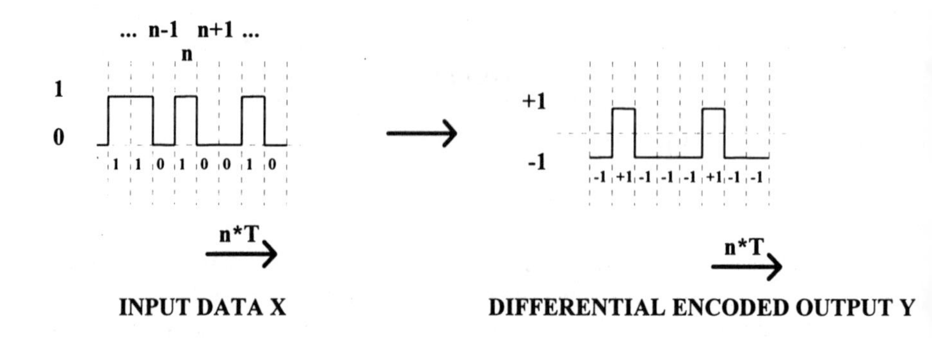

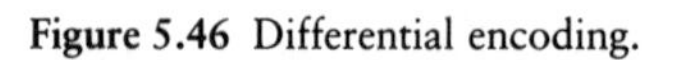

Figure 5.46 Differential encoding.

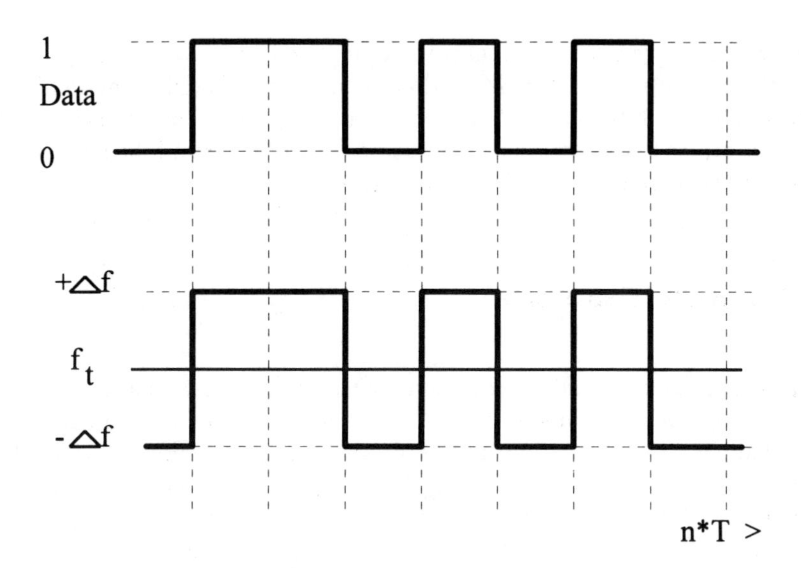

Figure 5.47 MSK frequency trajectory.

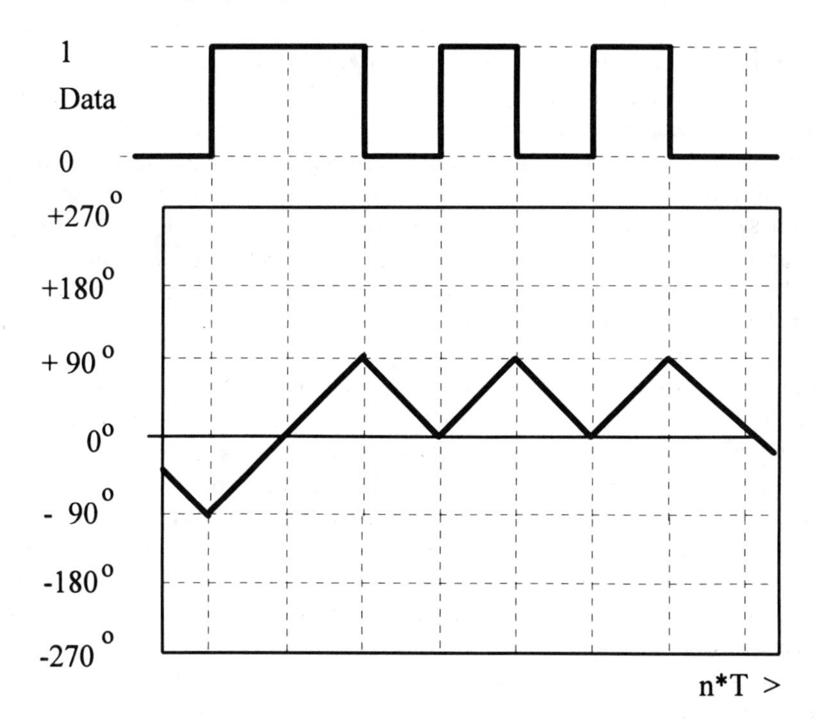

Figure 5.48 MSK phase trajectory.

is $\theta(t)$, which we can control to give the disciplined and gentle frequency shifts we need. We can consider the GMSK modulator as a type of OQPSK modulator, where we have not two bits per symbol, but less than one-half bit per symbol. We use phase transitions to cause frequency transitions, and we spread the phase transitions out to almost three bit times (Gaussian filter).

5.16 FREQUENCY HOPPING

The propagation of a radio signal and the corresponding quality of speech are influenced by the environment. Digital radio enhances the apparent quality of the radio channel through coding and error protection techniques, which hide the destruction of the information in the channel until the received signal strength is almost lost. A further improvement can be made to the radio channel on top of all the coding. The radio channel, as we have seen, is a frequency-selective fading channel, which means the *propagation conditions* are different for each individual radio frequency channel. Whereas channel number 1, for example, experiences problems when a mobile passes a large building, channel number 92 may not suffer any degradation in quality. As some channels degrade, others are improving as a mobile moves around in a cell. To average these conditions over all the base station's available frequencies in a cell, *slow frequency hopping* (SFH) is introduced. It is so called (to distinguish it from *fast frequency hopping*) because the operating frequency is changed only with every TDMA frame. Fast frequency hopping is used in spread spectrum systems and changes the operating frequency of a link many times per symbol, rather than only one time in each frame. The rather slow frequency changes in SFH are in deference to the synthesizers in the mobile station, which are required to alter their operating frequency even more often than once per frame so that they can monitor adjacent cells as well as hop around in the frequency domain. To perform well, a frequency synthesizer must be able to change its frequency, and settle quietly on a new one, within approximately one time slot (577 μs).

Frequency hopping reduces the SNR required for good communications. For a nonhopping link, the minimum required ratio is 11 dB, whereas frequency hopping reduces the requirement to only 9 dB, which is an additional increase of 2 dB of margin in the channel [8]. The hopping adds frequency diversity to the channel.

Frequency hopping is an option for each individual cell. A base station is not required to support this feature. A mobile, however, has no choice but to switch to a frequency-hopping mode when the base station tells it to; mobiles must support SFH. A mobile station has much less RF power to flood a channel with than a base station does; it has only a small antenna, is roughly handled, and is carried around to locations far from optimum for propagation conditions, such as parking structures.

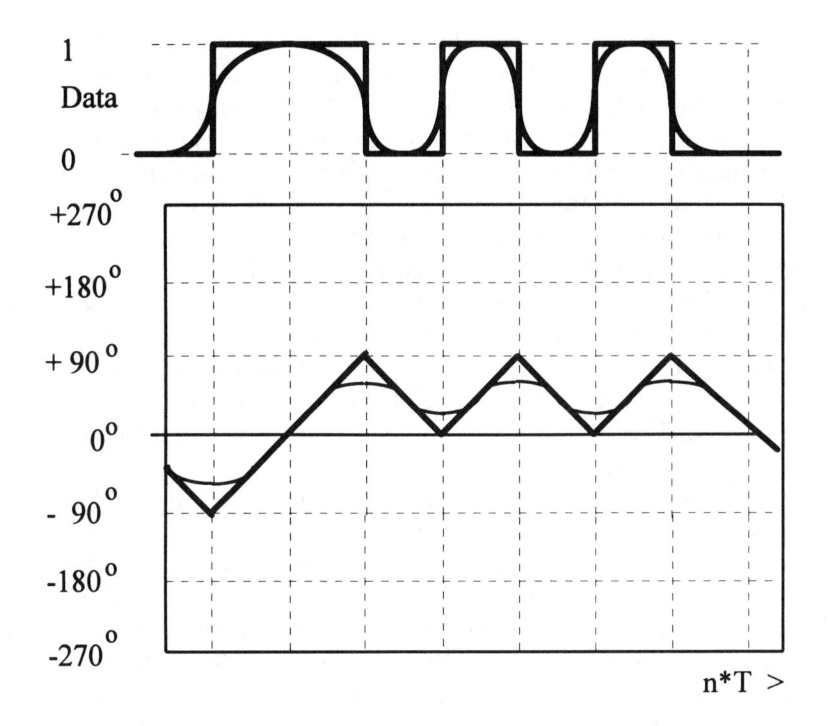

Figure 5.49 MSK versus GMSK phase trajectory.

The mobile needs to add frequency diversity to its transmissions as it moves toward the edge of a cell or as it enters an area of high interference, or at any other time the channel becomes marginal. When the BSC observes the failing channel from the mobile and decides to tell the mobile to turn on frequency hopping, it simply assigns the mobile a full set of RF channels rather than a single RF channel. The mobile performs a "dance" on the assigned set of frequencies to satisfy its SFH obligations. Different *hopping algorithms* can be assigned to the mobile station with the channel set. One is *cyclic hopping*, in which hopping is performed through the assigned frequency list from the first frequency, the second frequency, the third, and so on until the list is repeated. The other general algorithm is (pseudo) *random hopping*, in which hopping is performed in a random way through the frequency list. There are 63 different random dances that can be assigned to the mobile. When the mobile station has to assume SFH operation, it is advised of the channel assignment (a set of channels), and which one of the hopping algorithms it should use with an appropriate *frequency-hopping sequence number* (HSN). The base channel is not allowed to hop. The base channel, confined to time slot number 0—carrying the FCCH, the SCH and the BCCH—is the beacon upon which mobiles perform their periodic signal strength measurements on neighboring cells, and it is also the signal the mobile station uses to synchronize with a system as it initially seeks service or gets ready to move to another cell. The FCCH, SCH, BCCH exist only on the base channel on time slot 0, and all the other time slots on the base channel's frequency are filled with some kind of data to bring the base channel's power above all the other channels in the cell. A hopping traffic channel, however, can use time slot 0 on any BTS frequency channel not reserved for the base channel.

There are also two different implementations for frequency hopping in base stations. One of the implementations is *baseband hopping*, which is used if a base station has several transceivers available. The data flow is simply routed in the baseband to various transceivers, each of which operates on a fixed frequency in accordance with the assigned hopping sequence. The different transceivers receive a specific individual time slot in each TDMA frame, which contains information destined for different mobile stations (see Figure 5.50).

The other implementation is *synthesizer hopping*. One can find base stations in remote areas fitted with only one or two transceivers, but still want to use SFH (e.g., in hilly terrain). In this unusual case, the hopping is performed on the RF transceiver, which requires the transceiver to hop on the different frequencies itself.

Figure 5.51 shows the timing conditions for a mobile station during frequency hopping. It is SFH applied on three different frequencies. Since the mobile station first receives a message from the base station and then responds back to the base station three time slots later, there are four time slots of time during which the mobile station can accomplish different tasks, such as *monitoring adjacent cells*, before it has to hop to the next frequency, on which it will find the base station.

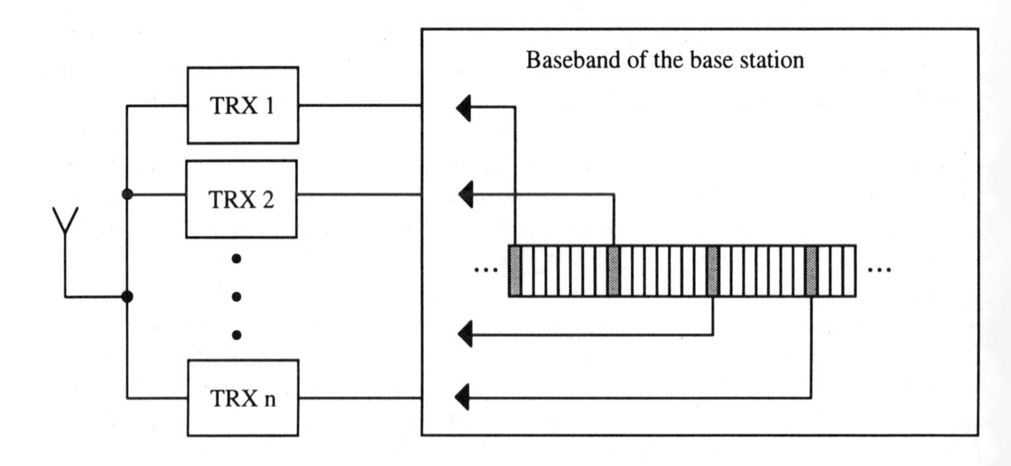

Figure 5.50 Implementing frequency hopping.

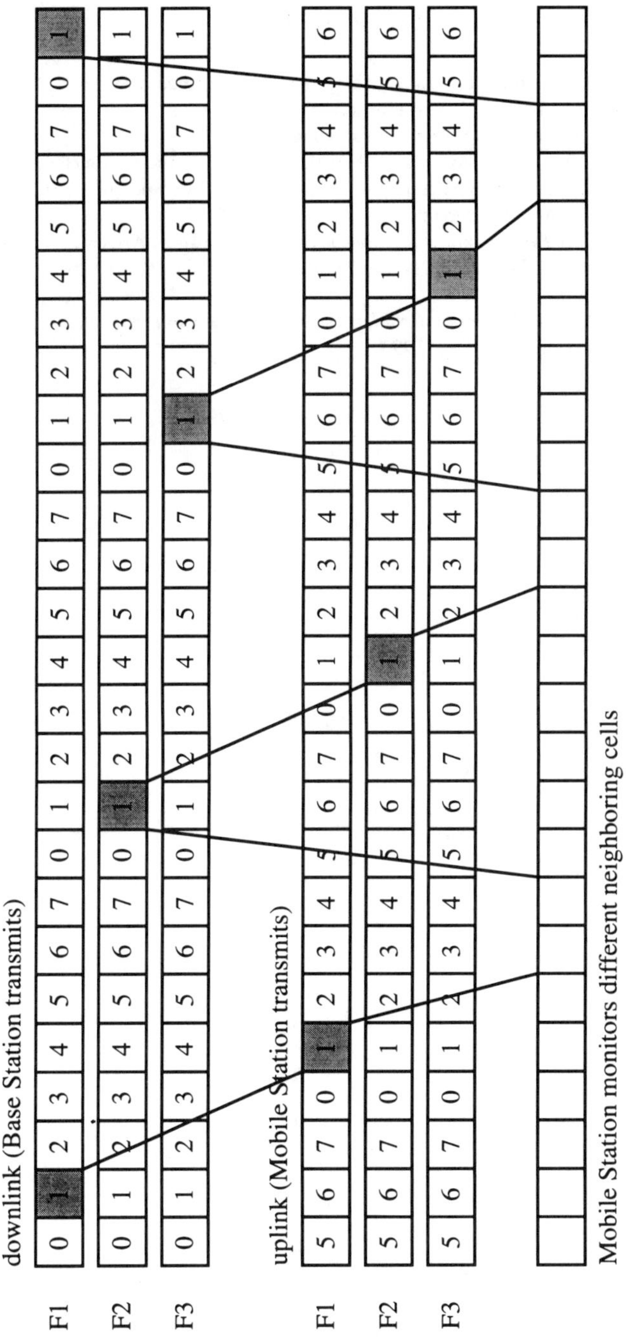

downlink (Base Station transmits)

F1 0 1 2 3 4 5 6 7 0 1 2 3 4 5 6 7 0 1 2 3 4 5 6 7 0 1

F2 0 1 2 3 4 5 6 7 0 1 2 3 4 5 6 7 0 1 2 3 4 5 6 7 0 1

F3 0 1 2 3 4 5 6 7 0 1 2 3 4 5 6 7 0 1 2 3 4 5 6 7 0 1

uplink (Mobile Station transmits)

F1 5 6 7 0 1 2 3 4 5 6 7 0 1 2 3 4 5 6 7 0 1 2 3 4 5 6

F2 5 6 7 0 1 2 3 4 5 6 7 0 1 2 3 4 5 6 7 0 1 2 3 4 5 6

F3 5 6 7 0 1 2 3 4 5 6 7 0 1 2 3 4 5 6 7 0 1 2 3 4 5 6

Mobile Station monitors different neighboring cells

Figure 5.51 SFH timing.

REFERENCES

[1] Watson, C., "Radio Equipment for GSM," Chap. 7 of *Cellular Radio Systems*, D. M. Balston and R. C. V. Macario, eds., Boston: Artech House, 1993.
[2] GSM TS 6.10, "GSM Full Rate Speech Transcoding."
[3] GSM TS 6.12, "Comfort Noise Aspects for Full Rate Speech Traffic Channels."
[4] GSM TS 6.31, "Discontinuous Transmission (DTX) for Full Rate Speech Traffic Channels."
[5] GSM TS 6.32, "Voice Activity Detection."
[6] Wiggert, D., *Codes for Error Control and Synchronisation*, Boston: Artech House, 1988.
[7] Abbott, E. A., *Flatland, A Romance of Many Dimensions*, 5th edition, New York: Barnes & Noble Books, 1963.
[8] Shostek, H., *Digital/Next Generation Cellular Technologies: A Competitive Analysis and Forecast*, 2nd edition, Silver Spring, MD, Dec. 1993.

CHAPTER 6

THE DATA LINK LAYER—LAYER 2

The protocol used in GSM for signaling transfer between a mobile station and the BTS is called the *link access protocol data mobile* (LAPDm) [1]. It is a mobile adaptation of the link access protocol data (LAPD), which is defined in the ISDN for fixed line networks [2]. LAPD is used in GSM on the *Abis interface* between the BTS and the BSC. This chapter provides a look at how *Layer 2* (LAPDm) works for GSM signaling transfers over the air. The tasks and functions of Layer 2 are listed and explained just as they were for the fictional ship's crew members in Chapter 4. Frames and field formats and their parameters are introduced, as well as an example of their function in a dynamic situation.

6.1 TASKS OF THE DATA LINK LAYER

The adaptation of the fixed network's LAPD for GSM's mobile radio applications was made because of there being no need to use frame limiters (flags) for synchronization in the LAPD protocol. This brought less development work and saved wasted bits on the air interface. Physical blocks of 184 bits (184 bits = 23 octets, 1 octet = 8 bits), which are compatible with LAPD, are generated by Layer 1 for signaling messages (see Section 5.14.4 and Figure 5.35). The 184 bits are channel-coded into 456 bits, which fit perfectly into 4 time slots. Special consideration was also given to the presence of different types of signaling channels that moving mobiles require. The modified and expanded data link layer is responsible for the correct and complete transfer of information blocks between Layer 3 entities over the GSM radio interface. The protocol contains the following functions:

- Organization of Layer 3 information into frames;
- Peer-to-peer transmission of signaling data in defined frame formats;
- Recognition of frame formats;
- Establishment, maintenance (supervision), and termination of one or more (parallel) data links on signaling channels;
- Acknowledgment of transmission and reception of numbered information frames (I-frames);
- Unacknowledged transmission and reception of unnumbered information frames (UI-frames).

The BTS passes signaling messages between the mobile station and the BSC or MSC. The BTS seldom takes part in the conversations except when it has to respond to commands for adjustments in its operation.

6.2 FRAME FORMATS

GSM defines four different frame formats used by Layer 2, named according to their formats: *A, B, Abis,* and *Bbis.* The *bis* designation is sometimes written as a prime mark, where A′ means the *other A* or the *A prime* frame format. There is also a fifth format defined for the RACH, *format* C, which is an exception to the rule that signaling channels use the LAPDm protocol; format C does not use real Layer 2 structures. A RACH contains only a tiny amount of information (8 bits) and should be considered a trivial case of signaling. The various frame formats are introduced in the following paragraphs. The frames are subdivided into fields and their structure is based on octets (1 octet = 8 bits).

6.2.1 Frame Formats A and B

Format A (see Figure 6.1) is used by DCCHs in those cases where no real Layer 3 information is to be transmitted. This can be, for example, on an SDCCH or on an FACCH. A frame is divided up into fields of eight bits each, and a field of eight bits is called an *octet.* The frame in Figure 6.1 is depicted as being made of four different kinds of fields: (1) the *address field,* (2) the *control field,* (3) the *length indicator field,* and (4) a field of *fill bits.* The control field has a fixed length of only one octet, where each bit in the octet is numbered 1 through 8. The other three fields have variable lengths as indicated by the dotted lines in their representations.

Now, why is such a silly thing necessary? Why do we need a frame format to transmit no information? In order to explain this, we have to imagine the case where a Layer 2 entity (in a mobile, for example) has to acknowledge frames coming from

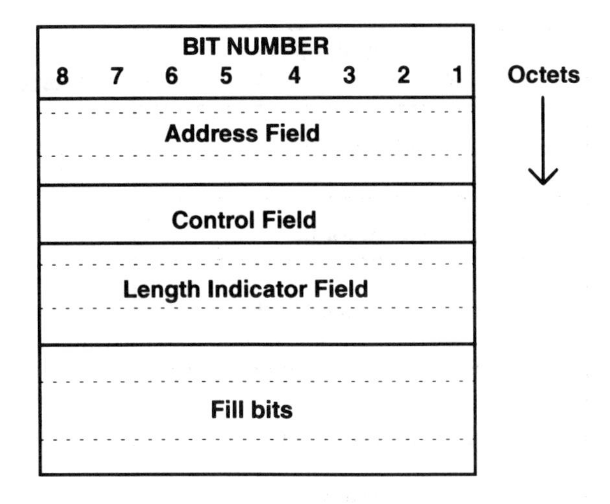

Figure 6.1 Frame format A.

the peer layer in another entity (in the BSS) during the presence of a signaling link. It often occurs that the receiving Layer 2 entity does not have a Layer 3 message to send after receiving a frame from its peer. In this case, it will simply transmit an empty frame to indicate that it has received the last frame, but does not yet have any information to send in response. The empty frame contains fill bits, which are coded with the hexadecimal value 2B (or not to be?) or FF. The binary representation is 00101011 for 2B and 11111111 for FF. The mobile station's Layer 2 can fill its empty frames with 2B or FF, and the base station's Layer 2 must use 2B.

While a signaling channel is active, there is no new Layer 3 information ready for transmission, and all previous frames have been acknowledged on both sides, then *fill frames* are exchanged. The fill frames indicate to the counterpart, the peer, that the signaling link is still active. Frame format A is used for this humble task. Frame format A is used when there is no signaling information to send but we do not want to give up the signaling channel.

The interesting situation occurs when real information has to be transferred over a DCCH. This can occur on either an SDCCH, an FACCH, or an SACCH. In the case where signaling information is ready for transmission, frame format B is used. In contrast to format A, format B contains an *Information field* (see Figure 6.2). Such a frame is called an *I-frame.* The mode in which numbered I-frames are transmitted or exchanged by Layer 2 entities is referred to as the *acknowledged mode,* and the procedure is called *multiple-frame operation.* The numbered and acknowledged mode is used to ensure a flawless transport of important information blocks. In doing so, the procedure numbers and counts the frames to make sure that no frames are lost and that all the frames arrive at the correct receiving Layer 3

destinations. To do this reliably, it needs mechanisms and parameters contained in the fields of these frames. Now we take a look at these fields, their parameters, and their meanings.

6.3 FIELDS AND PARAMETERS

Layer 2 uses the link access protocol to convey Layer 3 information. In other words, a Layer 2 entity (in a machine) receives information (signaling data) from its Layer 3 (the boss), sticks it in a numbered and properly addressed frame (envelope), and passes it on to Layer 1 (the post office) for transmission. Its peer, which is Layer 2 on the other side in another machine, receives the envelope from its own Layer 1 (mailbox), removes it from the envelope, and passes its contents up to its Layer 3 (another boss). Layer 2 acts as a clerk or a receptionist. Of course, the receiver of a letter is a clerk, who is conscientious and always checks whether the address on the envelope identifies the boss as the legitimate receiver of the letter, and whether the number on the envelope indicates that this letter is in the correct sequence and if the clerk should not have, perhaps, received and opened another letter earlier. In case the address and the sequence number are correct, the receiver will check whether the whole letter was received with no missing pages or obliterated words. The clerk would not dare deliver something incomplete to the supervisor (Layer 3), who is somewhat willful, impatient, and easily angered.

A length indication on the envelope helps the clerk find out how many lines or pages it contains. With this helpful information, the clerk can learn whether the envelope contains the whole letter, or whether the rest of the letter is in the next envelope to be expected. Perhaps the first envelope was not big enough to hold the whole letter, or there is a postage limitation imposed by the managers. In case there was some trouble with the mail, the receiver will have to indicate this to the sender. One clerk would have to ask the other clerk—by writing a short note—to send a certain letter again. The integrity of such a protocol can only be guaranteed when the senders of such letters (the clerks) hold strictly to the conventions listed in the protocol in all its details and without exception.

The practice of splitting up long letters, sending the parts in more than one envelope, and putting the parts together again after reception is called *segmentation and concatenation*. The protocol tools that enable the Layer 2 to perform all the functions described here are included in the fields of the frames.

6.3.1 Address Field

The address field holds its place as the first field in A- and B-formatted frames. As Figures 6.1 and 6.2 show, it has a variable length. In the case of the GSM control channels, however, the address field is fixed at one octet (eight bits).

The address field is primarily used to address the *service access point* (SAP), which is a defined interface at which Layer 2 offers and provides its services to a Layer 3 entity. GSM provides two such SAPs for use on the radio interface: one is for signaling and the other is for short messages, or SMS. There is a gray area between user traffic and SMS. The messages in SMS are, in a sense, user traffic, because they are read by the user and are not really signaling messages meant for network entities. They are transported as if they were signaling in much the same way pages over the public address system in an airport to a specific passenger are handled (i.e., though the public address mechanism was intended for calling airport personnel, it is convenient to use the same system to pass a message to a customer awaiting an arrival of a flight). A single Layer 2 entity can provide service at more than one SAP, in parallel, on the same signaling channel. The SAP address determines to which and from which Layer 3 entity a message is to be transported by Layer 2. Figure 6.5 shows the content of the address field and the meaning of each of the parameters in it.

Bit number 8 is unused. Bits numbered 7 and 6 are coded as zeros (link protocol discriminator (LPD)). Bits 5 to 3 identify the SAP, and thus their value is referred to as the *SAP identifier* (SAPI). SAPI = 000 is used for Layer 3 signaling, and SAPI = 011 is used for short messages (SMS). All other values are reserved. Bit number 2 is the command/response (C/R) bit, which indicates whether the Layer 2 frame contains a command or a response to a command. A mobile station would send a command with the C/R bit set to zero (C/R = 0) and a response with C/R = 1. The base station uses the opposite sense and sends a command with C/R = 1 and a response with C/R = 0. Finally, bit number 1 is the extension bit (E/A bit) in the address field. It is set to the value 1 when the address octet is the last one in the particular address field. A zero means that the next octet in the frame is part of the address field. When there is only one octet used for the address field, then the E/A bit will always be coded with 1.

6.3.2 Control Field

The control field consists of only one octet and is only found in A- or B-formatted frames. It identifies the type of frame, which can, again, be either a command or a response frame. There are three types of such frames.

BIT NUMBER

8	7	6	5	4	3	2	1
X	LPD		SAPI			C/R	E/A

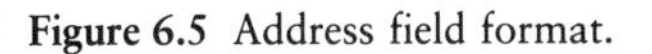

Figure 6.5 Address field format.

1. I-frame: This type of frame is numbered and carries numbered Layer 3 information. The number is included in counter variables contained in the control field.
2. S-frames: These kinds of frames (supervisory frames) are control functions like (a) I-frame acknowledgment with fill frames (format A), (b) requests for repetitions, and (c) intermittent suspension of numbered information transfers. The Layer 2 entities use S-frames to talk among themselves as they make use of the counter variables contained in the control field. The S-frames represent the conversations the radio operators in Chapter 4 might have in order to agree on codes and frequencies to use for reliable communications.
3. U-frame: The unnumbered type of frame offers additional data link functions for Layer 2 and also allows for the transfer of unnumbered and unacknowledged information frames on the control channels. Frames are not counted and no counter variables are used.

The three possible configurations of the control field for the three types of uses the A and B frames can assume are shown in the Figure 6.6(a–c).

The parameters shown in the Figure 6.6(a–c) have the following meanings and purposes.

N(S): Transmitter Send Sequence Number

The N(S) parameter is only found in I-frames and indicates the current number of the transmitted I-frame, including the one in which the N(S) parameter is contained.

BIT NUMBER

8	7	6	5	4	3	2	1
N(R)			P	N(S)			0

BIT NUMBER

8	7	6	5	4	3	2	1
N(R)			P/F	S	S	0	1

BIT NUMBER

8	7	6	5	4	3	2	1
U	U	U	P/F	U	U	1	1

Figure 6.6 (a) Control field format for I-frames; (b) Control field format for S-frames; (c) Control field format for U-frames.

It starts with the value 0. Three bits are reserved for this parameter, so that it counts from 0 (000) to 7 (111), and then starts over again with 0. This is also called a *modulo-8 counter.*

N(R): Transmitter Receive Sequence Number

The N(R) can be found in an I-frame or in an S-frame, and indicates the number of the next expected I-frame. It also holds three bits and can count from 0 (000) to 7 (111).

Poll/Final Bit

In command frames, the poll/final (P/F) bit is referred to as the *poll bit,* in which case it is set to 1 (P = 1), and indicates that a response is requested. This will occur when no response to a command frame has been received before. In response frames, the P/F bit is referred to as the *final bit.* It is set when a response to a command is sent back to a peer which had the poll bit set to 1.

S-Bits and U-Bits

These bits are called *supervisory* and *unnumbered function bits,* respectively, and are coded according to the different functions of the supervisory and unnumbered frames they are part of.

These functions are explained in the following paragraphs.

6.3.2.1 I-Frames

When transmitting information frames, Layer 2 has to set the transmitter send and receive sequence numbers accordingly.

6.3.2.2 S-Frames

For supervisory frames, there are three different functions defined, and each of them is distinguished from the other by a different coding of the S-bits.

Receive Ready: RR Frame, S S = 0 0

The RR frame has three functions:

- It indicates that a data link entity (Layer 2) is ready to receive an I-frame.

- It acknowledges the reception of up to N(R)-1 frames.
- It resets Layer 2 in case it was not in a state of being ready, which can occur after an indication of "receive not ready" (see below).

Receive Not Ready: RNR Frame, S S = 0 1

The RNR frame is used by a data link layer in case it cannot (temporarily) receive I-frames.

Reject: REJ frame, S S = 1 0

A REJ frame is sent when a repetition of one or more I-frames is required. This can happen, for instance, when the transmitter send sequence number N(S) of a received frame indicates that one or more frames have been lost. The REJ frame would, in this case, carry the number (in the N(R) field) from which the peer should start the retransmission of I-frames.

6.3.2.3 U-Frames

There are five different U-frame functions defined, each of which has a different coding of the U-bits.

SABM Command: U U U U U = 0 0 1 1 1 (bits 8-7-6-4-3 in the control field)

SABM (set asynchronous balanced mode) starts the modulo-8-counted I-frame transmissions (multiple-frame operation establishment). It is allowed to place a valid information field into an SABM frame. SABM is usually used in the uplink (i.e., from the mobile station to the base station).[1] It is also used for initial contention resolution; that is, in the case where more than one Layer 2 entity (multiple mobile stations) try to start multiple-frame operation on the same channel. The SABM containing an Information field with real information (an identification number is an example) specific to a single mobile station helps to identify the frame with which the other entity (base station) wants to establish multiple-frame operation (by unnumbered acknowledgment, see below).

DISC Command: U U U U U = 0 1 0 0 0 (bits 8-7-6-4-3 in the control field)

The DISC command (disconnect) ends a multiple-frame operation session. There is no information field in a DISC frame; fill bits are used instead (frame format A).

[1]One exception is for short messages coming from the network.

UA Response: U U U U U = 0 1 1 0 0 (bits 8-7-6-4-3 in the control field)

The UA response (unnumbered acknowledgment) is sent to acknowledge an SABM command or a DISC command. When sent after receiving an SABM frame, the UA frame contains the same information that was included in the SABM frame, thus allowing initial contention resolution (see above). It identifies the sender of the SABM it responds to. The UA response to an SABM command also implies a reset of the counter variables, N(S), N(R), which are internal variables within a Layer 2 entity.

UI Command: U U U U U = 0 0 0 0 0 (bits 8-7-6-4-3 in the control field)

The coding of the control field with the UI pattern (00000) appropriately indicates the transmission of unnumbered information within the frame. This can be, as we have seen, the measurement reports in the SACCH from the mobile station, or it can be fill frames.

DM Response: U U U U U = 0 1 0 0 0 (bits 8-7-6-4-3 in the control field)

The DM response (disconnect mode) is sent from a data link layer entity when it is in a state that does not allow multiple-frame operation. This can be the case after reception of a command it cannot respond to. There is no information field in a DM response frame.

6.3.3 Length Indicator Field

The length indicator field can consist of a variable number of octets. As with the address field, one octet is sufficient for the applications in GSM signaling channels. Figure 6.7 shows the format of the length indicator field.

The *L* in Figure 6.7 (bits 8 through 3) stands for the length of the Information field in the remainder of the frame. It is a binary representation of the number of octets. The *M* stands for the "more" data bit. It indicates whether the frame contains

BIT NUMBER							
8	7	6	5	4	3	2	1
L						M	EL =1

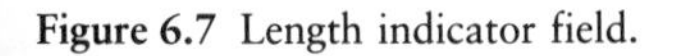
Figure 6.7 Length indicator field.

a whole message or whether another part of the message is contained in a subsequent frame. Again, we are talking about segmentation and concatenation here. When $M = 0$, then either a whole Layer 3 message is included in the frame's Information field, or the frame contains the last part of a segmented message. When $M = 1$, then the frame does not contain a whole Layer 3 message; another part of the message should follow in the next frame(s).

In accordance with the definition [2], the length indicator field can consist of more than one octet. The extension bit of the length indicator field (EL) indicates whether the particular octet is the last one in the length indicator field ($EL = 1$), or is not the last octet in the field ($EL = 0$).

6.3.4 Information Field

In case there is an Information field present in a frame, which would be the case for the B and Bbis formats, the length indicator field is followed by the information field. The information field carries a Layer 3 message, and its maximum length is dependent on the logical channel in which the frame appears. Table 6.1 shows the maximum lengths for each channel type.

When the Layer 3 information is too small to fill a whole frame, the remainder of the frame is filled with fill bits (2B or FF in hexadecimal notation). No information content would mean that the whole frame is filled with fill bits (thus making it a fill frame), and the length indicator L would be equal to zero.

6.4 EXAMPLE OF A LAYER 2 TRACE

To illustrate the data link layer's functions, we will now have a closer look at a trace or a "recording" of a real Layer 2 connection. This trace was generated by a GSM test system, which simulated a location update procedure with a GSM mobile station.

Table 6.1
Length of Information Field for Different Signaling Channel Types

Channel Type	*Number of Octets*	*Number of Bits*
SACCH	18	144
FACCH	20	160
SDCCH	20	160
BCCH	22	176
AGCH	22	176
PCH	22	176

The mobile station starts by transmitting on the RACH as it requests a channel. The RACH from the mobile station should be looked at as the mobile's first attempt to get the base station's attention; it has no information that could identify the mobile and puts the mobile station in the position of a student in a crowded classroom trying to attract the base station's attention by raising its hand. The base station, which is simulated by the test system, responds on the AGCH with an unacknowledged frame assigning a channel, the description of which is in the information (info =) field. An SDCCH comes back from the mobile station. We will not look into the message contents any further now, but rather will confine our attention to the Layer 2 dynamics. The messages themselves, which are confined to the Information fields, will be examined in Chapter 7. The mobile station transmits an SABM command on the new SDCCH channel in order to establish multiple-frame operation. The SABM frame already contains a Layer 3 message in the Information field (info = 05 08 00 . . .). It is acknowledged by a UA response from the base station on the SDCCH. Note that the UA frame carries the same information field content as the SABM frame from the mobile station (initial contention resolution, see Section 6.3.2.3), except that the fill bits are 2B instead of FF. From this point in time, we watch the two counter variables, N(S) and N(R), which were reset after the SABM/UA frame. Before the radio channel (Layer 1 connection) is released, the data link layer connection is terminated by a DISC command and a UA response.

The following additional information applies to the trace shown in Figure 6.8.

The frame number (FN) is indicated by the time stamp (T1:T2:T3) and counts the frames we saw in Section 5.11.5. It is our little pocket-sized schedule for the GSM physical layer. T2 counts from 0 to 25 and T3 counts from 0 to 50. The direction of transmission over the radio interface, Um, is away from the base station number 1 (dl1 = downlink, base station number 1) to the mobile station. The mobile station transmits in the uplink direction to base station number 1 (ul1 = uplink, mobile station to base station number 1). The mobile's Layer 2 messages appear on the left-hand side of the trace, and those of the base station on the right-hand side. Signaling messages occur on logical channels (RACH/AGCH/SDCCH). In order to get a better overview, all fill frames, except one example, were erased from the trace. Fill frames tend to litter these kinds of traces beyond the point where we can follow what is happening. Additional comments were added and are printed in a distinctive type face.

No voice traffic is moved here; the radios are merely getting ready to move the user's voice.

Turn to Part III, particularly Chapter 11, to see when someone would use an instrument capable of generating a Layer 2 trace. For now, the reader should relax in the knowledge that this kind of sophisticated capability is not required in general service testing or commissioning.

Layer 2 Trace

	MS transmits (uplink)	**BS transmits (downlink)**
Frame Number T1:T2:T3		
FN 0066178 0049:08:31	Umul1 rach Info = 04	
FN 0066204 0049:08:06		Umdl1 agch Length = 13, M = 0, EL = 1 Info = 06 3F 00 20 C0 1F 04 8B E8 00 08 F1 23 45 67 89 0A BC DE 2B 2B 2B
FN 0066235 0049:13:37	Umul1 sdcch Address: SAPI = 0, Command Control:SABM : N(R) = -, P/F = 1, N(S) = - Length = 0F, M = 0, EL = 1 Info = 05 08 00 62 F2 50 01 00 01 05 F4 00 00 09 41 FF FF FF FF FF	***Set Asynchronous Balanced Mode Command(SABM)***
FN 0066271 0049:23:22 ***UA response*** ***Info Field = SABM Info***		Umdl1 sdcch Address: SAPI = 0, Response Control:UA : N(R) = -, P/F = 1, (S) = - Length = 0F, M = 0, EL = 1 Info = 05 08 00 62 F2 50 01 00 01 05 F4 00 00 09 41 2B 2B 2B 2B 2B
FN 0066322 0050:22:22 ***BS transmits the first I frame N(S) number =0***		Umdl1 sdcch Address: SAPI = 0, Command Control:I : N(R) = 0, P/F = 0, N(S) = 0 Length = 13, M = 0, EL = 1 Info = 05 12 00 00 00 00 00 00 00 00 00 00 00 00 00 00 00 00 00 2B
FN 0066337 0050:11:37	Umul1 sdcch Address: SAPI = 0, Response Control:RR : N(R) = 1, P/F = 0, N(S) = - Length = 00, M = 0, EL = 1 Info = FF FF FF FF FF FF FF FF FF FF FF FF FF FF FF FF FF FF FF FF	***Acknowledgement by RR response (N(R) =1, one I frame received)***
FN 0066424 0050:20:22 ***Example of a fill frame (no Layer 3 message ready to be transmitted) Length = 0, fill bits= 2B hex***		Umdl1 sdcch Address: SAPI = 0, Command Control:UI : N(R) = -, P/F = 0, N(S) = - Length = 00, M = 0, EL = 1 Info = 2B 2B 2B 2B 2B 2B 2B 2B 2B 2B 2B 2B 2B 2B 2B 2B 2B 2B 2B 2B

Figure 6.8 Example of a Layer 2 trace.

FN 0066439
0050:09:37
Umul1 sdcch
Address: SAPI = 0, Command
Control:UI : N(R) = -, P/F = 0, N(S) = -
Length = 00, M = 0, EL = 1
Info = FF FF FF FF FF FF FF FF FF FF
FF FF FF FF FF FF FF FF FF FF

Fill frame from MS
Length = 0, fill bits = FF hex

FN 0066490
0050:08:37
Umul1 sdcch
Address: SAPI = 0, Command
Control:I : N(R) = 1, P/F = 0, N(S) = 0
Length = 06, M = 0, EL = 1
Info = 05 54 46 F5 17 73 FF FF FF FF
FF FF FF FF FF FF FF FF FF FF

MS transmits the first I frame
(Number 0)

FN 0066526
0050:18:22
BS transmits the second I frame
(Number 1) and acknowledges the first
frame received from the MS (N(R)=1)

Umdl1 sdcch
Address: SAPI = 0, Command
Control:I : N(R) = 1, P/F = 0, N(S) = 1
Length = 03, M = 0, EL = 1
Info = 06 35 00 2B 2B 2B 2B 2B 2B 2B
2B 2B 2B 2B 2B 2B 2B 2B 2B 2B

FN 0066541
0050:07:37
Umul1 sdcch
Address: SAPI = 0, Response
Control:RR : N(R) = 2, P/F = 0, N(S) = -
Length = 00, M = 0, EL = 1
Info = FF FF FF FF FF FF FF FF FF FF
FF FF FF FF FF FF FF FF FF FF

Acknowledgement by
RR response (N(R) =2)
(two I frames
received)

FN 0066592
0050:06:37
Umul1 sdcch
Address: SAPI = 0, Command
Control:I : N(R) = 2, P/F = 0, N(S) = 1
Length = 02, M = 0, EL = 1
Info = 06 32 FF FF FF FF FF FF FF FF
FF FF FF FF FF FF FF FF FF FF

MS transmits
its second I frame (No. 1)

FN 0066628
0050:16:22
BS transmits its third I frame (No 2)
and acknowledges the second received
from the MS (N(R)=2)

Umdl1 sdcch
Address: SAPI = 0, Command
Control:I : N(R) = 2, P/F = 0, N(S) = 2
Length = 07, M = 0, EL = 1
Info = 05 02 62 F2 50 01 00 2B 2B 2B
2B 2B 2B 2B 2B 2B 2B 2B 2B 2B

FN 0066643
0050:05:37
Umul1 sdcch
Address: SAPI = 0, Response
Control:RR : N(R) = 3, P/F = 0, N(S) = -
Length = 00, M = 0, EL = 1
Info = FF FF FF FF FF FF FF FF FF FF
FF FF FF FF FF FF FF FF FF FF

Acknowledgement by RR
response (N(R) =3)
(three I frames
received)

FN 0066679
0050:15:22
BS transmits its fourth I frame (No. 3)
(Note: BS's Layer 3 wants to
disconnect the channel)

Umdl1 sdcch
Address: SAPI = 0, Command
Control:I : N(R) = 2, P/F = 0, N(S) = 3
Length = 03, M = 0, EL = 1
Info = 06 0D 00 2B 2B 2B 2B 2B 2B 2B
2B 2B 2B 2B 2B 2B 2B 2B 2B 2B

Figure 6.8 (continued).

FN 0066694 Umul1 sdcch
0050:04:37 Address: SAPI = 0, Response

Control:RR : N(R) = 4, P/F = 0, N(S) = -
Length = 00, M = 0, EL = 1
Info = FF FF FF FF FF FF FF FF FF FF
FF FF FF FF FF FF FF FF FF FF

Acknowledgement by RR response (N(R) =4) (four I frames received)

FN 0066745 Umul1 sdcch
0050:03:37 Address: SAPI = 0, Command
Control:DISC : N(R) = -, P/F = 1, N(S) = -
Length = 00, M = 0, EL = 1
Info = FF FF FF FF FF FF FF FF FF FF
FF FF FF FF FF FF FF FF FF FF

MS disconnects the Layer 2 connection

FN 0066781
0050:13:22
BS acknowledges with a UA response

Umdl1 sdcch
Address: SAPI = 0, Response
Control:UA : N(R) = -, P/F = 1, N(S) = -
Length = 00, M = 0, EL = 1
Info = 2B 2B 2B 2B 2B 2B 2B 2B 2B 2B
2B 2B 2B 2B 2B 2B 2B 2B 2B 2B

The physical channel will finally be disconnected by the Layer 1.

Figure 6.8 (continued).

REFERENCES

[1] *GSM Technical Specifications,* ETSI, Sophia Antipolis, Vols. 04.05 and 04.06.
[2] CEPT TS 46-20 Recommendation and CCITT Q.920/921 Recommendations.

PART III

Testing GSM

CHAPTER 7

THE NETWORK LAYER—LAYER 3

Now it is time to hijack the GSM freight train, to see what is inside, to break into the time-slotted boxcars and spill the drums of unknown acids and solvents on the ground, to free the circus animals from their cages, and to pry loose the carefully marked and numbered boxes and bags from their neat and orderly framed pallets. Then we will paw through the letters, tear open the neatly addressed and numbered envelopes, and read through the private matters inside. It is time to dig deep into the freight and leave the carefully packed contents strewn on the side of the tracks so that we can regard the goods in all their details. In doing so, we will notice that everything was packed neatly and carefully, one box inside another. While we are engaged in our vandalism, we are careful to leave the passengers and their bags undisturbed; our interest lies only with the signaling freight, not the paying passenger traffic. The network layer (Layer 3) is the valuable freight in the GSM freight train, and the passengers are the traffic.

The network layer in the GSM architecture, also referred to as the *signaling layer,* uses a protocol that contains all the functions and details necessary to establish, maintain, and then terminate mobile connections for all the services offered within a GSM PLMN. The network layer also provides control functions to support additional services such as supplementary services and short message services. Due to the complex forms of signaling, particularly those introduced by a mobile user, the GSM signaling system needs thoroughly defined procedures and structures for the protocol to be implemented reliably in the network layer [1]. This chapter will describe the structure of Layer 3, which can be divided further into three *sublayers*. The general procedures of the Layer 3 protocol are introduced, as well as the parameters and the elements of a Layer 3 message. For better practical understanding, an example is offered on how Layer 3 signaling messages are exchanged during the process of setting up a call over the radio interface.

Layer 1 is the freight train, switches, lights, and tracks. Layer 2 is the pallets, boxes, drums, and carefully labeled envelopes in the train. Layer 3 is the valuable contents of all the containers and envelopes themselves. Please do not confuse the layered protocols with the speech traffic; the user's speech data do *not* move within the secure confines of the signaling layers.

7.1 SUBLAYERS OF LAYER 3

As mentioned above, there are three sublayers defined for Layer 3: (1) *radio resource management* (RR), (2) *mobility management* (MM), and (3) *connection management* (CM). A sublayer can be regarded as an entity unto itself, which handles the tasks and functions of a specific segment of signaling in the GSM PLMN. Its protocol consists of messages and procedures that allow each sublayer to fulfill the tasks assigned to it.

7.1.1 Radio Resource Management Sublayer

The tasks covered in this segment of the network layer are closely connected to the physical layer. The RR sublayer is responsible for the management of the frequency spectrum, the GSM system's reactions to the changing radio environment, and everything related to maintaining a clear channel between the PLMN and the mobile station. These responsibilities include channel assignment, power-level control, time alignment, and handover from one cell to another. The RR sublayer handles all the procedures necessary to establish, maintain, and release dedicated radio connections. Appropriately, the radio connections are also called RR connections. An RR connection is necessary in order to provide a path for further signaling traffic and then, eventually, a suitable traffic channel to carry the user's data.

Within the RR sublayer, we find some carefully defined procedures used to cover these tasks:

- Channel assignment procedures;
- Channel release;
- Channel change and handover procedures;
- Change of channel frequencies, hopping sequences (hopping algorithms), and frequency tables;
- Measurement reports from the mobile;

- Power control and timing advance;
- Modification of channel modes (speech and data);
- Cipher mode setting.

A fixed network has no equivalent of the RR sublayer, because it has permanently wired connections to each of its subscribers, who gain access to it by simply lifting the receiver on the phone.

7.1.2 Mobility Management Sublayer

The MM sublayer has to cope with all the effects of handling a mobile user that are not directly related to radio functions. These tasks include the kinds of things a fixed network would do to authorize and connect a user to a fixed network, which are modified to account for the fact that the user may not remain in the same place:

- Support of user mobility, registration, and management of mobility data;
- Checking the user and the equipment identity;
- Checking if the user is allowed to use the services and what kind of extra services are allowed;
- Support of user confidentiality (registering the user under a temporary mobile subscriber identity (TMSI));
- Provision of user security;
- Provision of an MM connection, based on an existing RR connection, to the CM sublayer (see Section 7.1.3).

The MM sublayer uses a large number of predefined procedures to meet the needs listed above:

- Location update procedure.
- Periodic updating.
- Authentication procedure.
- IMSI attach procedure. A mobile station will perform (when indicated by the *attach flag* in the base station's BCCH) a location update procedure after power-on, present its IMSI to the network, and get a TMSI in return.
- IMSI detach procedure. A mobile station will perform (when indicated by the attach flag in the base station's BCCH) a *detach* procedure just after power-off, telling the network that it is no longer in service.
- TMSI reallocation procedure.
- Identification procedure.

7.1.3 Connection Management Sublayer

The CM sublayer manages all the functions necessary for circuit-switched call control in the GSM PLMN. These functions are provided by the call control entity within the CM sublayer. There are other entities within the CM sublayer to cope with providing supplementary services and SMSs. The call control responsibilities are almost identical to those provided in a fixed network; the CM sublayer is virtually blind to the mobility of the user. The call control access schemes used in GSM are, to a large degree, inherited from the ISDN, with only a few simplifications and adaptations. The call control entity in a GSM network establishes, maintains, and releases call connections for communication links. Again, there are some specific procedures defined for this purpose:

- Call establishment procedures for *mobile-originated calls;*
- Call establishment procedures for *mobile-terminated calls;*
- Changes of transmission mode during an ongoing call (incall modification);
- Call reestablishment after interruption of an MM connection;
- Dual-tone multifrequency (DTMF) control procedure for DTMF transmissions.

Note: Two low-frequency tones within the audible frequency range from a 4 × 4 matrix can represent any one of 16 depressed dial keys. This DTMF dial method is used in the U.S. fixed telephone network, in contrast to the pulsed dialing used in so many other places. The analog transmission of such tone sequences can also be used to access certain supplementary services in the network (e.g., to check a mailbox or enable call forwarding to another number). GSM is a digital network, and digital values (bits) have to be transmitted instead of analog binary representations such as tones.

A call control connection is always based on an existing MM connection. The call control entity uses the MM connection to exchange information with its peer.

7.2 STRUCTURE OF A LAYER 3 MESSAGE

Before taking a closer look at a typical signaling sequence, let us see what the process should do. A signaling process may contain one or more signaling procedures, and each procedure consists of an exchange of signaling messages. A signaling message is, however, not the smallest unit we can consider when dividing signaling procedures into their constituent parts. A signaling message consists of elements, and each element has a specific purpose, function, and information content.

Figure 7.1 shows a schematic view of the structure of a Layer 3 message. Note the similarity to the Layer 2 envelope in which this message will eventually be placed. The information contained in the message, and its binary representations (each octet of eight bits), is the Layer 3 part that goes into the Information field of a Layer 2 frame.

Only the *optional* and *mandatory* information elements of the Layer 3 message have a variable length, indicated with the dotted lines in Figure 7.1. All the elements of the message are explained in the following four sections.

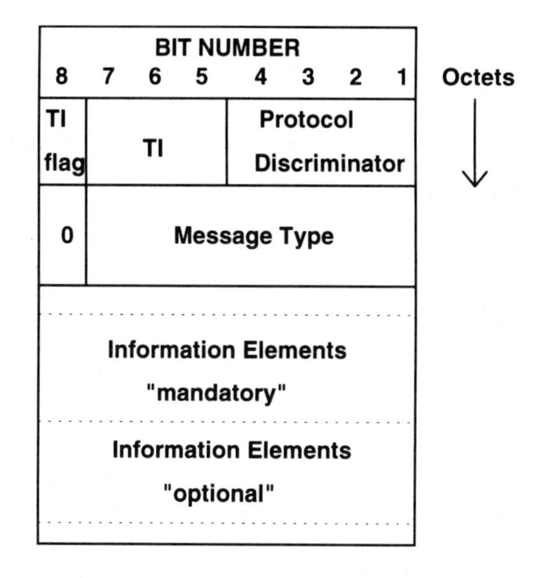

Figure 7.1 Structure and elements of a Layer 3 message.

7.2.1 Transaction Identifier

The transaction identifier (TI) is a pointer with a length of four bits. It is used to distinguish between (possible) multiple parallel CM connections and between the various transactions taking place over these simultaneous CM connections. For RR and MM connections, the TI is not relevant; the TI field is always coded with 0000 in these cases. The TI uses the bits 8-7-6-5 in the first octet of a Layer 3 message. Bit number 8 is called the *TI flag* and indicates the originator of a CM connection. It is set to 0 by the originator of a CM message, and when a message is sent back to the originator by its peer entity, it is coded with a 1. The originator also fills in the *TI value* (bits 7, 6, and 5). The TI value is kept for the duration of a transaction and serves as a label for it.

7.2.2 Protocol Discriminator

The protocol discriminator (PD) links the Layer 3 protocol to the entity the message is addressed to. It has a length of four bits, and it identifies the six protocols shown in Table 7.1.

Table 7.1
Protocol Discriminators Used in GSM Signaling

Protocol	*PD Binary*	*PD Hexadecimal*
Radio resources mgmt.	0 1 1 0	6
Mobility management	0 1 0 1	5
Call control	0 0 1 1	3
Short-message service	1 0 0 1	9
Supplementary service	1 0 1 1	B
Test procedure	1 1 1 1	F
All other values are reserved		

7.2.3 Message Type

The message type (MT) indicates the function of the Layer 3 message. It uses the lower six bits of the second octet in the Layer 3 message. Six bits are sufficient to address 64 different message types in a protocol indicated by the PD. The MT is part of a set of messages in a protocol. Bit 8 is reserved and is always set to 0. Bit 7 is a send sequence variable and may be used for MM and CM messages.

7.2.4 Information Elements

A Layer 3 message may contain one or more information elements (IE). Whether or not there are IEs, and how many IEs are contained in a message, depends on the MT. There are distinct definitions for the different MTs, each of which decides which IEs (if any) must or may be present in a message. GSM distinguishes mandatory IEs, which appear earlier in a message and have a reserved place in the structure of the message, from optional IEs. Optional IEs, if present, appear after the mandatory IEs. Optional IEs always have to carry an *information element identifier* (IEI), which tells the receiver of the message the purpose of the information contained therein.

Another distinction can be made between IEs with fixed and variable lengths. IEs with variable lengths also have to carry a length indicator, which is one octet. The length indicator is not to be confused with the length indicator field in a Layer 2 frame.

There are, then, four possible combinations of IE types:

1. Mandatory fixed length (MF);
2. Mandatory variable length (MV);
3. Optional fixed length (OF);
4. Optional variable length (OV).

7.3 EXAMPLE OF A LAYER 3 MESSAGE

The structure and functions within a message can be illustrated with a complete signaling message. The example chosen is an *immediate assignment* message, which is sent on the AGCH by a base station in response to an initial channel request (random access burst) from a mobile station. Its purpose is to assign a dedicated channel for a continuing signaling procedure. It is, itself, part of an initial channel assignment procedure, which is performed in order to establish an RR connection. Figure 7.2 shows the content of an immediate assignment message.

The immediate assignment message contains most of the information the mobile station needs to establish an RR connection with the network. This information is contained in the IEs of the message. The reader should note that the mandatory IEs

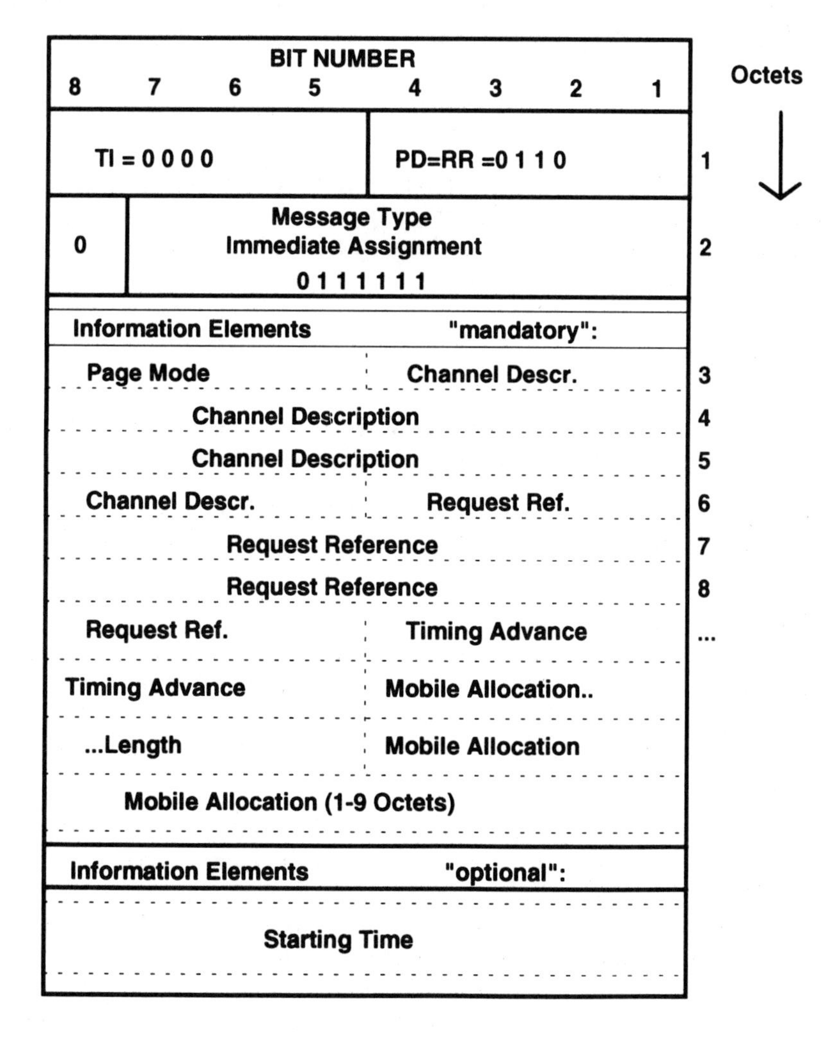

Figure 7.2 Example of a Layer 3 message, immediate assignment.

appear before the optional IEs, the ones with a fixed length before the ones with variable length. The message, including all the fixed-length fields at its start, are eventually mapped into the Information fields in an appropriate Layer 2 frame.

The mobile station may request a channel for various reasons. Some of the reasons are (1) establishing a call after being paged, (2) establishing a call after the user has dialed a number and pressed the SEND button, and (3) performing a location update procedure.

The TI has the value 0000, because the immediate assignment belongs to the RR protocol (the PD indicates this; see the first entry in Table 7.1). The MT is followed by the mandatory and optional IEs. Six IEs are examined here.

7.3.1 IE Page Mode

The page mode informs the mobile station how and with which paging blocks within the PCH it will still have to monitor the network. The PCH configuration on the broadcast channel is subject to change due to different loads on a base station. The IE page mode consists of only half of an octet (four bits).

7.3.2 IE Channel Description

The IE channel description contains all the necessary information about the physical dedicated channel identified in this initial assignment.

1. Channel type: Five bits are used to indicate the channel type, including a subchannel number, if applicable (e.g., FACCH/TCH, SDCCH/4, SDCCH/8). The applicable bit patterns are:

00001	for FACCH/TCH;
0001T	for half-rate FACCH/TCH (when half-rate is introduced);
001TT	for SDCCH/4 or CBCH (cell broadcast for short messages) on SDCCH/4;
01TTT	for SDCCH/8 or CBCH (cell broadcast for short messages) on SDCCH/8.

   The T-bits indicate the subchannel number. The bit pattern 01010 would indicate an SDCCH/8 (01 binary = the fourth channel type), subchannel 2 (010 binary = 2 decimal).
2. The time slot number: Three bits address the time slot number to be used on the assigned channel, and the range is from 0 to 7.
3. The training sequence code (TSC): Three bits indicate the TSC to be used on the assigned channel. There are eight different TSCs defined, and they range in from 0 to 7.

4. The hopping bit: The hopping bit indicates whether frequency hopping should be used.
5. The mobile allocation index offset (MAIO) and the HSN: In the case of frequency hopping, these two parameters inform the mobile stations of the channels to hop among and which hopping sequence to use.
6. The ARFCN: In the case where no frequency hopping is to be used, this parameter defines the channel number of the newly assigned channel.

7.3.3 IE Request Reference

This IE consists of three octets. It contains the five-bit random number, which was sent by the mobile station in the channel request (random access burst), and the three-bit coded cause of the channel request (eight bits total). Two more octets are used to tell the mobile station at which frame number the channel request (in the mobile's random access burst) was received by the base station. The mobile's channel request is answered by this particular immediate assignment message. This little trick makes it possible for the mobile station to identify its own unique immediate assignment message after sending only eight information bits. The mobile station knows the random number, and it knows the frame number in which it sent the channel request. The problem of more than one mobile station accessing the same channel because of reading the same immediate assignment message is solved with a rather high probability of success with this trick.

7.3.4 Timing Advance Value

The first correction value for the mobile station's burst timing is sent in the first dedicated message to a terminal. The base station derives this value from the timing error in the random access burst it receives from the mobile station. The first timing correction makes sure that the next burst from the mobile station fits into the assigned time slot (see Section 7.3.2, item 2). Additional timing advance adjustments are contained in the SACCH, which is sent along with any dedicated channel (SDCCH or TCH/FACCH).

7.3.5 IE Mobile Allocation

When it comes time to assign a dedicated channel with frequency hopping, this IE assigns the frequency channel numbers to be used in the hopping sequence. The selection is a subset (maximum: all) of the frequency channels used by the base station in a cell or a cell sector. This IE has a variable length of from one to nine octets.

7.3.6 IE Starting Time

The only optional IE in the immediate assignment message has a length of three octets, including the IEI. It is used to tell the mobile station a frame number. The mobile has to wait until the indicated frame number appears (in the T1, T2, and T3 timers) before it can start further transmissions. The maximum time covered by this IE is a little more than three minutes.

7.4 EXAMPLE OF A SIGNALING PROCESS: CALL ESTABLISHMENT

A number of signaling messages have to be exchanged between a mobile station and a base station in order to establish a call connection over the radio interface. The base station is the element in the GSM network the mobile station *talks* to. However, most of the signaling messages in the downlink (from the base station to the mobile station) do not originate in the base station (BTS) itself, but originate further back in the network, usually from the BSC or the MSC. Some signaling messages originate in the mobile, and most uplink signaling messages are directed to the entities behind the BTS, as well. The BTS only transports the messages between the data link layer and the physical layer; it acts as a bit mapping device as it transforms the Um bit sequences to the Abis bit patterns. The BTS can make the reverse transformation as well. Figure 7.3 shows the signaling flow exchanged during a *mobile-originated call setup* in the form of MTs. Mobile users dial the number they want to reach and then push the SEND button on the phone. The mobile station, of course, has to be registered to and synchronized with the network before it can indicate service availability on its handset display. Figure 7.3 shows that 14 messages are required to activate a traffic channel, but this signaling flow takes place within just a few seconds. The actual duration of this signaling sequence is strongly dependent on the logical channel used for information transfer.

The mobile station starts with the transmission of a *channel request* message (random access channel) for the remainder of its signaling needs. The immediate assignment message assigns the requested signaling channel (see Section 7.3 for some of the details). Once it is assigned to a dedicated channel, the mobile station tells the network what it desires, in this case the services of the CM entity. It transmits a *CM service request* message to make its wishes known to the network. Before this service can be granted, some RR matters and some MM procedures have to be sorted out. We have already learned that a call control connection or a CM connection will be based on existing RR connections and MM connections. The necessary connections are established by the authentication procedure (*authentication request* and *response*) and the subsequent RR procedure, which sets the cipher mode (*cipher mode command* and *cipher mode complete*). The authentication response will contain a *key* that the network uses to verify the mobile station's authorization to obtain

Mobile Station transmits		Network transmits
Channel request	→	
	←	Immediate Assignment
Connection Management Service Request	→	
	←	Authentication Request
Authentication Response	→	
	←	Ciphering Mode Command
Ciphering Mode Complete	→	
Setup	→	
	←	Call Proceeding
	←	Assignment Command
Assignment Complete	→	
	←	Alerting
	←	Connect
Connect Acknowledge	→	

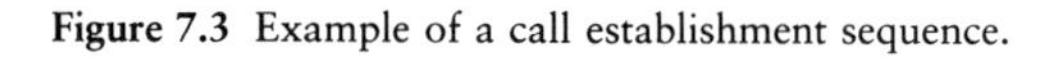

Figure 7.3 Example of a call establishment sequence.

service. The cipher mode command is used by default in this procedure in order to switch on an encryption process in Layer 1. It is, however, also possible that the cipher mode command indicates that *no ciphering* is used. Ciphering would start with the transmission of the cipher mode complete message, which will already be encrypted. Then the mobile station will transmit the *setup* message indicating the type of service it requires from the CM entity, the user number it would like to reach, and (optionally) identification of itself and its capabilities. The network starts a routing process in order to connect the addressed user. In the meantime, the network assigns a traffic channel for the transmission of user data (speech). This *assignment command* is answered by an *assignment complete* message from the mobile, which is already on the new channel (TCH/FACCH) and is using the FACCH for the remaining signaling messages. The old signaling channel is released and this resource is made available for other signaling links. When the other party is reached and alerted (the phone rings), the network transmits the *alert message* to the mobile. The mobile station generates an audible ring-back tone to tell its user that the other party has been reached and is being alerted. There are also some procedures defined in case the other party is busy or cannot be reached at all. In the successful case, the other party will accept the call by lifting the receiver, which is indicated by the *connect message* to the mobile. The acknowledgment by the mobile station is accomplished with the *connect acknowledge* message it sends back to the network. The FACCH, which was used for the final signaling messages, is replaced by user data.

In order to disconnect the communication link, such as would happen if the land party hangs up, the network will initiate a call release procedure by sending a *disconnect* message (see Figure 7.4). This message will be followed by a *release* message from the mobile station. The network acknowledges this with a *release complete* message. Then comes the transmission of the *channel release* message after which the Layer 3 connection is terminated. The data link layer connection will also

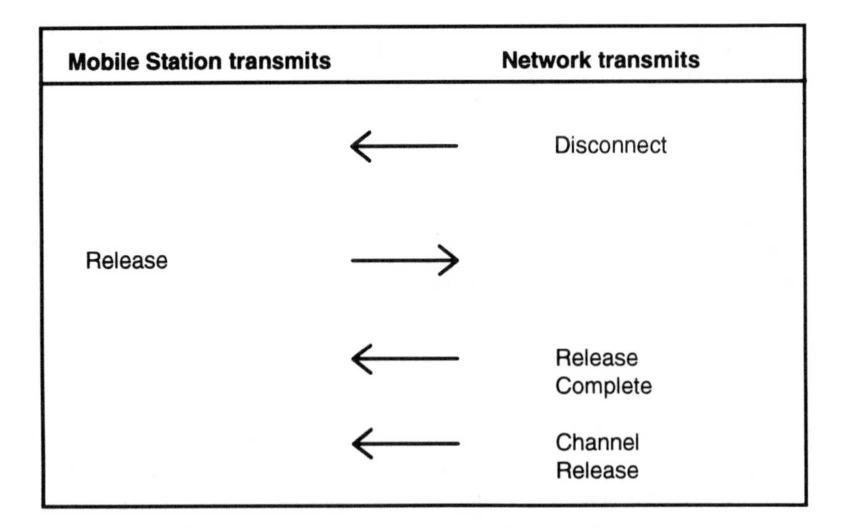

Figure 7.4 Example of a call release, network-initiated.

be disconnected (see Chapter 6), and, finally, the physical radio link will vanish. Figure 7.4 shows the procedure of a network-initiated call release.

There are some defined procedures for a call setup originated by the network (paging) and a mobile-originated call release. There are also mechanisms to be called upon when problems arise during the signaling processes. These mechanisms generally bring the signaling entities back into a defined state in order to avoid tied-up network resources, wasted capacity, and bad service. Such mechanisms include the repetition of messages that were not answered or ignoring bad or incomplete messages, or they may simply release the whole signaling link.

7.5 A LAYER 3 SIGNALING TRACE

A call setup recorded by the same GSM test system we used in Chapter 6 will illustrate the exchange of signaling data over the radio interface (Figure 7.5). This time we ignore the Layer 2 envelopes and look only at their contents. The procedure is a mobile-originated call setup that a GSM mobile station has performed in a network simulated by the test system.

In a manner similar to the Layer 2 trace (in Section 6.4), the test system has added some information that is useful when examined in combination with the exchanged messages. The information includes frame numbers, time stamps, the direction of transmission (uplink = ul, downlink = dl), the logical channel (RACH/AGCH/SDCCH/FACCH), time slot number, TI, PD, as well as MTs, IE names, and their contents. The T1, T2, and T3 counters appear just as they did in Chapter 6.

The trace used an SDCCH/8 signaling channel. We can also look at the TDMA frame numbers to determine how long the whole procedure lasted.

FN start (channel request):	4,610
FN connect:	5,386
Difference:	776

Since the duration of one TDMA frame is 4.615 ms, the duration of the whole signaling sequence is 4.615 ms/frame × 776 frames = 3.58 seconds.

7.6 TIMERS

In order to ensure a smooth flow of signaling messages and to avoid tied-up links or the excess use of processor time and network (radio channel) capacity, there have to be restrictions on the time used for the procedures. There also have to be restrictions on the number of attempts made to perform certain signaling procedures. A mobile station that continuously transmits channel requests, even when a base station tempo-

Layer 3 Trace

MS transmits (uplink) **BS transmits (downlink)**

FN : Frame Number

```
FN:0004610: Umul1,rach, TS(0),TI(0),pd(rr),
0003:08:20 chan_req,
        chan_req(= 7h,0Ah);

                                        FN:0004622: Umdl1,agch,TS(0),TI(0),pd(rr),
                                        0003:20:32                imm_ass,
                                                  pag_mod(= 0h),
                                                  chan_desc(= 8h,3h,0h,0i,0h,02h),
                                                  req_ref(= EAh,03h,14h,08h),
                                                  tim_adv(= 00h),
                                                  mob_alloc(= );

FN:0004656: Umul1,sdcch(0), TS(3),TI(0),pd(mm),
0003:02:15 cm_serv_req,
        cm_serv(= 1h),
        ciph_key_seq(= 0h),
        mob_class2(= 0h,0h,2h,0i,0h),
        mob_id(= 4h,0i,12345678h);

                                        FN:0004743: Umdl1,sdcch(0),TS(3),TI(0),pd(mm),
                                        0003:11:00                auth_req,
                                             ciph_key_seq(= 0h),
                                             rand(= 0123456789ABCDEF
                                             0123456789ABCDEFh);

FN:0004809: Umul1,sdcch(0), TS(3),TI(0),pd(mm),
0003:25:15 auth_res,
        sres(= 90DFD9F0h);

                                        FN:0004845: Umdl1,sdcch(0),TS(3),TI(0),pd(rr),
                                        0003:09:00                ciph_mode_cmd,
                                                     ciph_mode_set(= 0i);

FN:0004911: Umul1,sdcch(0), TS(3),TI(0),pd(rr),
0003:23:15 ciph_mode_com;

FN:0004962: Umul1,sdcch(0), TS(3),TI(0),pd(cc),
0003:22:15 setup,
        bear_cap(= 1i,1h,0i,0i,0h),
        cd_p_bcd(= 0h,1h,089996410h);

                                        FN:0004998: Umdl1,sdcch(0), TS(3),TI(8),pd(cc),
                                        0003:06:00                call_proc;

                                        FN:0005100: Umdl1,sdcch(0), TS(3),TI(0),pd(rr),
                                        0003:04:00                assign_cmd,
                                                  chan_desc(=01h,4h,0h,1i,00h,00h),
                                                  pow_cmd(= 0Fh),
```

Figure 7.5 A Layer 3 trace, call establishment.

```
FN:0005304: Umul1,facch, TS(4),TI(0),pd(rr),
0004:00:00  assign_com,
            rr_cause(= 00h);

                                            FN:0005334: Umdl1,facch, TS(4),TI(8),pd(cc),
                                            0004:04:30                alert;

                                            FN:0005369: Umdl1,facch, TS(4),TI(8),pd(cc),
                                            0004:13:14                conn;

FN:0005386: Umul1,facch, TS(4),TI(0),pd(cc),
0004:04:31  conn_ack;
```

Figure 7.5 (continued).

rarily has no capacity to serve it, would disturb the network and generate unnecessary radio traffic. This negative effect can be controlled and managed by the use of additional timers. The GSM network and its entities work with a number of timers in the signaling processes. A timer is started at the transmission of a message that requires an answer within a defined period. Upon receiving the answer, the timer is stopped or reset. For example, the network side starts a timer called *T3101* at the transmission of an immediate assignment message. It is reset after the mobile station transmits on the newly assigned channel. If the mobile station does not seize the new channel within this time (in T3101), the channel request will be ignored and the reserved channel will be released for another communication link. The spurned mobile station would have to request a channel again.

For the data link layer, Layer 2, a timer called *T200* is defined. It specifies how long a Layer 2 entity can wait before repetition of a Layer 2 frame. An example is an SABM command not answered by a UA response. The possible response times are a function of the data rates of the different logical channels; the FACCH is much faster than the SDCCH. There are, therefore, different timer values for different logical channels. They range from 155 ms to approximately 1 second.

7.7 SIGNALING ARCHITECTURE IN A GSM PLMN

A GSM PLMN uses complex and sophisticated signaling processes to provide service and still be flexible enough to interface with a variety of communication entities and partner systems. Although a multitude of thick specifications exist, all GSM networks installed in the world and all those yet to be installed in the future will differ from each other in their architectures. This is because of the different services offered in different markets and the different access and interfaces to the fixed networks. The signaling architectures still have some common ground on which we can work. One of the major objectives in designing the GSM system was, and remains, to maintain an open and flexible character in the architecture, leaving it open to adaptations, different interfaces, and new service implementations. The chief infrastructure components interface physically with each other (radio interface Um, Abis interface, A interface) as well as logically with each other (OSI layers and CCITT SSN7). The logical entities in a GSM PLMN have their assigned tasks and functions. They communicate with each other with protocols and procedures. They all represent a giant communications system: a huge machine, which connects and disconnects, switches and relays, routes and registers, counts and bills. Taken together with all the connecting fixed networks, it becomes the largest machine ever made.

This chapter will give a general view on how a GSM PLMN logically interconnects from the exchange (MSC) to the user terminal (mobile station), since this subject does not belong anywhere else in the book. A complete analysis and discussion of the structures and architectures inherent in the definition of a GSM PLMN is far beyond the scope of this book. The interested reader should refer to the applicable *GSM Technical Specifications* for further details. Another, far less painful approach to discussing the subject in all its variety can be found in [2], which is recommended as an intermediate step between this book and the official standards and recommendations.

Figure 7.6 shows the GSM signaling environment in a simplified diagram with logical blocks, functional planes, and interfaces.

The are four entities (devices or machines):

- Mobile station;
- Base transceiver station;
- Base station controller;
- Mobile switching center.

The three interfaces that connect them have names:

- Um (between the mobile station and BTS);
- Abis (between the BTS and BSC);
- A (between the BSC and MSC).

7.7.1 Mobile Station and Base Transceiver Station Connection

The GSM mobile station includes all Layer 3 functions and its sublayers (RR, MM, CM). It interfaces with the BTS via the radio interface, Um (Layer 1), and the LAPDm (Layer 2). LAPDm ends in the BTS. The BTS itself only needs very restricted Layer 3 capabilities. There are only a few RR functions in the BTS, where some examples are (1) to control transmission timing (timing advance) and (2) power control. The BTS is transparent to the much larger amount of signaling that comes with MM and CM.

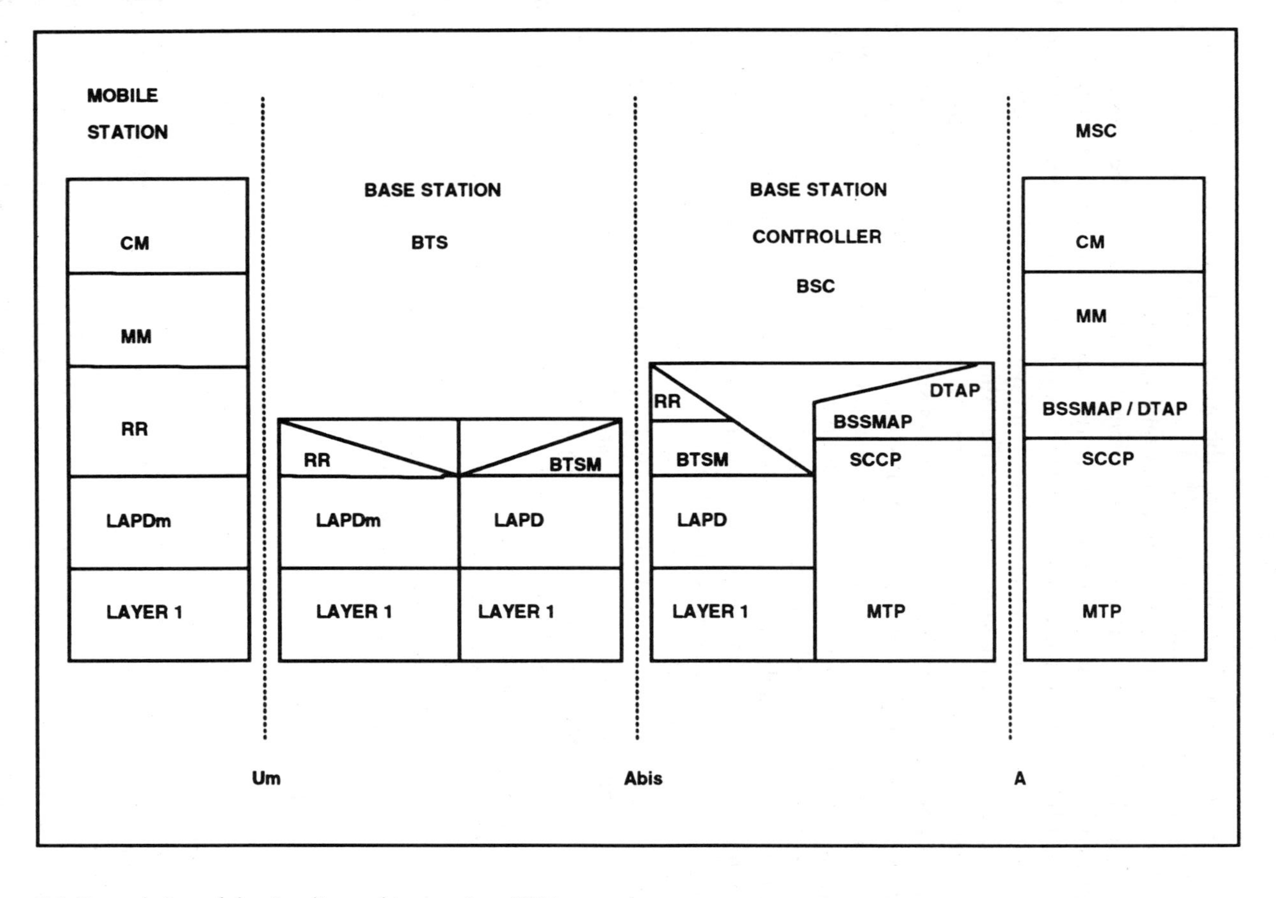

Figure 7.6 General view of the signaling architecture in a GSM network.

7.7.2 The BTS-to-BSC Connection

The BTS interfaces with the BSC on the Abis interface via a PCM-30 link, which is a 2.048-Mb (32 subchannels (or slots) × 64 kbps = 2.048 Mbps) fixed-line standard [3]. See Chapter 11 in Part III on base station testing for lots of details on the Abis interface. Due to its late and initially fragmented standardization, the Abis interface appeared in a variety of different interpretations and implementations. This led to incompatibilities among network components from different manufacturers. So, if network operators decided to buy a BSC from one supplier, they had little choice but to buy BTSs from the same supplier. The Abis interface uses 64-kbps signaling subchannels in order to carry signaling data and submultiplexed 16-kbps channels (one 64-kbps subchannel has four 16-kbps *sub*-subchannels) for the transfer of user data (speech). For the transformation of speech in and out of the 64-kbps links, which are used as a standard in the ISDN and GSM networks as the only entities that can be switched, a transcoder rate adapter unit (TRAU) is used. The TRAUs can be physically located in the BSC, in the BTS, or even in the MSC. This leads to a wide variety of installations. Rate adaptation is also an important issue for non-speech user data transmissions. In general, it is more economic to have the TRAU/rate adaptation as close as possible to the switch (MSC), because the traffic adapted to the rates used in GSM radio transmission (speech: 13 kbps easily fits into a 16-kbps subchannel) is substantially lower (and thus cheaper to transmit) than the 64 kbps of fixed ISDN lines. In the case of speech traffic taking place between two users of GSM mobile stations, the signals will always be transformed twice: first from 13 to 64 kbps, and second from 64 back to 13 kbps, even if they are both connected to the same BSC or BTS.

The BSC controls the BTSs connected to it and maintains contact with the MSC. A BSC and the connected BTSs are, together, also referred to as a *base station subsystem* (BSS). The BSC has to perform the remainder of the RR functions and relay information, such as random access, paging, and ciphering messages.

7.7.3 The BSC-to-MSC Connection

The BSC is connected via a PCM-30 link to the MSC; this interface is called an *A-interface.* The switching entity of the GSM PLMN, the MSC, is organized with a CCITT SSN7 protocol. Consequently, it transmits and receives user data (speech), and signaling data in 64-kbps subchannels of the PCM-30 link (CCITT G.703, ISDN standard).

The A-interface also transmits the CM and MM protocols to and from the MSC. The BSC and BTS are transparent to these messages, which is to say they merely hand them back and forth using their own protocols and various means of physical transmission (LAPD, LAPDm, and Layer 1). The functions within the MSC that are used to handle and convey this kind of information (meant for a specific mobile station), as well as general BSS control information, is allocated to the base station system management part (BSSMAP) and the direct transfer application part (DTAP). They make use of the lower planes' signaling connection control part (SCCP) and message transfer parts (MTP), which are equivalent to the lower three layers in the OSI model. The SCCP is responsible for the establishment of signaling nodes, and the MTPs transfer the information between the nodes.

REFERENCES

[1] *GSM Technical Specifications,* ETSI, Sophia Antipolis, Vols. 04.07 to 04.88.

[2] Mouly, M., and B. Pautet, *The GSM System for Mobile Communications,* Palaiseau, 1992.

[3] CEPT T/S 46-20 Recommendation and CCITT Q.920/921/G.703 Recommendations.

CHAPTER 8

TESTING—ANALOG VERSUS DIGITAL GSM SYSTEMS

We have come a long way, and have arrived at the pinnacle and purpose of our work: we are ready to test GSM radios. Our path to competence and certification was interesting and full of adventure. We moved to a new town and landed a job with the local tram system, we probed into the details of the GSM Freight Train Company and its demanding but eccentric customers, we listened in on some private and trivial radio correspondence on the high seas, and we finally visited the criminal's realm as we hijacked and robbed the GSM freight train. But we live in an orderly and civilized society, and it seems time we turn from our childish behavior, that we stop hopping from job to job, that we cease listening in on other's radio conversations, and stop robbing trains and reading other people's mail. It is time to put the GSM train back together and test the system and all its components.

8.1 TRANSITION FROM ANALOG TO DIGITAL TESTING

The transition from analog to *digital radio testing* confounds many people, and often seems mysterious at first. When we use instruments to examine the characteristics of analog systems, we are never far from the physical reality of the *device under test* (DUT). Analog radios offer a particularly friendly test environment: (1) we can *hear* the recovered audio and signaling tones, (2) we can *feel* the heat of the transmitter's power in the attached dummy load, (3) we can *speak* into the handset and *watch* the reaction of the transmitter's modulator on a radio test set's oscilloscope or other readouts, and (4) we can *twist* knobs and make adjustments to circuits with immediate and comforting results.

This very physical activity with all its sensory feedback seems to be missing when we try to test digital radios. We often hear nothing at all when we test digital radios, and if we do, it seems to be only the sound of inscrutable noise. We make relatively few connections (usually just one to the radio's antenna connector) between the DUT and our test instruments, which are a mystery unto themselves as they display only numbers and confusing terms on CRTs. We speak into the handset with no apparent reaction from the transmitter. There is a distinct lack of knobs to twist, levels to set, and adjustments to make. The whole testing process seems to be nothing more than an arbitrary computer exercise.

Power is not measured in mere watts; we have to take into account whether the radio is bursting its power or not, and the readings we get are sharply dependent on the type of modulation employed. Receiver testing seems to be particularly mysterious. The relatively intuitive SINAD method is replaced by mere numbers. What do they mean? What is the difference between BER and RBER?

The radios themselves often seem to defy testing. We cannot seem to get them to transmit or receive anything at all without attaching a mysterious test system of some kind. We search frantically in the test documentation for a test mode that will bring the radio back to the real analog world, and we are almost always disappointed.

The authors of this book are keenly sensitive to the trials and frustrations that so often accompany those just entering the world of digital radio testing from the comforting and sane world of analog radio. This chapter, then, represents a careful transition from the analog to the digital test domains for the reader. Starting with Section 8.2, we will compare each and every characteristic we test in a digital radio with its corresponding intuitive analog task. On the way, we will offer a comforting insight and remark where appropriate. We will see that digital testing is not as mysterious as it first seems. With the new insights gathered in Chapter 8, we look at the details of testing GSM mobile stations in Chapter 9. Chapter 10 does the same for base stations, which, we will see, have lots of things to test even though they do not seem to do much. Chapter 11 concludes this important part with some applications and a survey of all the testing situations we will find in GSM. We will also have a look at some of the instruments available to us. Let us begin!

8.2 INTRODUCTION TO THE MEASUREMENTS

New digital radio measurement techniques add some challenges for manufacturers of radio test sets and general instruments, supplementing the normal metrological challenges that come with any new technology. Pulsed transmissions require special *peak power detection* for accurate power level measurements. Precise timing references are enlisted in order to measure the proper power-up and power-down ramping times. A high dynamic measuring range is required to examine and display all the characteristics of on and off switching transients. The criteria used to evaluate modulation quality are specifically adapted to GMSK modulation in its bursting mode. The use of a speech codec makes the traditional and familiar methods for measuring receiver quality obsolete; the fixed wire line's world of digital data transmissions offers some handy techniques for radio receiver testing. Finally, the huge amount of signaling transactions in a digital air interface requires a greater test effort than was known for the corresponding and rather simple analog functions.

8.2.1 A GSM Test Set

In Figure 8.1, we see a block diagram of a state-of-the-art *GSM radio test set*, which is able to perform all the necessary tests to fully evaluate a GSM mobile station or base station. It not only performs RF measurements, but also sorts through all the signaling tests to verify the correct behavior of a mobile station. The tester has three major parts: (1) a personal computer (PC), (2) a digital unit, and (3) an RF unit. For simplicity, the modules required for BTS testing are not shown in Figure 8.1; they appear later, in Chapter 10.

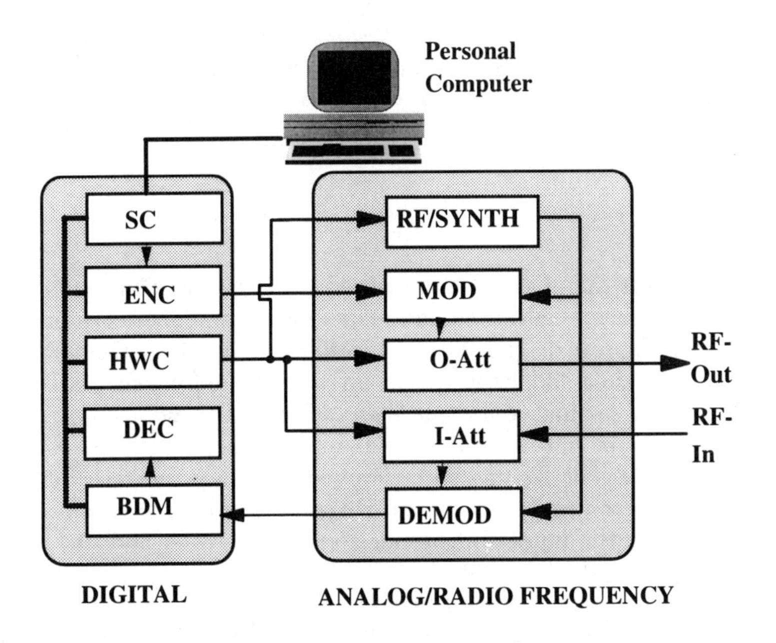

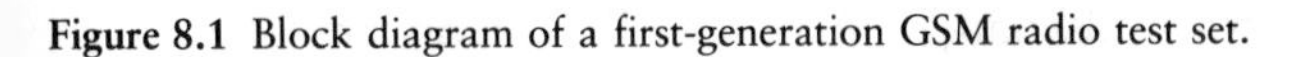
Figure 8.1 Block diagram of a first-generation GSM radio test set.

8.2.1.1 PC Part

The PC provides an instrument interface for the user; it is the instrument's front panel. It controls the whole test set such that the user has access to configuration data, different system parameters, and test limits. With an appropriate program running on the PC, different tests and measurements are initialized and run; the results are displayed on the screen or stored on the hard disk in the PC.

8.2.1.2 Digital Unit Part

The digital unit contains the circuits needed to process the digital information for the air interface and execute some control functions for the RF unit. The *signaling controller* (SC) inside the digital unit provides all the signaling functions on Layers 1, 2, and 3, and controls the different logical channels that may appear. Both uplink and downlink Layer 3 sequences are taken from the PC and are executed in parallel with measurement functions to check for the correct signaling messages coming back from the DUT. Once they are understood in all their details, Layer 2 functions tend to appear as rather bone-headed automatic chores. Layer 2, therefore, is managed automatically in the test set's digital unit, but it may be manipulated for error simulation. The *encoder* (ENC) performs the proper channel coding of the signaling and user data. The *training sequence*, the *stealing flags*, and the *tail bits* are inserted here. If *ciphering* is selected within a signaling sequence, it is performed within the encoder with a preselected sequence. The digital data finally appear from the encoder in Layer 1 form and are passed on to the *modulator* (MOD) in the RF unit.

8.2.1.3 RF Unit Part

The modulator alters the phase of an intermediate frequency (IF) signal to yield a Um simulation of either a GSM base station or a mobile station, depending on what type of DUT is to be tested. The IF comes from a frequency synthesizer in the RF unit. The GMSK-modulated IF is converted up to the radio frequency channel under test with traditional methods. The *hardware controller* (HWC) inside the digital unit manipulates all the hardware in the RF unit in concert with the data domain activities in the tester. The RF unit's functions include such things as adjusting the input and the output to the correct channel frequency, making different power level adjustments through the input and output attenuators (I-Att and O-Att), and, if selected, implementing the frequency-hopping details.

Back in the digital unit, the *bit demodulator* (BDM) accepts changing phase information from the *demodulator* (DEMOD) in the RF unit, which has converted the received RF signal back to a baseband signal. The BDM converts the phase information into a Layer 1 bit stream. The demodulated bit stream is passed on to the *decoder* (DEC), which recovers the messages as any channel decoder might do.

8.2.2 Testing Base Stations and Mobiles

The theory and techniques used to make measurements on base stations and mobiles are almost identical; only some of the parameters and test limits differ slightly. It is for this reason they are treated equally in this chapter. The practical details surrounding BTS and mobile station testing, however, are quite different for exactly the same reasons they are different in analog situations. The big difference is found in the base station, which has two ports rather than only one. One port is for the Um interface and the other is for the connection to the rest of the network through the BSC.

8.3 TRANSMITTER MEASUREMENTS

Transmitter tests on GSM radios, both base stations and mobiles, are almost exactly the same as those we perform on analog transmitters. Only the details of the methods are modified to account for the lack of analog signals and the peculiarities of the modulation.

Frequency error and *RF power* are the most common measurements performed on transmitters. The frequency error and the power of an analog transmitter should be measured in the absence of modulation or, as we say, in the CW mode. It is quite easy to measure the frequency error, from an assigned standard, of an unmodulated transmitter's output signal; we do it with a frequency counter. It is likewise rather easy to determine the transmitter's power with a suitable power meter. In the case of analog CW angle-modulated transmitters (FM or PM), such as those common in analog cellular systems, the frequency error can be determined in the presence of modulation by simply demodulating the transmitter's output in a calibrated receiver and measuring the frequency of the IF in that test receiver. Similarly, RF power can be measured with considerable certainty as long as we know the modulation is not imparting any amplitude variations to the carrier we cannot follow and account for; no AM is allowed.

8.3.1 RF Power—Myth and Reality

Perhaps no measurement process in radio causes more confusion and spawns more stories and false theories than RF power measurements do. The proper way to measure RF power is to couple the transmitter's output (with no loss—a real challenge at 900 MHz) into a device that converts *all* the power into heat, and then calculate the power from the change in temperature of the device making the conversion. One of the common techniques uses a thermocouple. A thermocouple is nothing more than two tiny pieces of dissimilar metals joined together at a junction, which, when heated, generates a small voltage at the junction. The voltage is a very precise function of the junction's temperature. The problem with this method is that it is very slow. Even if we take extraordinary measures to make everything small, it still takes *time* to heat up whatever thermal device we choose to make the conversion from power to voltage (or to current or to resistance—all variations exist). The other problem with this accurate method is that it tends to be very expensive, since the thermal conversion devices tend to be fragile and expensive, particularly at higher frequencies. Thermal power meters are the most accurate way to measure RF power, but they are far more trouble than they are worth, particularly in production and field-testing environments.

8.3.1.1 Traditional Power Meters

How do we avoid using thermal power meters? The popular solution is to use a diode. If we detect the transmitter's output in a diode and amplify the detected voltage, we have an indication of the transmitter's RF power. What kind of indication? A diode in this application is a square-law device, which means we have to perform a little arithmetic on the detected voltage in order to find the power. If the voltage is called V, then the RF power (P) is found with the square of the voltage multiplied by some conversion factor (e.g., $P = V^2(1.25)$). This is the method widely used in portable power meters and radio test sets designed for analog service, and it works well as long as we are careful to make good connections between the transmitter and the power meter. Establishing a good connection between a transmitter and a power meter at 900 MHz is not an easy task; mismatches caused by connectors and dummy loads quickly convert diode power meters to random number generators. The diode is, you see, only sensitive to the transmitter's output voltage (not the power), and we make the conversion to power by assuming the output's current is always in phase with the voltage, which will only be the case if the transmitter's output sees a perfectly resistive load exactly equal to the transmitter's own output impedance.

8.3.1.2 Digital Radios and RF Power

An even bigger problem arises when we attempt to measure the RF output from a digitally modulated transmitter using the TDMA technique. A GSM mobile's RF output is burst so quickly that the readouts attached to common linear power meters cannot follow the changing power output, and if they could, we would not know what the conversion formula back to power would be; we are looking at a complicated AM waveform. Our only hope for success with linear processes is, perhaps, to confine our RF power measurements to the base station that fulfills the duties of the base channel, providing we make sure the transmitter's output power does not change. Even in this case, the measurements will not be particularly accurate, because GMSK, as elegant as it is, has some small amount of AM on it in order to keep the transmitter's emissions confined to its assigned channel. Chapters 12 and 13 describe some competing digital cellular systems whose transmitters, owing to the highly AM character of their modulation, call on extraordinary methods to determine their RF power. Our goal is to find a way to look at the transmitter's output, and convert it somehow into what an imaginary super-fast thermal power meter would measure. The solution in digital radio is to simply abandon traditional power meters and turn the whole process over to test sets that employ powerful signal processing techniques to calculate the RF power directly from the recovered modulation. The tempting and continued use of low-cost traditional power meters is pure folly in the world of digital radio. Life is too complicated to further confuse it with traditional power meters pressed into digital service. There are lots of things we should not do: we should not steal things from our neighbors, we should not allow our children to eat too many sweets, and we should not use traditional power meters on digital radios.

Mobile stations and base stations in the GSM system are not normally capable of transmitting unmodulated signals. A tiny few mobile stations have a service mode that allows the transmission of unmodulated carriers (CW). For this reason, new methods had to be found to measure the frequency error. The same techniques, thanks to digital signal processors, can be adapted to RF power measurements, which take into account the bursting modes of the transmitters and the residual AM associated with GMSK.

8.3.2 Phase Error and Frequency Error Measurements

Digital cellular base site radios, of all types and protocols, usually get a reference clock from the fixed telephone network, and it seems odd and somewhat silly, therefore, to test the base transmitter's frequency error. We do so from time to time, since apparent frequency error is usually a symptom of something amiss somewhere else, usually with the modulation. The mobiles, on the other hand, synchronize themselves with the base station, and we have to test the mobile's ability to perform this critical task accurately.

Phase error is a new criterion for evaluating the quality of a modulator. It is an indication of how accurately the individual bits modulate the RF carrier. The frequency error is an indication of the mobile's ability to synchronize itself onto the frequency correction channel of the base channel and maintain this frequency. For the base station, it is an indication of the accuracy of its own frequency reference, or if the base station synchronizes itself with its host BSC, an indication of the accuracy of the BSC.

Both measurements, phase error and frequency error, are performed together. The frequency error can be calculated from the phase error. To perform the measurements, the *phase trajectory* of an individual burst is recorded. From this trajectory, the radio test set recovers the individual bits to yield a bit stream. Since the modulation is based on an MSK technique, the information at any instant, whether it is a 0 or 1, lies within the phase changes. This means that the bit stream may be derived from the incoming phase information. It is not necessary to record the information of one complete burst before the actual data recovery can start; the underlying data are merely a function of the changing phase of the carrier (see Section 5.15). To perform phase and frequency error measurements with significant accuracy and resolution, the RF signal has to be sampled at least at twice the bit clock, which is 540 kHz (2 · 270 kHz = 540 kHz). In Figure 8.2, we see the different steps as they are described in the text.

From the derived bit stream, the ideal phase trajectory is calculated, which is the one an ideal modulator would provide under theoretically ideal conditions. But since everything in this world is real, not ideal, there is a difference between the measured phase trajectory and the one calculated from the recovered bit stream. To get the difference between the ideal and the actual trajectories, one is subtracted from the other, and the difference is plotted in time together with the reference and the actual trajectories. The resulting differential curve contains not only the phase error, but also the frequency error of the sampled signal. To separate both effects and treat them separately, a regression line is fitted onto the curve. This line traces the minimum average distance from each point of the curve.

The gradient of this curve is an indication of the frequency error. The sign of the slope of the curve tells us the sense of the frequency error, too high or too low. The distance from the regression line at any point is the phase error for each instant. Two different values describe the phase error: (1) the *peak phase error* and (2) the *RMS phase error*. The peak phase error indicates the maximum deviation from the ideal phase trajectory at a certain instant within one single burst (e.g., at the change of a bit). The RMS phase error considers all the values and gives an average value for the phase error according to the formula

$$x_{\mathrm{RMS}} = \sqrt{\frac{\sum_{i=1}^{N} x_i^2}{N}}$$

It does not display the modulator quality for a single instant as the peak phase error does, but it does provide a long-term *average* value—an indication of the trend. A digital radio system's performance does not hang on the proper reception of any single

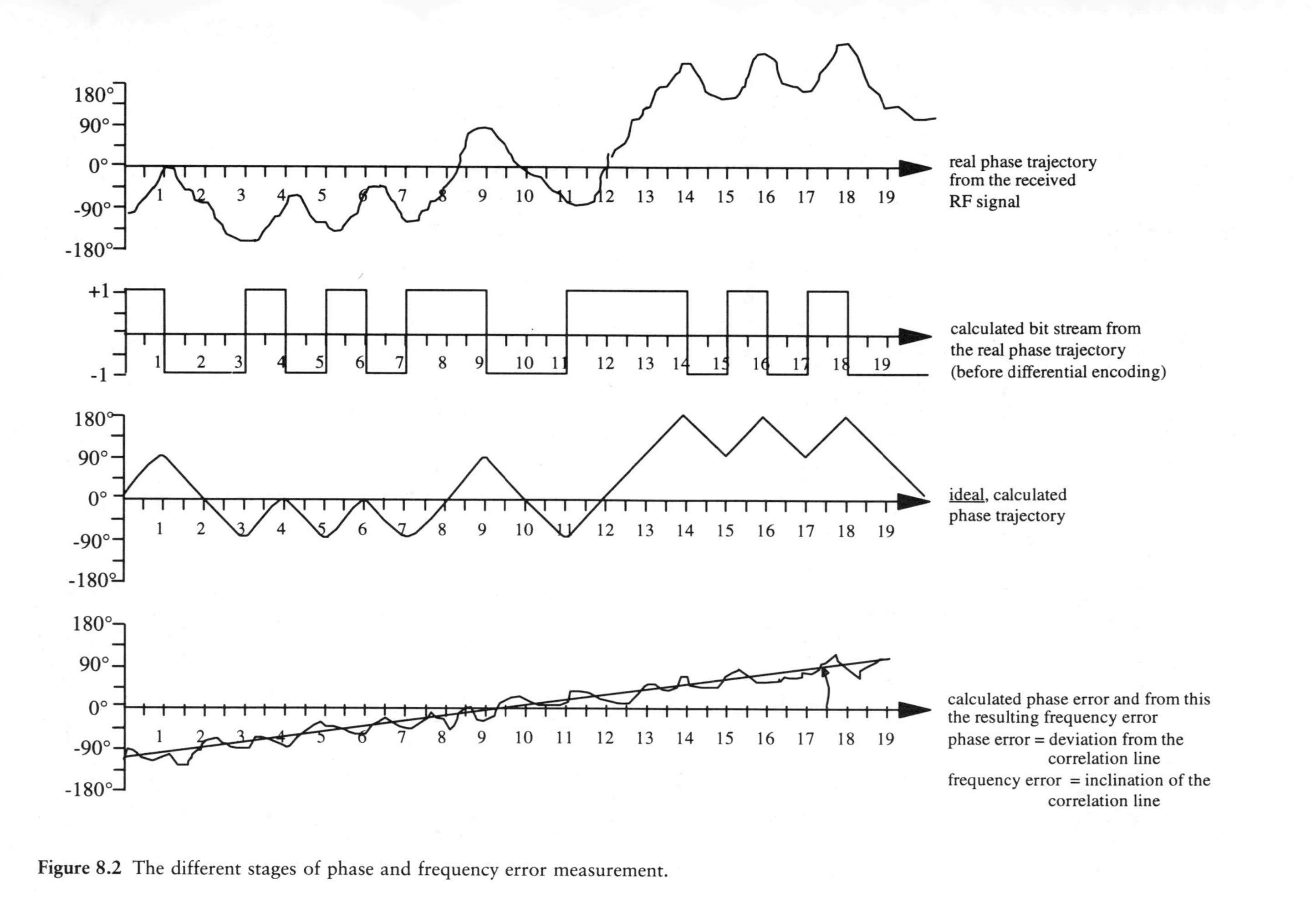

Figure 8.2 The different stages of phase and frequency error measurement.

bit; the receiver's channel decoder can recover data from some rather spectacular catastrophes on the Um interface. An average indication of the modulator's performance makes sense in this operating environment. The modulator, however, should deliver undamaged goods to the antenna, and the peak readings and plots are appropriate for this consideration. Together, both measurements provide a good indication of the modulator's quality.

8.3.3 Special Transmitter Power Measurements

Along with the measurement of power in the sense we normally think of it, there are some peculiarly digital and GSM aspects of the transmitter's output that should be examined: (1) *peak power*, (2) *power-time template*, and (3) *burst timing*. Most people would think that burst timing is not really a power measurement, since it seems to belong to the *time domain*. We include it in our *power measurements*, because we use the same equipment and techniques to measure burst timing as we use for routine RF power measurements. This tendency of different processes in a digital radio (and its test equipment) to share a small pool of resources is common.

8.3.3.1 Peak Power

The peak power is the RF power we measure after the transmitter is full on, which is the power level the transmitter assumes while it sends data on the air interface. Both the mobile station and the base station can set themselves to different power levels, and some (or perhaps all) of these should have their accuracy verified (see, for example, Tables 5.1 through 5.3 in Section 5.4). We have to make sure that the mobile accepts and responds properly to power adjustment commands from a base station and that the power levels are properly calibrated. Likewise, we need to check the calibration of the various power levels a base station can assume. A base station transmits a certain RF power level as a balance between the optimum SNR at the mobile's receiver while maintaining a minimum amount of interference in other cells.

8.3.3.2 Power-Time Template

The power-time template displays the power-up and power-down ramping of either the mobile or the base station. If the carrier is switched on too early or is switched off too late, it might disturb a mobile station on an *adjacent time slot*. If, on the other hand, it is switched on too late or switched off too early, it might disturb its own transmission, since some of the transmitted data will be cut off. Small deviations within the power-time template are allowed, since the tail bits ensure a certain margin of error-free transmission. If the deviations from the optimum become too large, then they start to disturb the transmissions on the radio path. From the power-time template, one gets a first indication of switching transient anomalies that may appear in the spectrum. If the power-up or power-down ramping is too steep, then the energy outside the assigned channel, or even outside the assigned band, will probably be greater than is allowed. This will disturb transmissions in neighboring bands. The reference for the power-time template is the training sequence. After this sequence is identified, the power within each individual bit may be related to any other bit starting with bit 0, the first bit.

8.3.3.3 Burst Timing

Burst timing only affects the mobile station and is an indication of *when* the burst appears in its assigned time slot. It seems that burst timing is related to the power-time template, but this is not the case. Burst timing is associated with the time slot structure or schedule, and is an indication of how accurately the mobile station can maintain a timing advance value within defined limits. The base station transmits *timing advance* adjustment commands to the mobile as it moves around in the cell. To verify the accuracy of burst timing, the radio test set simulates a base station that sends timing advance commands to the mobile station and checks the proper timing of the returned bursts.

For the purposes of the measurements, the base station is the arbitrary master regarding time slot structure. It makes no sense to perform these kinds of measurements on base stations in the same way we make them on mobiles. A base station can be fitted with up to 16 transceivers. We must at some time assure ourselves that the timing between these transceivers is precise enough to make sure there are no timing offsets between the base channel and all the other channels. To verify proper timing among transceivers, the test system has to measure the timing on the base channel and compare this with all the other installed frequencies.

8.3.3.4 Some Power Measurement Techniques

All these different aspects of a *power measurement* are performed together in one step. The signal coming from the unit under test (UUT) is actually examined and analyzed at the IF of the test set's calibrated receiver. This signal is sampled at each transmitted bit, the measurements are stored, and then the maximum power level is determined for a defined series of bits. In order to evaluate this signal in its power-time template, the phase or timing indication of each sample is also stored. From the two columns of stored measurements (power and time), we have the timing information as well as the bit stream that created the modulation.

We are making measurements and not looking at data from the UUT. This being the case, for what purpose do we need the bit stream's contents? We need to recover the bit stream so that we can find the training sequence, and we need to find the training sequence in order to make timing measurements. The bit stream is derived from the same recorded data we used to calculate the phase and frequency error. The precise timing and level values of all the bits remain available in memory for further study and manipulation. From the training sequence, the timing within the power-time template can be evaluated. If the training sequence is found, then we can count backwards to bit 0, at which point the power should have already reached its maximum value. With the power and timing indications for each bit safely stored away in the tester for analysis, we can put this information into any kind of presentation we like, including a power-time template. There can be more than one power-time template. The much shorter *random access burst* (only from a mobile) uses a shorter template with only 87 bits, and it also has a different *synchronization sequence.*

For the burst timing, the exact reception time of the burst from the UUT is compared with the time slot structure generated within the tester. The timing between the uplink and the downlink is precise and well defined. The tester supplies its own time line reference against which it determines the starting time of a burst from the mobile station. From the starting time, the test equipment can also measure whether the mobile's burst fits exactly within its specified time slot and whether the mobile's bursts track the timing advance commands it receives from the tester.

With the single exception of the random access burst, the transmitter measurements evaluate only 147 bits of a burst. The time period of the first and the final half bits are not taken into consideration for the measurements. Figure 8.3 graphs the different steps taken for the presentation of the power-time template. The figure demonstrates that if we know the power and timing of a series of samples, and if the samples are accumulated fast enough (twice the symbol rate), we can manipulate the data into impressive displays such as the power-time template.

8.3.4 Measurements in the Spectrum

The transmitter measurements discussed to this point treated only the behavior of the signal within a channel itself. Sadly, mobile stations and base stations can disturb other users within the GSM system, or even other services outside the GSM bands, when they emit signals above certain power levels outside their assigned channels. This is a particularly common hazard with digital radios, as we saw in Chapter 5. To prevent this, special care is taken in the design of a transmitter to make sure it retains its band-limited character. Moreover, type approval requires that transmitters must be designed in such a way that there is no reasonable possibility of creating a disturbance on other frequencies. There are, then, a whole series of spectral tests to check the orderly behavior of transmitters with regard to their off-channel emissions. Different causes and influences require several different types of measurements, and

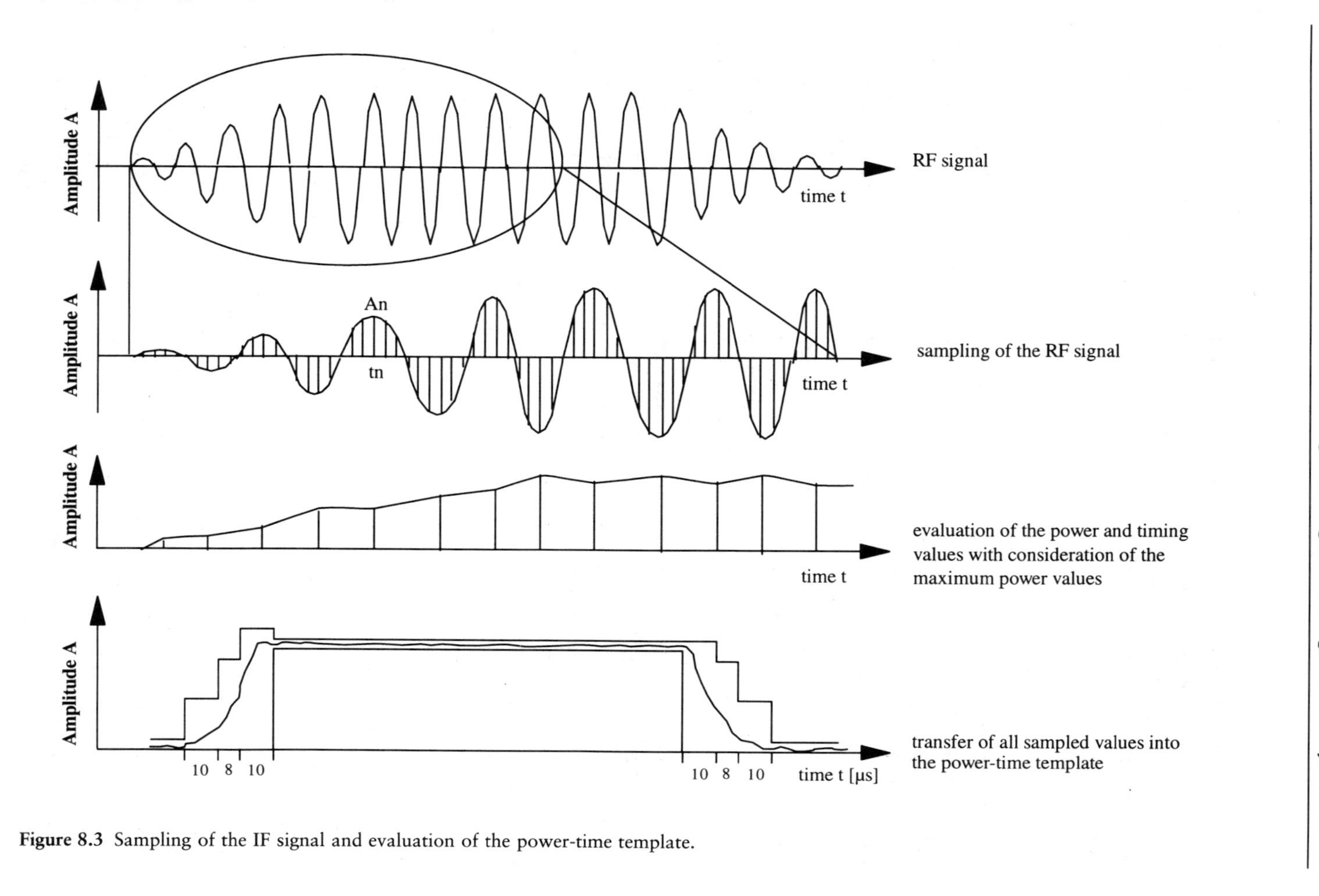

Figure 8.3 Sampling of the IF signal and evaluation of the power-time template.

all of them are identical to similar measurements performed on analog transmitters; only the reasons for doing the tests are different. Spectral examinations are generally not performed in the field, because their failure is almost always caused by a design fault rather than something that is likely to fail in an after-market field testing situation.

8.3.4.1 Spectrum Due to Modulation

As the name implies, the *modulation spectrum* should be evaluated with this measurement. Digital modulation and power bursting makes this measurement a popular and important one. To get the most significant measurement, the test is performed only with coded data which give the Layer 1 bit pattern a decidedly random character. The reason for this requirement is that the results of off-channel measurements on digital transmitters are a function of the information content in the transmissions at the time the measurements are taken; it is best to look at the worst case, which will occur when all possible transitions (in phase and power) are included in the signal under test. Witness, for example, the different spectral content we find in a base channel and a traffic channel. Random transmission data are often a part of digital transmitter measurements. The test data and the resulting spectrum should not contain the tail bits and the training sequence, because the content of these stays the same and reduces the random character of the signal under test. The evaluation is done for a transmitter operated in slow frequency-hopping mode on three different frequencies, where one of the frequencies is on the bottom end of the frequency band, one is in the middle of the band, and one is on the upper end of the band. This is done to drive the test conditions as far into the worst case realm as possible.

As the reader is likely to guess, spectral measurements are performed with a spectrum analyzer. To determine the modulation's spectral content, the spectrum analyzer is operated in its zero span mode, which causes all the measurements to be performed with a fixed and defined filter bandwidth. The spectrum analyzer's x-axis (frequency) no longer represents the frequency domain in the zero span mode; it becomes the time axis. Figure 8.4 shows a typical spectrum display for a single burst as it would be seen on a spectrum analyzer, with a resolution bandwidth setting of 30 kHz, and the center frequency offset from the carrier. The spectrum analyzer displays the power-time relationship for a certain fixed frequency. In this figure, the impact of the switching transients can easily be distinguished from the modulation component. It is important for this measurement that the spectrum analyzer is equipped with an input for a *gated sweep*. A fast sweep is required to resolve such a short-lived event such as a burst ($<600\ \mu s$). The triggered sweep lets the analyzer start its sweep (and its measurements) when the burst starts transmitting in its ramp-up sequence. Either the user has to evaluate where the data bits are positioned within the display or an intelligent system performs an automatic analysis. Several consecutive bursts are averaged to get a representative measurement.

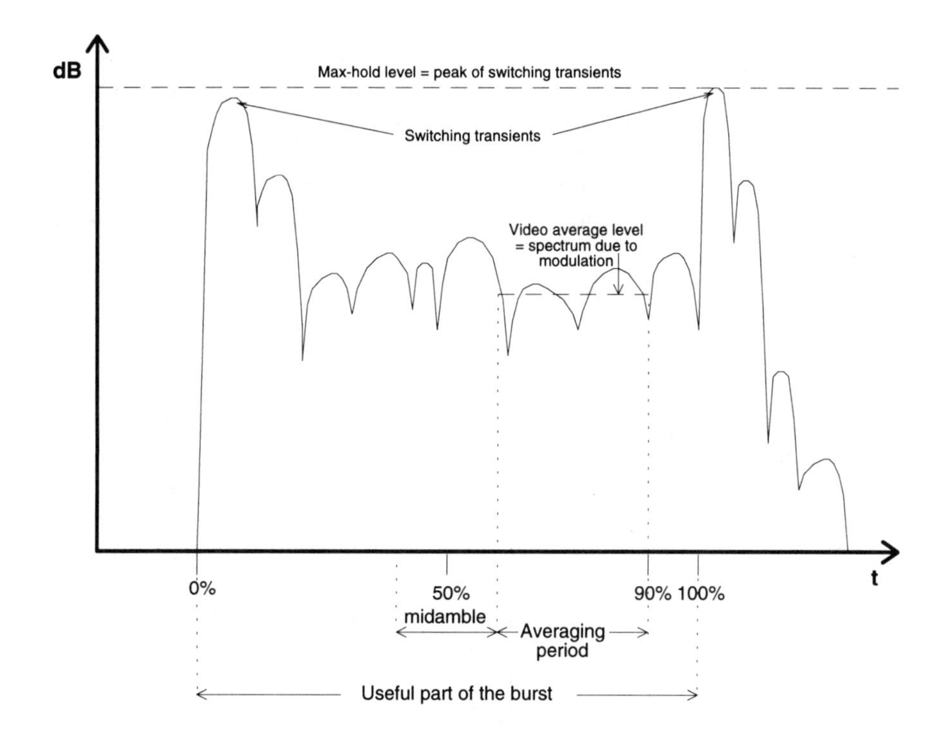

Figure 8.4 Display of the spectrum due to modulation (picture from the specifications).

If the sweep cannot be triggered with the burst event, it would be difficult to distinguish the effects due to modulation on the spectrum analyzer's display from those caused by switching transients. The requirement for a gated sweep already narrows the selection of test equipment manufacturers. The shopper may wish to look first at oscilloscope manufacturers who supply this kind of feature on their scopes; they are much more likely to include a triggered sweep in their spectrum analyzers as well.

8.3.4.2 Spectrum Due to Switching Transients

Switching transients are the chief cause of excess RF power appearing on adjacent channels or even further away from the assigned transmitting frequency, perhaps invading spectrum space reserved for other services. This phenomenon was reviewed in several discussions in Chapter 5, in which we learned that we must take precautions to make sure no aspect of a carrier changes too quickly as we impart information to it. Sudden amplitude changes add especially troublesome off-channel components to the transmitter's output, which tend to be much stronger than those derived from faulty modulation. We can evaluate the switching characteristics with this measurement, and owing to the potential for system trouble, it should be evaluated. The transmitter is, again, operated in its slow frequency-hopping mode over three different frequencies, but only one of the channels is evaluated at a time. The spectrum analyzer must not be triggered for this measurement, because the contribution of the modulation at the observed and measured frequencies is smaller than that caused by the switching transients.

8.3.4.3 Spurious Emissions From a Transmitter

Spurious emissions are frequency components derived from intermediate or clock frequencies within the equipment itself, which are radiated to the outside world via the antenna, power leads, or the cabinet. The effects of these emissions are almost always troublesome, because they can disturb external equipment such as car radios, audio equipment, and just about anything electronic. The measurement is performed on the mobile station at the transceiver's antenna connector. Base station measurements are done separately on the transmitter port and on the receiver port. Next, all the measurements are performed again in a shielded chamber with antennas pointed at likely radiation sources. In the course of performing the tests, it is most efficient to exercise some care in looking for the source and root cause of any spurious radiation rather than to simply document the measurements and go back at some later time to discover the reasons for the test's failure. The means of radiation should be grouped into three possible cases: (1) through the antenna, (2) from the cabinet, and (3) from the power leads.

8.3.4.4 Intermodulation Attenuation

As is true for analog transmitters, GSM base stations require an additional measurement of their *intermodulation attenuation*. The test seeks an indication of the overall linearity of the transmitter, any attached amplifiers, and any filters found within the transmit path. A chance interfering signal would, together with the desired signal, intermodulate at some point of nonlinearity within the transmitter. The simulated interfering signal is injected into the transmitter's output port at a level 30 dB lower than the transmitter's normal output level and at different frequencies offset from the desired output signal such that intermodulation products appear at the receiver's input frequency. Special care is taken for the intermodulation products of the third, fifth, and seventh orders in the receiver's input band (see Figure 8.5). We should mention that although the specifications require the investigation of the seventh-order products, these are very hard to find and are not likely to cause problems. Intermodulation products of the third order are, by far, the most likely to cause trouble, and they should be investigated first.

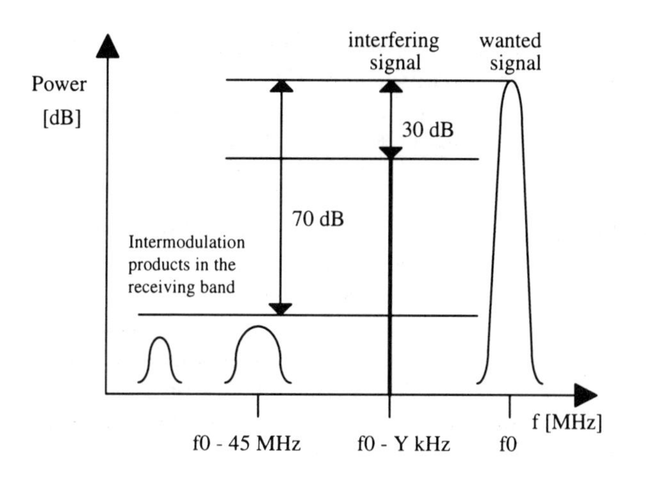

Figure 8.5 Intermodulation of a transmitter caused by an interferer.

Further intermodulation may occur in base stations in which several transmitters are installed. Bad linearity in an output amplifier common to all the transmitters may cause them to disturb each other without an external interferer being present. This case is investigated for several channel spacings, and special care is, again, exercised on the receiving band. Exact requirements and a block diagram of these measurements are found in Chapter 11. Intermodulation is a problem confined to base stations, because they are located in fixed sites with other powerful RF sources only a few centimeters away.

8.4 RECEIVER MEASUREMENTS

Receiver measurements fall into two classes along lines similar to those that separate transmitter tests. One is a collection of on-channel operational tests, and the other investigates behavior in the frequency domain for both on-channel and off-channel specifications. The measurements we perform on GSM receivers are almost identical to those we perform on analog receivers, where the differences, when they occur, are in the procedural details.

8.4.1 Measurements on Analog Receivers

Receivers in analog cellular systems tend to be FM, and testing them is quite intuitive. The most common method is the *SINAD* procedure, in which we inject a low-level tone-modulated (FM) signal into a receiver's RF input port, and recover the demodulated result from the receiver's audio output terminals. We quantify the receiver's performance as we consider the ratio

$$\frac{\text{signal + noise + distortion}}{\text{noise + distortion}}$$

If we reduce the injected RF level to a very low level, and if this ratio is considered at the receiver's audio output, then it is a comparison of the desired signal (signal) plus all the undesired products (which are (1) the *noise* accompanying the weak signal (noise) plus (2) any *distortion* the receiver adds to the recovery process (distortion)) to only the undesired outputs (the noise plus the distortion). If we measure, say, 4V for the value in the numerator and then filter the desired modulating tone from this signal and measure the resultant signal as 1V (the denominator), then we can say that the receiver's shortcomings (noise plus distortion) represent 25% of the composite output. A distortion analyzer would indicate the receiver has 25% distortion. The term SINAD is an acronym for the phrase

signal + noise and distortion

As is common in radio work, ratios are expressed in decibels. SINAD is expressed as the ratio of two voltages: the *good plus the bad* ($V1$) divided by *only the bad* ($V2$).

$$\text{SINAD(dB)} = 20 \log(V1/V2)$$

which would be, in this example,

$$20 \log(4/1) = 20\ (0.6) = 12 \text{ dB}$$

The nice thing about the SINAD method over many other similar receiver tests (such as the simpler SNR method) is that it uses a modulated test signal. It evaluates, therefore, the whole receiver: the RF stages, IF stages, demodulator, and all the audio processing stages up to the point where the voltages are measured.

The determination of SINAD has some standard conditions. The most common ones recommend a 1-kHz modulating tone at a deviation of 60% of the peak deviation normally specified for the receiver under test. The injected RF level is decreased until the SINAD value reduces to some specified ratio, usually 12 or 20 dB, and then the level of the injected RF is noted.

8.4.2 Measurements on Digital Receivers

The SINAD technique cannot be used on a GSM radio, because the speech codec does not process sine wave signals, be they pure tones or noise. Testing the quality of a receiver used in digital services such as GSM is, however, very similar to the analog methods; we consider a *good part–bad part* ratio.

The most popular technique is borrowed from the fixed wire-line services: the BER technique. The principle is that a known bit sequence modulates a carrier in the appropriate way and is injected into the receiver's RF input port. The receiver has to demodulate and decode the bit sequence, and then somehow pass this back to the test equipment where both the sent (the *good* part) and received bit sequence (the *good* part plus the *bad* part) are compared. The error rate (sometimes called the *error ratio*) is evaluated at some low RF level that creates a certain proportion of questionable bits at the receiver's output. For the mobile station, the bit sequence is looped back to the test equipment on the duplex radio channel. Specifically, the recovered data, including the bits decoded in error, are fed into the transceiver's encoder, modulated on the RF carrier, and sent back to the test instrument on some logical channel enlisted for the purpose. The method used in base stations is slightly modified from the mobile method, and it will be thoroughly explained in Chapter 10. Depending on the channel type simulated during the test and which bit classes are considered if a traffic channel is simulated, different BERs are distinguished. Higher class traffic bits will be decoded more accurately than lower class bits under otherwise identical test conditions. Each type of BER test, then, has a different measurement procedure.

8.4.2.1 Bit Error Rate

The BER takes every single bit of a speech or data frame into consideration when the BER is calculated. If, for example, we use a full-rate traffic channel for the test, there will be 260 bits sent and compared in each speech frame passed to the receiver; no distinction is made between their classes: class Ia, class Ib, or class II bits. Signaling data, and the channels in which they find themselves, are not evaluated with this type of measurement.

8.4.2.2 Residual Bit Error Rate

The *residual bit error rate* (RBER) is only performed for speech frames that are not marked as bad or corrupted frames. Such frames, we should recall from Chapter 5, are discarded if they cannot be recovered properly in the receiver. A frame is labeled as a bad one if, for the class Ia bits, the *cyclic redundancy check* (CRC) (test on the parity bits) is not successful. This will occur if an error in the class Ia bits has not been detected and corrected by the channel decoder which "undoes" the convolutional code. The residual rate is calculated only on frames that make their way to the speech codec. Two RBERs are distinguished. One is reserved for the class Ib bits and the other for the class II bits. We write "RBER Ib," which evaluates all

132 class Ib bits of a good full-rate speech frame, and "RBER II," which does the same for all 78 bits of class II.

8.4.2.3 Frame Erasure Rate

The *frame erasure rate* (FER) indicates how many of the received frames are bad. The FER is not only performed for the speech and data frames, but also for signaling frames such as the FACCH. The receiver evaluates the parity bits to determine whether a frame should be erased, independently of whether it is done for a speech frame for the class Ia bits (3 parity bits) or for a signaling channel (40 parity bits). If a frame is erased from a speech channel, the bits are all set to a specific value to indicate the erasure. In the signaling channel case, the L2 acknowledgment is not sent back to the source expecting an acknowledgment, and this becomes an indication to the test system that this frame has been erased, or was not, for whatever reason, understood by the receiver.

8.4.3 Sensitivity Measurement

If we confine our examination of a receiver to its *sensitivity*, then we are concerned with the receiver's ability to recover very weak signals and still be able to demodulate and decode the information modulating those signals within a certain margin of safety. The measurement is performed for the different logical channels using RBER and FER measurements under various *propagation conditions*. Propagation conditions are selected to simulate the operating environments GSM radios are most likely to be asked to deal with, such as urban areas, hilly terrain, or remote areas in the countryside. The equipment used to simulate the propagation characteristics in the different environments is explained in Chapter 9.

8.4.4 Cochannel Rejection

Just as was the case with transmitters earlier in this chapter, there are some receiver tests that are in no way peculiar to digital radio or GSM. The first of these is *cochannel rejection*, which is an indication of the receiver's robustness against the reception of a signal on the same channel. This would be the case under peculiar propagation anomalies when the receiver has to reject a weak signal on the same frequency from a cell far away, perhaps from another network. To perform the measurement, the desired on-channel signal is injected together with an interfering signal, which is also GMSK-modulated at a power level below the desired signal (see Figure 8.6). The receiver should be able to distinguish between the signals and demodulate the stronger desired signal with no errors.

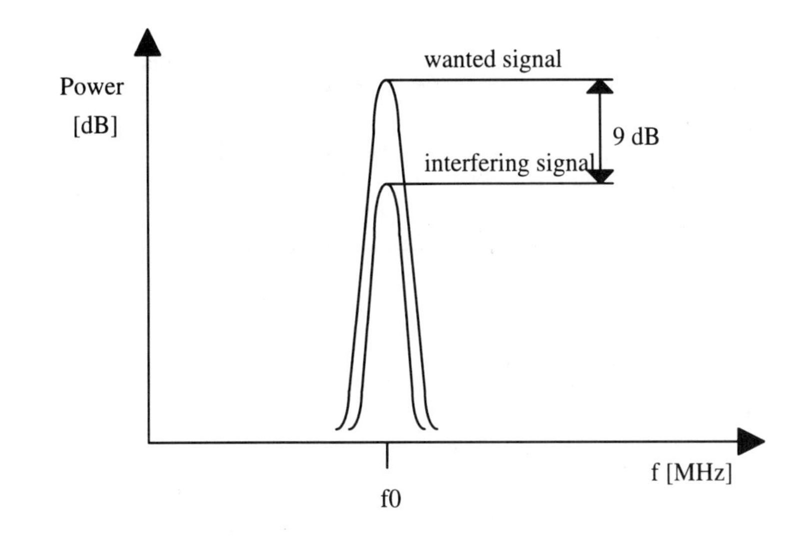

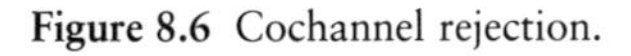
Figure 8.6 Cochannel rejection.

8.4.5 Adjacent Channel Rejection

Adjacent channel rejection is similar to the cochannel case and is an indication of the receiver's ability to receive the desired signal even in the presence of a stronger signal in an adjacent channel. The adjacent channels tested in this case are the first and second ones at an offset of 200 and 400 kHz. It is not likely that adjacent channels are used within one cell or an area of neighboring cells. The common case is within a cell as several transmitters are transmitting at the same time at different power levels on different channels. The interfering signal is, again, GMSK-modulated to simulate another GSM transmitter (see Figure 8.7). An adjacent channel in a TDMA system is not only found on an adjacent frequency, but also on an *adjacent time slot*. The impact on the receiver's sensitivity to a stronger signal in an adjacent channel transmitting in both adjacent time slots is measured.

8.4.6 Intermodulation Rejection

Receivers can generate intermodulation products just as transmitters can. We can judge the apparent linearity of a receiver by measuring its ability to reject intermodulation products caused by nonlinear elements in the receiver. Along with the desired signal, which carries important information, two additional signals are fed into the receiver: one is GMSK-modulated, and the other one is a pure CW signal. The frequencies of the two invading signals are selected in such a way that the intermodulation products of the third order of both interfering signals is the same as the desired

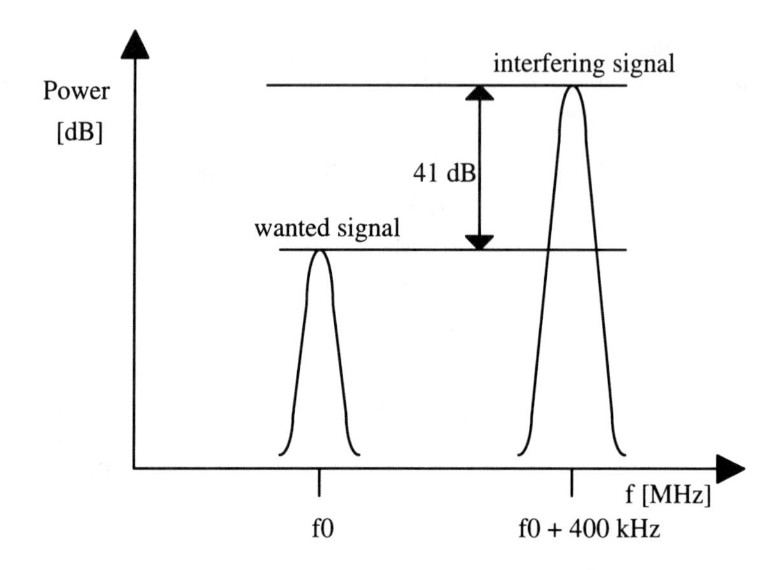

Figure 8.7 Adjacent channel rejection.

input signal (see Figure 8.8). The impact of the third-order intermodulation products is always the greatest, and these are the ones that should be investigated. *Intermodulation rejection* is good if the receiver—under these hostile conditions—still has a low BER according to the specifications.

8.4.7 Blocking and Spurious Response Rejection

The *blocking and spurious response rejection* measurement investigates the behavior of the receiver when a strong transmitter is present on *any* frequency, such as would be the case at a cell site located with a broadcast TV transmitter. The interfering transmitter may be on a very different frequency from that of the receiver under test, and may even be modulated completely differently, but it has a very high output power. The frequency bands that were used to investigate the adjacent channel rejection characteristics are spared from this test, since using them would be redundant. An RBER measurement is used to perform this test on the receiver.

8.4.8 Conclusion

We have dealt with the common tests for both the mobile station and base station receiver measurements. They look similar to each other, but their realization is different for the mobile and base stations. This is due to the fact that the latter

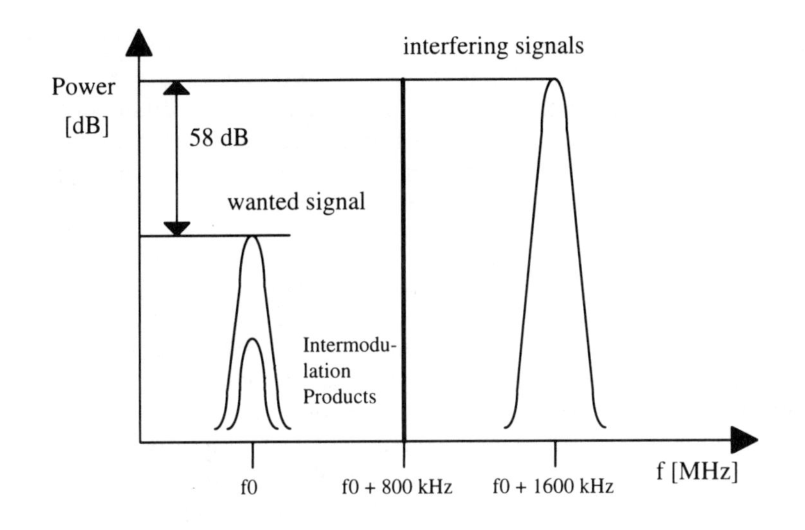

Figure 8.8 Intermodulation rejection.

device not only provides a Um interface through which the tests are performed, but also an Abis (network) interface, which makes the measurements more complicated. In Chapters 9 and 10 we will see the practical aspects of both types of measurements, and there will be more elaborate and specific descriptions of them.

8.5 GSM SIGNALING TESTS

The complex signaling environment employed in GSM, described in a great pile of pages in the specifications, implies an enormous amount of test effort. A first glance into the signaling complexity may give an impression of a giant software monster (another interpretation of the three letters GSM) created by the standardization bodies. A closer look, however, will reveal that there is a structure behind it, a structure that has been defined down to the bit level and that leaves little room for misinterpretations. This structure makes it easy for the design engineer to develop network entities, both hardware and software, in a highly organized manner. Eventually, the defined structure yields guidelines for testing the devices that employ them. The signaling environment, with a focus on the radio interface, has been described in some earlier chapters. The remainder of this chapter gives an overview of all the aspects that have to be considered when testing the signaling performance of both GSM mobile phones and BTSs.

Analog cellular systems employ signaling functions, too, but they are so crude and simple that it is hard to draw an analogy between analog and digital signaling tests, particularly in the GSM case. The comparison is one of scale rather than function; the signaling in an analog system is trivial compared to GSM. Perhaps we can only say that some analog cellular systems require explicit signaling tests, because some protocols use tones on the voice channels, and tones can wander off frequency, they can assume the wrong level, their detectors can stop working, or they can be missing altogether. Explicit signaling tests are never performed in digital cellular systems except during design and verification functions.

8.5.1 Signaling Tests—When Do We Perform Them?

Signaling tests are part of *approval procedures* and thus have to be taken seriously by design engineers who implement all the necessary signaling functions in a product on its way to commercial reality. Signaling functions are realized as software programs (running on microprocessors), logic arrays, and customized application-specific integrated circuits (ASIC). Manufacturers of GSM equipment, the operators who buy it, and the regulatory bodies that watch the whole process unfold have a common set of signaling goals. All of them want to make sure that a particular product performs according to the rules laid down in the *GSM Technical Specifications*, and is thus resistant to systematic and transmission-related errors, as well as to any problems introduced by other external network entities. It has to be clearly verified that when problems arise, the particular entity under development performs as defined, that it eventually enters a defined state and does not "hang up" the system. Because signaling functions are implemented in the form of software or integrated circuits, they are not explicitly part of test protocols in production or service environments; they are thoroughly tested and then never examined again. Indeed, a good test engineer can increase testing efficiency in a production or service situation by carefully avoiding explicit signaling tests. Performing signaling tests in production and after-market environments makes as much sense as proofreading a book already in print. The single exception to the rule is the case where a certain manufacturer's mobile phone, for example, is suspected of exhibiting incorrect responses to some specific signaling conditions. This is caused by a misinterpretation of the signaling specifications at design time.

8.5.2 Tests

In order to verify the defined performance of the network components, such as a GSM mobile terminal, procedures similar to the more common radio tests are consulted. Detailed specifications of signaling procedures have been developed. These procedures systematically check *signaling capabilities* by stimulating the DUT in numerous and extremely specific ways, and then examining the response for a series of precise responses: if *A* happens, then *B* should happen, but not *C* or *D*, unless *E* occurred two steps earlier, and so on. These signaling procedures, when used during type approval tests, assume the correct RF performance of the mobile phone or base station. They are carried out in a logical sequence, checking the lower layers before the upper layers. Dedicated GSM test equipment is necessary for design and type approval. This test equipment must be able to follow the protocols and specified radio parameters as defined in *Technical Specifications*. It must also be able to simulate certain scenarios, introduce protocol errors on purpose, and trace all the information exchanged with the DUT. An indication of a pass/fail result is desirable rather than sorting through the often complex responses for a judgment.

8.5.2.1 Tests of Layer 1–Related Signaling Functions

Signaling functions related to Layer 1 are defined to support the establishment, maintenance, and termination of physical links, as well as to control and monitor the radio resource. They are routine facilities, which the higher signaling functions use for everything they do. These functions are also referred to as *radio link management* functions. They are employed and controlled by the RR sublayer of the network layer (Layer 3). As such, RR procedures (see also Section 7.1.1) are always based on *physical facts* related to Layer 1 influences. The physical facts can be such things as (1) the availability of and access to channels, (2) power levels, (3) timing of signals, and (4) interference conditions.

The procedures and the required performance of the physical functions within a GSM network entity, be it a mobile terminal or a base station, are listed below.

1. Synchronization and timing:

- *Initial synchronization* of a mobile station to the network (FCCH).
- Synchronization during a dedicated radio link: a base station has to measure burst timing and give correct timing advance values to the mobile station (SACCH and channel assignment), and the mobile station has to correctly process the timing advance and adjust its burst timing.
- Synchronization of a base station to the BSC, synchronization of all transceivers within a single base station (*timing tolerance*), and synchronization to adjacent base stations.

2. Power control:

- *Downlink power control or mobile station power control*: A mobile station has to set its radio output power as commanded by the base station (SACCH and channel assignment). A base station has to monitor the dedicated radio links with mobile stations and evaluate the power levels. Downlink power control is performed in 16 levels separated by 2 dB each (mobile station power control levels 0 to 15). This function is checked in combination with the transmitter tests (during the power tests). Also, the base station optionally may control its output power according to the distance to a connected mobile station.

- *Measurement report*: A mobile station has to continuously report back to the network values corresponding to received-signal level (RX-LEV) (similar to received-signal strength indication (RSSI) used in analog networks) and receiver quality (RX-QUAL) by the determination of a BER (number of decoder errors during one multiframe) on a particular channel. It therefore has to measure the received signal strength and quality. A distinction is made for the monitored base channel (carrying the BCCH) and the monitored active dedicated channel: (1) RX-LEV-FULL and RX-QUAL-FULL (base channel), and (2) RX-LEV-SUB and RX-QUAL-SUB (dedicated active channel). When monitoring adjacent cells, the measurements are only made for the particular base channels, since there is no dedicated link established.
- A base station reports signal quality parameters to the BSC, which are measured on a particular mobile station's signal when a dedicated link is active. These parameters are named RX-LEV and RX-QUAL, too, after the mobile's measurement reports.

3. Radio channel management:

- A GSM base station is checked for the appropriate management of channel resources: activation, modification, and deactivation of control channels and dedicated channels.
- A GSM mobile station is checked for the correct activation of dedicated channels, channel changes (assignment procedures, handover, frequency redefinition), and radio link maintenance and termination.
- A very important function of both the base and mobile stations must be verified as well: in the case of interruptions in the radio link that do not allow the mobile or base station (or both) to keep up the link (it cannot read the counterpart anymore), the dedicated link has to be terminated. As a criterion for the termination upon such a *radio-link failure*, a counter has been introduced. It is used by both the base and mobile stations. This counter, named *S*, is initially (at the establishment of a dedicated channel) set to a certain maximum value decided by the network side. It is decreased by 1 when an SACCH block has not been read correctly, and it is increased by 2 for each SACCH block that has been received correctly (without unrecoverable errors). Of course, the S-counter will not be increased above the initial maximum value. The S-counter's contents is the hysteresis left when the radio channel fails, thus preventing an immediate release of the channel. Such a function is necessary in order to terminate unrecoverable interrupted radio links on both sides in an orderly way. The maximum time for a *radio-link timeout* can be derived from the maximum number of SACCH blocks that can be coded for the counter: 64 (approximately 30 seconds).

8.5.2.2 Test of the Data Link Layer (Layer 2) Protocol Implementation

The data link layer has to ensure the correct transfer of signaling information over the radio path. The functionality defined for this layer and its terminology are explained in Chapter 6. The verification of the protocol used on this layer—LAPDm—has found its place in the test specifications and test procedures. In order to check for the correct implementation in a GSM mobile or base station, not only the basic procedures of acknowledged and unacknowledged block transfer have to be tested, but also the performance in the case of errors. *Technical Specifications* [1,2] define procedures for the following tests:

- Establishment of an acknowledged data link (SABM);
- Error control during data link establishment (e.g., loss of a UA frame);
- Sequence counting during acknowledged block transfer (N(S)/N(R));
- Segmentation and concatenation;
- Loss of information frames (I-frames);
- Reaction to erroneous C/R values in the address field;
- Reaction to erroneous values in the control field;
- Reaction to invalid frames.

Since basic Layer 1 functions are a prerequisite for the verification of Layer 2 performance, the data link layer tests are described (and in type approval sessions also performed) after the Layer 1 tests.

8.5.2.3 Test of the Network Layer (Layer 3) Protocol Implementation

The network layer, as described in Chapter 7, comprises tasks, messages, and procedures to control the processes necessary for the establishment, maintenance, and termination of dedicated communication links within a GSM network. There are three sublayers defined within Layer 3: RR, MM, and CM. The signaling architecture in the GSM network (see Figure 7.5) defines the implementation of all three network sublayers within a GSM mobile station, whereas a BTS only takes care of basic RR procedures. It is *transparent* to the rest of the Layer 3 signaling functions. It only transports these messages between the mobile phone and the other network entities without paying any attention to their content. Since the establishment of CM sublayer signaling links are, in general, based on the presence of MM connections, and those, in turn, are based on RR connections, the functionality of the three sublayers within a mobile station are logically tested in the sequence of their establishment: First RR, then MM, and finally CM. Another important part of the test procedure is dedicated to the verification of Layer 3 timer handling. The timers and the appropriate actions (on expiration) specified for the transmission and reception of Layer 3 messages or the completion of Layer 3 procedures are checked for their compliance within given tolerances.

An indication of the importance of network layer signaling tests can be determined from the effort spent on GSM mobile phone type approval: *Technical Specification* for approval (TS 11.10) contains approximately 170 pages of Layer 3 test procedures. In comparison, the verification of Layer 2 functions is covered in 35 pages, and the radio link management tests in approximately 30 pages.

REFERENCES

[1] *GSM Technical Specifications*, ETSI, Sophia Antipolis, Vol. 11.10, Phase 1.
[2] *GSM Technical Specifications*, ETSI, Sophia Antipolis, Vol. 11.20, Phase 1.

CHAPTER 9

Testing a GSM Mobile Station

With the GSM system spreading all over the world, not only in Europe, but also in Asia and Australia, the number of GSM phones in use will increase steadily. They are popular wherever they appear, and their success is assured. This means that there will be a high demand for the production and service of GSM phones. At the same time all this growth runs rampant across the GSM landscape, care has to be taken that these phones keep to their required specifications, because they may be used anytime at any place in the GSM world. We have to guarantee the quality of the equipment used in the networks. To accomplish this, R&D activities, production, and service facilities have to maintain the phone's compliance with the test specifications through the use of appropriate test equipment.

This chapter explores the most important tests of GSM mobile stations. It will focus particularly on the RF tests with a broad look at signaling tests at the end of the chapter. RF tests are most sensitive to the parts of phones that are under stress and are likely to fail in a network. Signaling tests, on the other hand, are usually confined to R&D and type approval functions, and test design accuracy rather than specific serviceability. Some basic signaling activity is needed for the RF measurements to work, but this is used only as a tool to gain access to the tested functions, not to exercise the signaling that supports them. The signaling, in this case, is invisible to the test engineer or the service technician. Besides descriptions of the most common RF tests, the requirements for passing the individual tests are also listed so that it is not necessary to flip through the technical specifications looking for them.

9.1 STRUCTURE OF A MOBILE STATION

Before the tests are described in detail, we take a closer look at the internal structure of a GSM mobile phone. We see the internal workings in block diagram form in Figure 9.1, where the different entities are shown as they are described in this chapter. It must be stated that though different functional entities may be distinguished from each other in all phones, these appear in the physical realm on only two or often even one printed circuit board. Most portable phones are divided into two parts: (1) an *RF part*, which handles the receiving, transmitting, and modulation tasks, and (2) a *digital part*, which takes care of the data processing, control, and signaling functions. In the newest hand-portable phones all these functions are integrated into one printed circuit board on which we find a number of ASICs.

The antenna combiner couples the receiving and the transmitting paths onto the single antenna connector or a fixed antenna. The receiver contains the front end, a receiving filter network, and a mixer to down-convert the input signal onto an IF that is eventually converted into the data domain by the ADC. The IF is examined, or demodulated, for its phase information. The IF's level is measured for the signal strength of the host base station, as well as transmitters in adjacent cells when they are monitored.

Due to *multipath propagation* and other reflections, the signals arriving at the receiver are distorted, since they take several paths of various lengths to the receiver. It is the role of the *equalizer* to compensate for these distortions and multiple signals. It can compensate for a delay spread of up to 16 μs. The *demodulator* extracts the bit stream from the IF, and the *demultiplexer*, thanks to the frame numbering in GSM, sorts the received information from the different time slots and frames into their appropriate individual logical channels.

The *channel codec* decodes or encodes a bit sequence coming from the demultiplexer or going to the *multiplexer*, respectively. It is the "packing, shipping, and receiving department" of the radio. Not only does it process signaling channels, such as the SDCCH, FACCH, or SACCH, but it also handles speech channels. If the channel codec discovers that a signaling frame has arrived for processing, it passes it on to the *control and signaling unit*. A speech frame is passed to the *speech codec*. From a stream of speech data, a signaling channel can appear as the FACCH. This case is indicated by the *stealing flag* status.

The speech codec either rebuilds specific human language sounds out of the 260-bit speech blocks received from the channel codec and passes the digitized speech to the DAC, or compresses digitized speech coming from the ADC so that the data are represented by blocks of 260 bits before being encoded.

The control and signaling unit performs all the control functions of the mobile station. These functions include *power control*, the selection of different channels that have to be used, and a host of other functions built into the mobile station. Depending on the currently executed function, different signaling messages have to

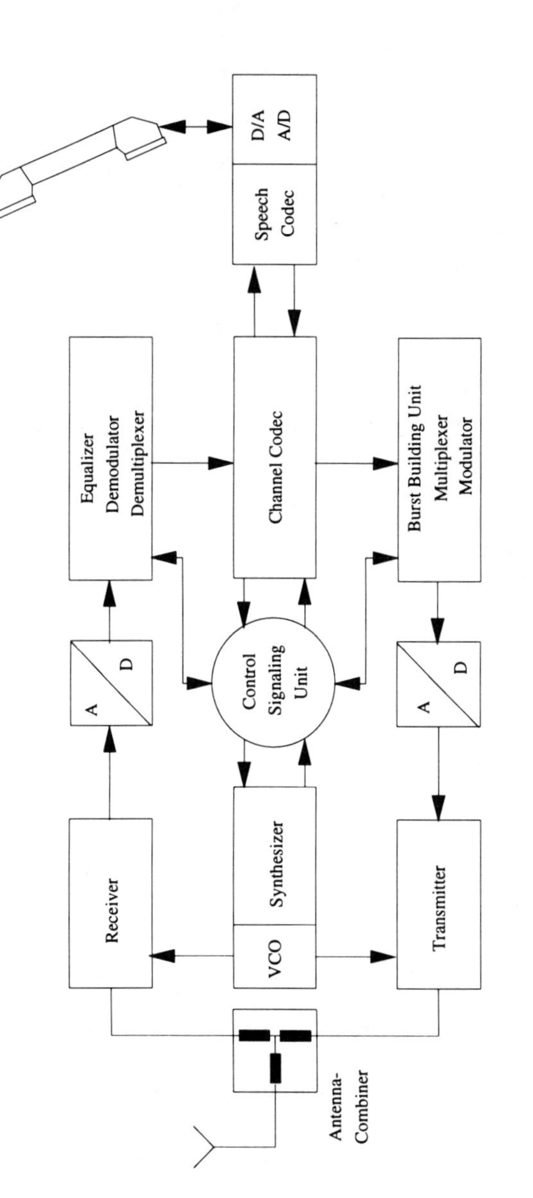

Figure 9.1 Block diagram of a mobile station.

be exchanged with the network, which are prepared or processed in this unit and are passed to or received from the channel codec.

The *burst-building* unit places the coded bits in the transmitter path into the appropriate burst structure and then adds the training sequence bits, the tail bits, and the stealing flags. The multiplexer assigns each individual burst to a time slot within a numbered frame in which it should be transmitted. After this sorting and ordering has been done, the modulator imparts this information onto the IF carrier.

The *transmitter* contains a mixer to up-convert the modulated IF signals from the modulator to the final radio frequency in the 900-MHz band. An amplifier increases the level of the signal to whatever value is commanded by the base station, and output filters limit the bandwidth of the output to its assigned channel to ensure that the transmitter does not disturb other channels in the GSM network or radio services on other bands.

The *synthesizer* provides the internal timing references for the bit and frame clock as well as for the RF sources in the transmitter and the receiver. A *voltage-controlled oscillator* (VCO) assumes a stable operating frequency in accordance with commands from the control and signaling unit. The accuracy of the mobile station's transmitter frequency is maintained within a network standard with the help of the FCCH the mobile finds on the base channel.

This is just an example of how a GSM mobile phone might appear if we pry it open for examination. It is also possible that a mobile station does not use an IF at all, or that the control structure is slightly different or buried within an ASIC with other functions.

9.2 PREPARATION FOR TESTING

It is, in general, easier to test a GSM mobile station than it is to test an analog mobile station. Just as we do with analog terminals, there are a few things we should consider at test time.

An *external power supply* is usually used for the mobiles under test rather than a battery, even if the terminal is a hand-portable phone with a built-in battery. A battery will influence the quality of the measurements. A weak battery may cause the mobile station to stop transmitting in the middle of the measurements, or there may not be enough energy in the battery to properly test output power; the measurements may be misleading or very confusing. An external power supply should be used if the battery itself is not checked at the start of the tests.

As of the date of this writing (mid-1994), there is no test equipment that allows off-the-air measurements of a GSM mobile station through its antenna. There is not much call for such equipment, since measurements directly through the antenna connector are more accurate and cheaper. If we do not use hard cable connections, then reference measurements would have to be made to measure the coupling factor between the transmitting and receiving antennas of the mobile and the test set. To keep measurements simple and reliable, an appropriate connecting cable between the mobile station and the test equipment has to be provided. The cable should not be longer than 1m in order to limit the attenuation the signals will suffer. This is no problem for car-mounted or portable mobile stations, which are fitted with standard antenna connectors. Experience shows that there is a major problem with many of the smaller hand-portable phones. The little phones have special adapters that replace the antenna during testing, or they have very unusual antenna connectors that can only be used with a fragile and expensive service kit. Stubborn problems appear with certain service kits of some hand-portable mobile stations. Some of these connectors cannot be screwed firmly to each other during testing, but are only pushed together with a complex mechanism of some kind. A fixed loss of up to 20 dB appears between the mobile station and the test set, and the loss is difficult to duplicate from phone to phone. These special connecting mechanisms and contraptions can eventually be mastered with patience and diligent practice. Even the slightest hint of frustration on the part of the attending technician will usually destroy these special connectors and test adapters.

The most important thing we need before we can test a GSM mobile station is the *SIM card*. The chip integrated within the card holds all the necessary data to identify the mobile station to the network and the tester, and, even more important, get the phone to simply transmit something. On one hand, customers might not leave their card with the service center, fearing an additional service charge on their phone bill, and on the other hand, the customers' commercial cards are not sufficient for testing mobile stations. Some of the measurements require specific service commands to be sent to the mobile station, which are executed only if the inserted SIM card is positively identified as a *test card* that contains special test data. A test SIM holds, for example, a known *ciphering sequence* for the test equipment, which allows us to thoroughly test the ciphering feature. Ciphering is something we exercise and test in R&D or type approval activities rather than in production or service situations.

Now the questions is where a service shop might get such a test card, which we need if we want to perform all tests needed to properly service a mobile station. These cards are usually provided by the network operators or they may come from test equipment suppliers. Depending on the size of the mobile station, two different SIM card sizes are possible. For normal-sized mobiles, it is the *credit card–sized SIM card*, and for the small hand-portable phones, which are smaller than the normal-sized SIM itself, it is the much smaller *plug-in SIM card*. Independently of the size of a SIM card, the chip containing all the data is always the same.

9.2.1 Some Calculations

In the *GSM Technical Specifications*, the unit for power levels commonly used or specified is in decibels over 1 μV (dBμV) or decibels over 1 mW (dBm). When talking about power classes, the watt (W) is used. Here are some formulas for converting measurements between the different units.

The calculation from watts into dBm is a basic one. The "m" indicates that everything is related to a milliwatt. Knowing this, (9.1) applies.

$$P(x\text{W}) = 10 \cdot \log\left(\frac{x\text{W}}{1\text{ mW}}\right)\ [\text{dBm}] \tag{9.1}$$

The transformation of dBμV into dBm is a simple linear one with a constant added value of 113 dB, as shown in (9.2):

$$P(x\ \text{dB}\mu\text{V}) = -113\ \text{dBm} + x\ \text{dB}\mu\text{V}\ [\text{dBm}] \tag{9.2}$$

This is an easy relationship, but it helps to know where the value of 113 dB comes from. To do this, we make the assertion that the power of 0 dBm (1 mW) shall be expressed in dBμV. Figure 9.2 shows the relationship of the powers and voltages in a source and its load. With these, the following assumptions are valid and the resulting calculation follow logically.

If

$$Ri = R_L = 50\Omega \tag{9.3}$$

then

$$U_L = \frac{Ui}{2} \tag{9.4}$$

With this, the power is calculated:

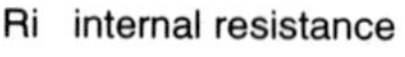

Figure 9.2 Relationship between power and voltage.

$$P_L = \frac{U_i^2}{4Ri} \tag{9.5}$$

$$\begin{aligned} P(1\ \text{mW}) &= 10 \cdot \log\left(\frac{1\ \text{mW}}{P_L(1\ \mu\text{V})}\right) \\ &= 10 \cdot \log\left(\frac{1\ \text{mW}}{\frac{(1\ \mu\text{V})}{4 \cdot 50\Omega}}\right) \\ &= 10 \cdot \log(2 \cdot 10^{11}) \\ &= 113\ \text{dB}\mu\text{V} \end{aligned} \tag{9.6}$$

9.3 TRANSMITTER TESTS

Even in the case where only RF tests have to be performed, there is still a need to perform some signaling with the mobile to get the mobile station into an active call state where it is transmitting continuously. The usual method is to bring the mobile onto a traffic channel. It does not matter whether the call establishment procedure is mobile-originated or mobile-terminated.

Modern test equipment allows us to perform a call setup automatically with a simple selection from a menu on a display. Users do not have to concern themselves with the different messages that have to be exchanged. Furthermore, it is possible to interactively change the channels and the power levels of the test equipment and the mobile phone during testing. The test equipment starts the necessary signaling automatically without the user observing the protocol exchange in the background.

The descriptions of the tests that follow assume that a call has been established and a traffic channel is active. The tests merely have to be executed one after another.

9.3.1 Phase Error and Frequency Error Measurement

The *phase error and frequency error measurement* is split into two separate tests, depending on the simulated propagation conditions on the RF path. The first part covers the static tests with ideal propagation, and the second one covers the propagation under real conditions with interference in a multipath propagation environment. The static case is the suggested method in service and in some manufacturing circumstances.

For the static propagation case, the phase error and frequency error are measured together. This is a test of the *modulator's quality*. It is performed under ideal conditions in order to avoid disturbing the mobile phone and to examine the modulator itself. The frequency error indicates the mobile station's ability to lock onto the network's master frequency from the base station as transmitted on the FCCH. The phase error is expressed with the *peak phase error*, which for a mobile station must not exceed 20 deg. If the error is expressed as the root-mean-square (rms) average, then the phase error must not exceed 5 deg. The frequency error should not be more than 90 Hz for the statistic propagation case.

Figure 9.3 illustrates the evaluation of the phase error and frequency error with a test system. The test equipment not only performs the different phase error measurements (rms and peak) and frequency error measurements, but also gives an indication of the transmission time delay, which shows how accurately the transmitted burst appears within the time slot structure. For each measurement, the results are given in a statistical form with their current, average (over all measurements), minimum, and maximum values. To make the evaluation easier, the results are checked against the required test limits and a simple pass/fail indication is given. Something that is not specified but very helpful for R&D purposes is the display of the phase error of a complete burst, with the burst content in terms of bits and symbols. This allows engineers to precisely evaluate a modulator and to determine at which bit position a reading is above the required limits and whether the reading is, for example, at the transition between two symbols.

In addition to the usual static test condition just described, multipath propagation is simulated so that the performance under real and typical operating conditions can be checked with reflections and delays on the RF path. Propagation conditions affect the signals at a mobile's antenna, and they are a function of the physical environment that a mobile may find itself in at any given instant. The mobile's reaction to them is primarily a design consideration rather than a service one. Under such circumstances the quality of the modulator should not get worse, even though the mobile station's ability to synchronize with the FCCH decreases. To make conditions even more difficult but also more real, up to six additional base stations may be simulated, just as would be the case within a typical cell. This leads to further interferences. These additional simulated adjacent base stations require two test constraints: (1) no simulated base station is allowed within five channels, in the frequency domain, of the base station frequency maintaining the link with the mobile station, and (2) one transmitter should be simulated at each end of the GSM frequency band.

The different propagation conditions and their limits are given in Table 9.1.

A *fading simulator* is used to simulate multipath propagation. These simulators usually incorporate two signal paths, where each path provides six poles. One pole produces one reflected original signal with a certain user-defined characteristic. For the tests using these simulators, only the signal path for the mobile station's receiver is altered to test the mobile's capabilities; it makes no sense to fade the mobile's transmitting path, since the test equipment is not fitted with equalizers to cope with

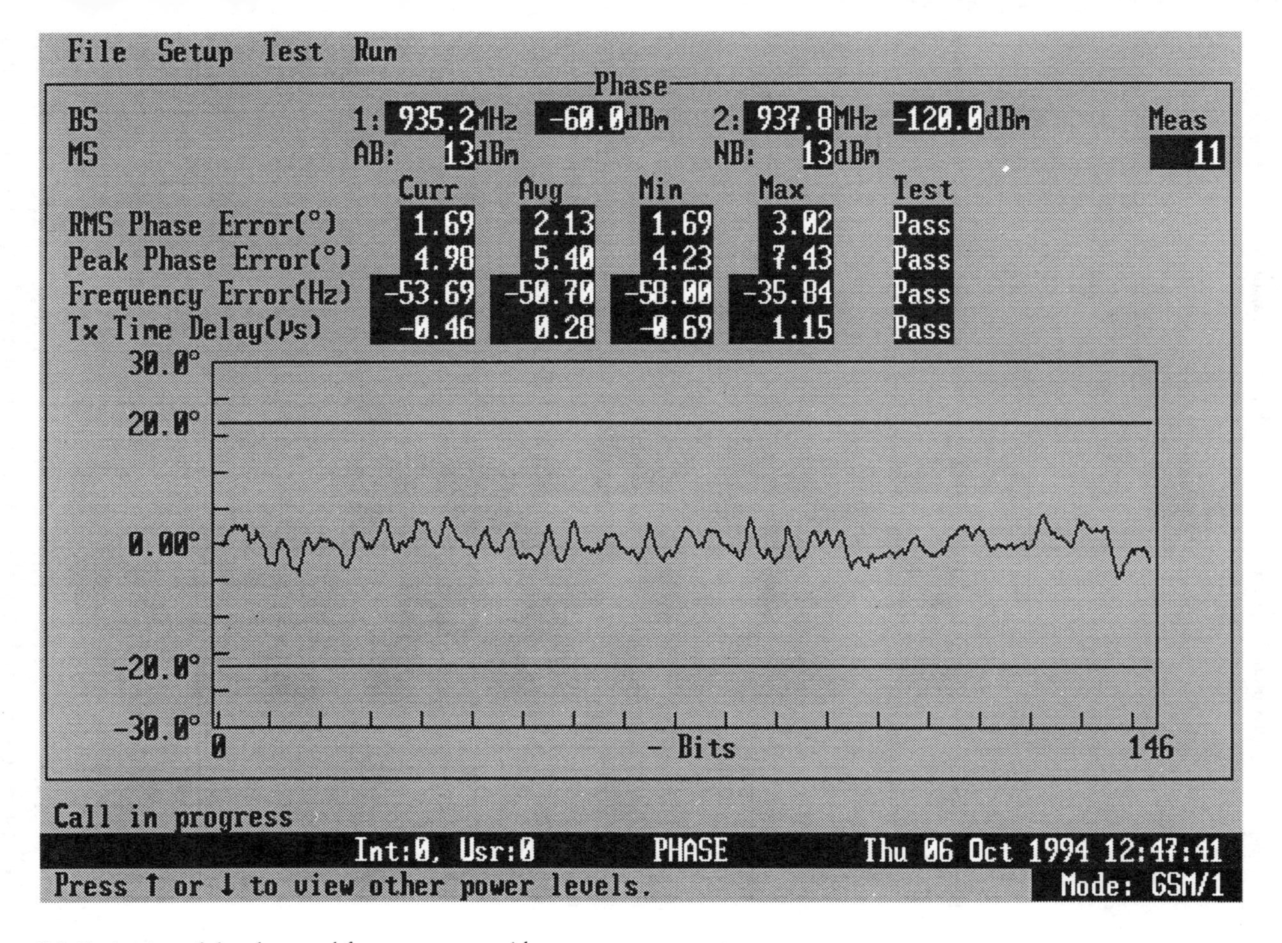

Figure 9.3 Evaluation of the phase and frequency error with a test system.

Table 9.1
Maximum Frequency Errors for Multipath Propagation

Propagation Conditions	*Maximal Frequency Error (Hz)*
RA250	±300
HT100	±175
TU50	±135
TU3	±95

multipath propagation. We are, after all, testing the mobile station, not the test equipment.

Most fading simulators have predefined simulation characteristics for the different environments used in the mobile or base station tests:

- Rural area (RA);
- Hilly terrain (HT);
- Typical urban (TU);
- Equalizer (EQ) (used to test the equalizer).

In addition to the selected multipath environment, the speed parameter is simulated. This adds *doppler shift* to the signals applied to the mobile under test. The term RA250 in Table 9.1 stands for the use of a fading profile found in a rural area which is passed through at a speed of 250 km/h (155 mi/h). This is the maximum speed the GSM system and its equalizer have been designed to accommodate. This rather high speed would apply to high-speed trains like the French TGV or the German ICE, and to the German highways (Autobahn).

The fading profiles are applied to the transmitting path within the test equipment, on which special connectors are provided on the front panel for the insertion of a fading simulator into the path. If an interferer should also be faded, then two times six poles are provided by the fading simulator. If, instead, only the on-channel signal is faded, it is possible to use all twelve poles for this signal. Figure 9.4 shows a test setup using a fading simulator with a test set.

9.3.2 Power Measurement

The power measurement evaluates the compliance with the standards of (1) the different *power levels*, (2) the *power-time template*, and (3) the *burst timing* of a mobile station.

To make the testing of the different power levels easier for the user, the test set makes use of the normal downlink power control signaling, which commands the mobile station to use different power levels. These commands are sent to the

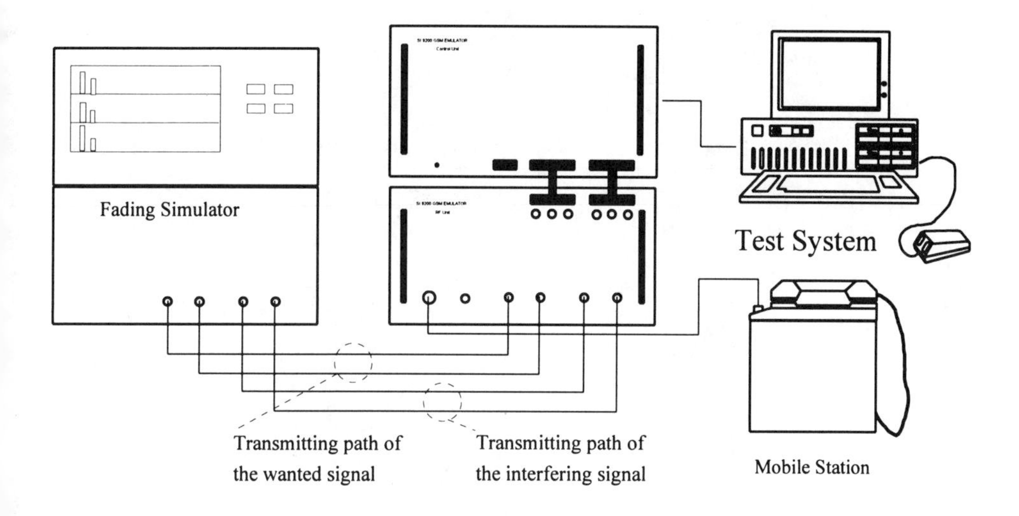

Figure 9.4 Test setup for mobile stations with multipath propagation.

mobile station on the SACCH, which, in addition to the assigned power level, also contains a *timing advance value*. If a testing routine requires either the power level or the timing advance to be altered, the user simply enters the new value into the test set, which then encodes these altered values into the Layer 1 header of the SACCH (see Chapter 5). The mobile reads the values from the SACCH and changes its output power level accordingly. Using the power control feature makes it easy to perform automatic test sequences for the complete set of power levels used by a particular mobile station. Table 9.2 shows the different limits for the power levels and classes. Some additional conditions have to be met to pass a power test successfully.

- The tolerance for the highest power level within one power class is more stringent when compared to the other levels; it must fall within ±2 dB, whereas all the other levels require a more relaxed ±3-dB window. Different power classes can be identified from Table 9.2. The peak power values that might apply to the narrower limit are indicated with an asterisk (*). This means that a hand-portable with power class 5 has to meet the power level of 29 dBm with a tolerance of ±2 dB, while a car-mounted mobile station with power class 1 has to meet the power level of 29 dBm with a tolerance of ±3 dB.
- Two consecutive power level values must have a difference of at least 0.5 dB but no more than 3.5 dB from each other. In addition, the values must be in an ascending sequence.
- The power-time template has to meet the constraints in Figure 5.6.
- All power levels starting from the power class up to level 15 have to be implemented. For example, a mobile station with power class 3 must support the power levels from 3 up to 15.

Table 9.2
Limits for the Power Measurement

Power Class	*Power Level*	*Peak Power (dB)*
1	0	43 dBm ± 2
1	1	41 dBm ± 3
1 2	2	39 dBm ± 3*
1 2 3	3	37 dBm ± 3*
1 2 3	4	35 dBm ± 3
1 2 3 4	5	33 dBm ± 3
1 2 3 4	6	31 dBm ± 3
1 2 3 4 5	7	29 dBm ± 3*
1 2 3 4 5	8	27 dBm ± 3
1 2 3 4 5	9	25 dBm ± 3
1 2 3 4 5	10	23 dBm ± 3
1 2 3 4 5	11	21 dBm ± 3
1 2 3 4 5	12	19 dBm ± 3
1 2 3 4 5	13	17 dBm ± 3
1 2 3 4 5	14	15 dBm ± 3
1 2 3 4 5	15	13 dBm ± 3

*If this power level is also the power class of the mobile station, then the tolerance for this level is ± 2 dB.

- The burst must be located within its assigned time slot with a tolerance of ±1 bit (3.69 μs), which is referred to as the burst timing.

Figure 9.5 shows the display of a mobile's power measurement on a GSM radio test set. For the peak power, different statistical values are displayed (current, average, minimum, and maximum). The same applies for the burst timing and the transmission time delay. The power-time template result is just a simple pass/fail criterion as the test system verifies the mobile's compliance with the technical specifications. Again, a graphic display helps the user to identify errors occurring during the power measurement. A power test is passed if, for all measurements, a PASS indication is given. If a single measured burst from the mobile fails one of the above criteria, the test system will give a FAIL indication.

9.3.3 Adjacent Channel Power Measurement

To perform the *adjacent channel power measurement*, which is actually a simple examination of the mobile's performance outside its assigned channels, the phone is operated in the frequency-hopping mode over three different frequencies. This should cause the frequency spectrum to show the worst cases of spurious emissions appearing in adjacent channels, or areas of the spectrum where energy from the

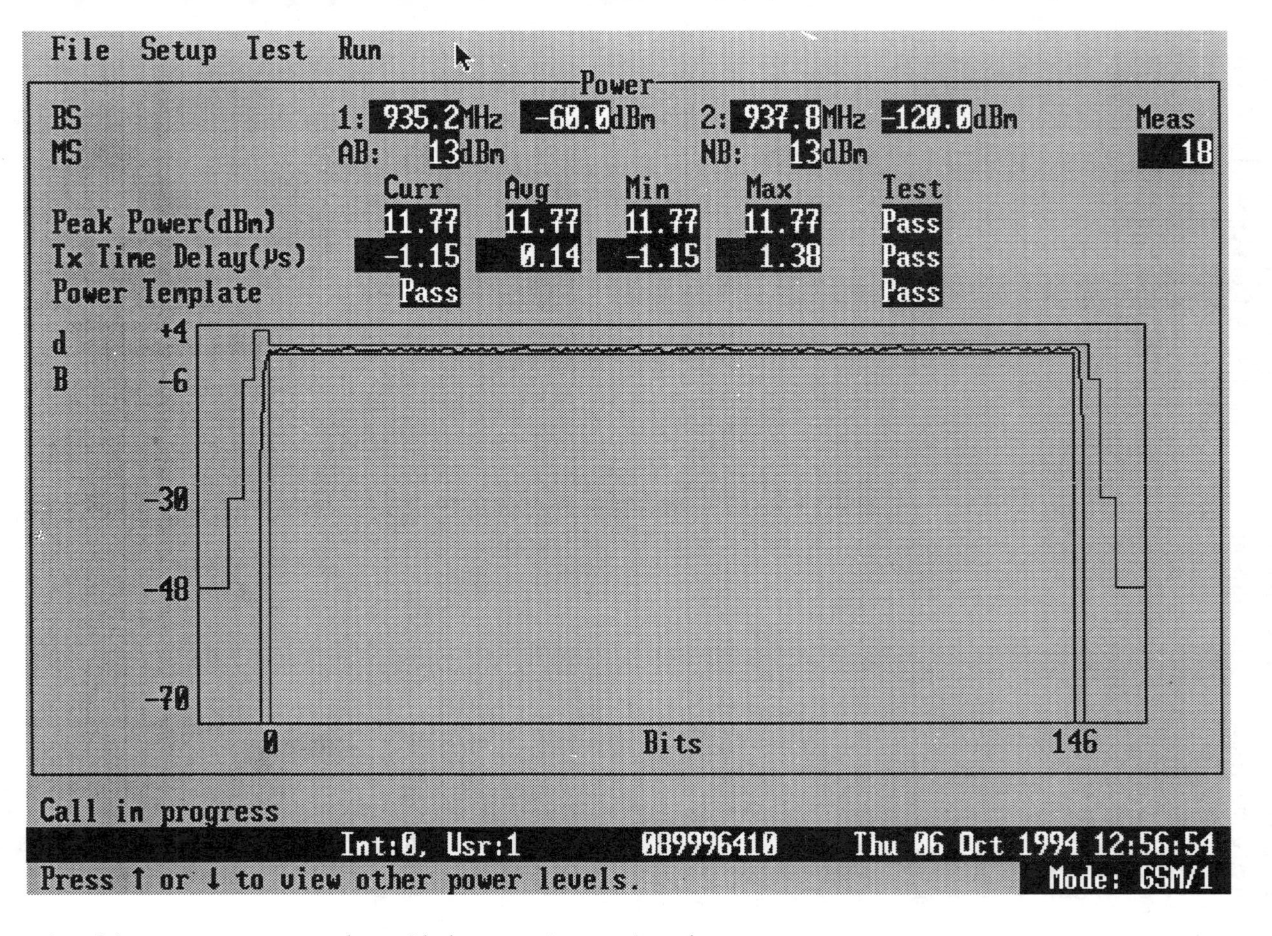

Figure 9.5 Display of the power measurement along with the power-time template of a test system.

mobile should be considerably limited. The three frequencies used for hopping should be selected from each of the following frequency ranges: from channel numbers 1 to 5, from 60 to 65, and from 120 to 124. While the mobile station is hopping over the frequencies in a simple sequence, the three frequencies are evaluated one after the other as we determine the proportionate power falling into the adjacent channels. Two different spectrum measurements are distinguished. One is the spectral content due to the modulation, and the other is the spectral characteristic due to the switching transients. The modulation spectrum measurements evaluate only the influence of that part of the burst containing the actual transmitted data, which should be random data during the test (see also Section 8.3.4.1). The spectrum is examined at a certain operating frequency (e.g., in the range of channels 1 to 5). The measured level within a channel is referred to as the *reference power level.* The other measurements at specified offsets from the channel frequency must have a diminished level expressed in decibels below the carrier's power (dBc). Table 9.3 shows the requirements for this measurement. Due to the fact that this part of the test analyzes the impact of the modulation on the spectrum, there are different measurements up to an offset frequency of only 400 kHz.

The spectrum analyzer is set to a resolution bandwidth of 30 kHz, a video bandwidth of 30 kHz, and a frequency span of 0 Hz. To make sure that only the contribution of the data bits is evaluated, a burst trigger has to be provided for the spectrum analyzer. This trigger might come from the radio test set or it may be generated internally in the spectrum analyzer.

The *switching-transient spectrum* result evaluates the impact of the power switching of a mobile (bursts), which we referred to as AM splash in Chapter 5. In a manner similar to the modulation spectrum, the proportionate power within adjacent channels is measured. Since the frequency components coming from this switching activity confine their influence to higher frequency offsets from the mobile's transmitter than the modulation effects do, the measurements start where the modulation measurements ended, at a 400-kHz offset, and is measured up to 1,800 kHz

Table 9.3
Relative Levels for the Measurement of the Modulation Spectrum

Power Level (dBm)	*Maximum relative distance of the measured level from the peak Power [dBc] Depending on the Distance From the Channel Frequency [kHz] (in the Range from 600 to 1800 in Steps of 200 kHz)*					
	0	100	200	250	400	600 – 1,800
43	0	±0.5	–30	–33	–60	–70
39	0	+0.5	–30	–33	–60	–66
37	0	+0.5	–30	–33	–60	–64
≤33	0	+0.5	–30	–33	–60	–60

on either side of the center frequency. Table 9.4 shows the requirements for this measurement.

The spectrum analyzer is set to a resolution bandwidth of 30 kHz, a video bandwidth of 100 kHz, a frequency span of 0 Hz, and the display is set to peak hold mode.

If we compare the allowed impact from both phenomena, it can be seen that the modulation will have a higher influence on the frequencies closer to the channel frequencies, while the switching transients contribute a higher level at larger frequency offsets.

9.4 RECEIVER TESTS

As we already mentioned in the introduction to this part of the book, Chapter 8, the receiver measurements for radios in the GSM system make use of BER techniques. A known pseudorandom bit sequence is transmitted in a precise way to the mobile station which loops the received bits, after channel decoding, back to the test equipment through its own transmitter. The BER is evaluated within the tester. The general principle is shown in Figure 9.6.

To generate a pseudorandom bit sequence, or bit string, the test equipment uses ciphering. Since part of the ciphering sequence is made up from the current frame number, the bit sequence modulated onto the RF injected into the receiver under test will change with each burst, even if the transmitted data are always the same. Therefore, a simple 010101 . . . bit sequence is specified to yield a pseudoran-

Table 9.4
Relative Levels for the Measurement of the Switching-Transients Spectrum

Power Level (dBm)	*Maximum Level (dBm) in the Defined Distance From the Wanted Frequency [kHz]*			
	400	600	1,200	1,800
43	−9	−21	−21	−24
41	−11	−21	−21	−24
39	−13	−21	−21	−24
37	−15	−21	−21	−24
35	−17	−21	−21	−24
33	−19	−21	−21	−24
31	−21	−23	−23	−26
29	−23	−25	−25	−28
27	−23	−26	−27	−30
25	−23	−26	−29	−32
23	−23	−26	−31	−34
≤21	−23	−26	−32	−36

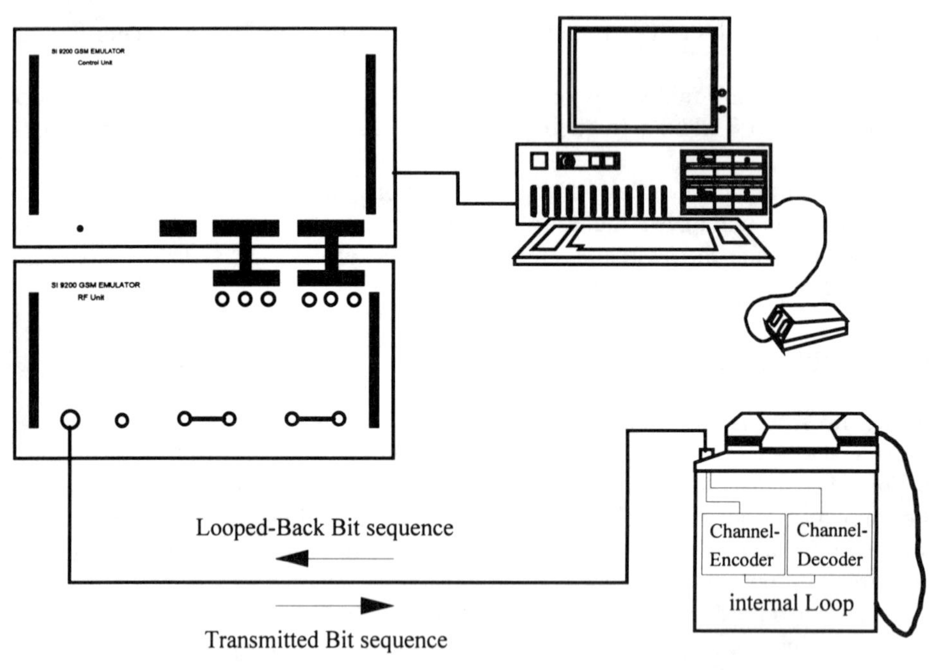

Figure 9.6 Test setup for the receiver measurement.

dom test sequence. The orderly 010101 . . . sequence is ciphered and the resulting pseudorandom bit stream (PRBS) modulates the tester's calibrated RF output to the radio under test. It is apparent that this solution makes it easy for the test equipment to keep track of and evaluate the BER, even without a synchronization frame.

An *internal loop* between the mobile's channel decoder and channel encoder has to be closed to get the mobile to loop the data back to the test equipment. A special command is sent to the mobile phone to close this loop through the signaling resources. Upon reception of this command, the SIM card is checked by the mobile to see if it is a test card, and, if so, the test loop within the mobile is allowed to close. After it closes the loop, the mobile sends an acknowledgment back to the test set. If the card is not a test SIM, then the loop is simply not closed, and no acknowledgment is returned to the tester. Since this condition is not necessarily indicated to the operator of the test set, it might cause trouble for the user to find out why a BER test is not executed properly. The test equipment usually transmits this loop-back command automatically when a receiver measurement is started.

9.4.1 Sensitivity

The BER technique is used to judge the sensitivity of a mobile phone. Different BERs are performed at a fixed power level, which is 11 dBμV (–102 dBm) for hand-portable phones and 9 dBμV (–104 dBm) for all other phones. For these levels, the mobile phones have to keep to the limits specified in Table 9.5. For the traffic channel, the sensitivity changes with different multipath propagation profiles. The table also contains values for the *adjacent time slot rejection test*, for which the power level from the test system in the two neighboring time slots is transmitted at 28 dBμV (–85 dB).

The decision to declare a speech frame a bad frame depends very much on the specific implementation of the equalizer and the channel codec by the various manufacturers of GSM mobile phones. To account for this variability, the correctional value α was introduced. It is related to all types of BER measurements that depend on the bad-frame decision, namely FER and RBER Ib. If the threshold to indicate a frame as bad is very low, then the residual BER for the good class Ib bits should be significantly lower. This is the basic relationship underlying the α-parameter. The values for this parameter may vary between 1 and 1.6. If one value has been determined, it must stay the same within an individual propagation profile.

For the control and data channels, the sensitivity is measured only for the typical urban propagation condition of 50 km/h (TU50). The FACCH is selected as a control channel representing the other channels because the coding scheme is the same for them. Table 9.6 shows the limits of the receiver's sensitivity as measured on these channels. The only possible measurement is the FER, since the bits are not differently coded or distinguished one from another. The significance of the adjacent time slot rejection measurement will be explained in further detail in Section 9.4.4.

Figure 9.7 shows the display of a BER test on a GSM test system. The test limits here are displayed in percentage values even though the recommendations give absolute values; that is, the number of bits that have to be evaluated is specified, and then for each single channel and BER type, the number of bits or frames that

Table 9.5
Test Limits for the Sensitivity Measurement on the Traffic Channel

BER Type	*Multipath Propagation Condition*			*Adjacent Timeslot Rejection (%) (Statistic/FH)*
	TU50 (%)	*RA250 (%)*	*HT100 (%)*	
FER	$3.14 \cdot \alpha$			$0.122 \cdot \alpha$
RBER Ib	$2.5/\alpha$			$0.410/\alpha$
RBER II	7.67	7.5	8.67	2.439

Table 9.6
Test Limits for the Sensitivity Measurement on the Control and Data Channels

Channel Type	*FER Rate (%)*
FACCH	14.63
TCH/FS9.6	0.778
TCH/FS4.8	0.0112
TCH/HS4.8	0.778
TCH/FS2.4	0.00126
TCH/HS4.8	0.0112

may be wrong is given. To make the receiver's operation clearer for the test technician, a percentage is calculated from the absolute values. Care has to be exercised that enough samples are taken to give the results some significance. This means that for each digit beyond the decimal point, the number of samples has to be increased by a factor of 10. For example, if we want to make a measurement with a resolution of 0.1%, then the number of samples used in the test should be at least 1,000 because each bit contributes 0.1% to the result.

9.4.2 Usable Receiver Input

Besides very weak signals, on the order of those we use to make sensitivity measurements, a mobile station must also be able to cope with very strong signals without any degradation in the quality due to distortion within the receiver or the demodulator. This condition might be the case when the mobile is very close to a base station. The measurement evaluates, then, only the residual bits of the class II bits (RBER II). At the beginning of the test, the power level at the test system is set to –85 dBm (28 dBμV) and the RBER II is evaluated. The same test is performed for a higher power level of –40 dBm (73 dBμV) and then an even higher –16-dBm (97 dBμV) RF level. The selected propagation conditions are statistical with no use of a fading simulator, and the other selected condition is EQ50, a special profile to test the equalizer at these high power levels into the receiver. Table 9.7 indicates the test limits for this test.

9.4.3 Cochannel Rejection

The *cochannel rejection* test verifies the ability of the receiver to demodulate the test signals properly even if a GMSK modulated signal is present at a power level of only 8 dB below the on-channel signal. A call is established at a power level of

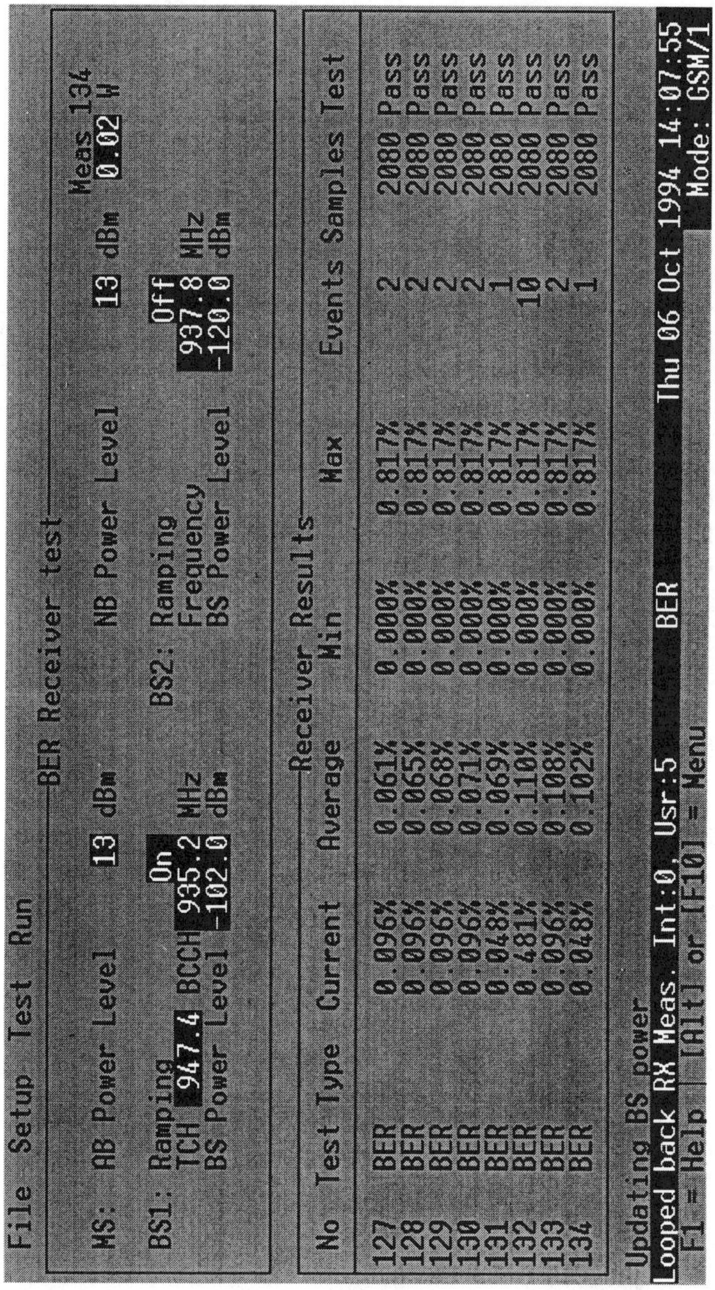

Figure 9.7 Measurements on the mobile receiver by a test system.

Table 9.7
Test Limits for the Usable Receiver Input

Propagation Condition	*RBER II Rate (%)*
Statistic	0.0122
EQ50	3.25

–85 dBm. At the same time, an interfering signal is injected at a power level of –93 dBm. This test is performed for the FACCH and the TCH/FS under the typical urban propagation profile at 3 km/h (TU3). The test for the traffic channel is performed first without SFH, and then it is performed again with SFH over 10 different frequencies within a range of 5 MHz. The requirements are shown in Table 9.8. From this table we see the helpful influence of frequency hopping. It was stated earlier that frequency hopping improves the quality we see in the radio channel by about 2 dB, which is, of course, also reflected in the requirements that apply to this test.

The α-parameter applies the same way as stated in Section 9.4.1, and it is, again, somewhere in the range from 1 to 1.6, but must be the same value within onset of propagation conditions.

9.4.4 Adjacent Channel Rejection

The mobile station's behavior in the presence of a stronger signal in an *adjacent channel* is examined with the adjacent channel rejection test. In a TDMA system, a neighboring channel is either located on an adjacent frequency channel or on an adjacent time slot.

The adjacent time slot rejection is measured with the active time slot transmitted from the tester at –102 dBm (11 dBμV) for handportable phones and –104 dBm

Table 9.8
Test Limits for Cochannel Rejection

Channel-/BER-Type	*Propagation Condition*	*Error Rate (%)*
FACCH/FER	TU3/no SFH	24
TCH/FS/FER	TU3/no SFH	$24 \cdot \alpha$
TCH/FS/RBER Ib	TU3/no SFH	$2.091/\alpha$
TCH/FS/RBER II	TU3/no SFH	4.3
TCH/FS/FER	TU3/SFH	$3.37 \cdot \alpha$
TCH/FS/RBER Ib	TU3/SFH	$0.215/\alpha$
TCH/FS/RBER II	TU3/SFH	8.33

(9 dBμV) for all other phones and the neighboring time slots at a level 19 dB higher at –83 dBm (30 dBμV)/–85 dBm (28 dBμV). Since this measurement is intended to be performed along with the sensitivity measurement, the test limits may be found in Section 9.4.1. This measurement does not use any propagation simulation, but the SFH feature is applied.

For the *adjacent frequency channel rejection* case, a stronger GMSK-modulated interfering signal is transmitted on a neighboring channel. Depending on the frequency offset from the active channel, different power levels are applied. If the offset is only one channel, then the interferer is transmitted at a level 9 dB higher. With an offset of two channels, the interferer is transmitted at a level 41 dB higher. The active channel uses a power level of –85 dBm (28 dBμV). With a typical urban propagation profile at a simulated speed of 50 km/h (TU50), the limits specified in Table 9.9 should not be exceeded for a TCH/FS. Again, the α-parameter is within the range 1 to 1.6 and should maintain a fixed value for the different BER types.

9.5 ADDITIONAL RF TESTS ON A MOBILE STATION

In addition to the tests described up to now, there are other tests defined in the specifications. Though these are also important, the effort to perform them is much greater and it is not likely that they will be performed in a service or production environment. They are more reflective of design parameters than conditions that may arise in manufacturing or after-market situations. To give an idea of these tests, they are listed below.

- The *bad-frame indication performance* verifies the mobile station's ability to detect bad frames. For this purpose, the test system has to send the mobile station bursts with a pseudorandom bit sequence that cannot be decoded as a traffic channel; that is, the mobile station must identify and indicate each speech frame as a bad frame. The test set evaluates the frames not indicated as bad frames.
- The *intermodulation rejection* measures the impact of interfering signals on the nonlinear parts in the mobile station's receiver (e.g., amplifiers). Besides the

Table 9.9
Error Rates for Adjacent Channel Rejection

BER Type	*Error Rate (%)*
FER	$6.74 \cdot \alpha$
RBER Ib	$4.2 \cdot \alpha$
RBER II	8.33

active channel, two additional signals are transmitted. The frequency relationship of these two signals is of such a nature that they produce intermodulation products exactly on the receiving frequency of the active channel, and their presence should not decrease the quality of the link.
- The *transceiver tests* mark the *spurious emissions measurement* for a mobile phone. They confirm that the transmitter is only operating in the appropriate frequency band and not on other frequencies that might disturb other mobile stations or radio services. The transceiver tests are split in two different states: (1) the active state, during which a call is performed (i.e., the mobile station is connected to a base station and is transmitting signals), and (2) the idle state, in which the mobile station is only listening to the base channel of the simulated base station and is not transmitting at all. The evaluation is performed in combination with a spectrum analyzer.

9.6 GSM MOBILE STATION SIGNALING TESTS

Considering the signaling complexity and capabilities built into a GSM network, it is obvious that a certain degree of attention has to be paid to the verification of the compliance with the specified protocols. Signaling tests for GSM mobile stations have to verify this compliance, to make sure that the device can cope with the signaling requirements and does not disturb or annoy the network, its services, and its users. The equipment necessary to perform signaling tests must be capable of simulating the protocols used by the network side in a way that allows the test engineer to check different scenarios, procedures, and messages in different constellations and combinations. The test equipment must enable the user to define or design such scenarios, execute them, and trace all the events and data (e.g., by online "recording"). An additional display with interpretations and checks (for errors or certain events) is helpful.

For the greatest number of specified protocol tests, it is enough to simulate two base stations and the signaling entities behind them. There are, however, a smaller number of conformance tests that require up to eight base stations to be simulated (most of them without any signaling dynamics behind them).

For use in design laboratories, another type of test equipment or another function incorporated in the test equipment has proven to be useful, especially for Layer 2 and Layer 3 protocol tests. The useful feature is *baseband* simulations. Most of the time there are no finished products to be tested in a design laboratory. It is very common that digital and analog radio components are developed in parallel so that they have to be tested separately during the design and verification processes. Therefore, using test equipment that is able to perform the signaling protocol verification without going over a finished RF part is appropriate. This indicates that test equipment that can test the radio parameters without having to deal with signaling functions is also necessary.

Signaling tests without the radio part can be realized by additional *digital data interfaces*, which, for example, make use of transistor-transistor logic (TTL) standardized signals in the data baseband just before the burst building and/or the modulation functions. The signaling data to and from the DUT can be transmitted with very little additional hardware effort and without having to deal with the complex radio parameters.

For type approval efforts, there are very tight and often tedious descriptions of how the tests have to be performed. The reader should understand that this tedium is not only found in signaling tests. The descriptions include the initial state, scope, and purpose of the test; the parameters to be considered; the method or sequence in which the test has to be performed; and, finally, the requirements that have to be met by the DUT in order to pass the test.

9.6.1 Test of Layer 1–Related Signaling Functions in a GSM Mobile Station

The verification of the *radio link management* performance of a mobile station, which is its capability to support the maintenance of Layer 1 links, is tested for the following functions:

1. The mobile's ability to comply with *downlink power control.* This is an important function, which is used to reduce the interference and battery drain imposed by high output power levels.
2. The mobile's correction of burst timing according to *timing advance* commands from a base station. This function will minimize incidences of adjacent time slot interference.
3. The mobile station's report of actual timing advance settings and power control levels (uplink SACCH, Layer 1 header).
4. The correct reporting of power levels and signal quality in the actual cell and in adjacent cells (*measurement report*).

The following test procedures are applied in order to check the mobile station's conformance to the specifications.

1. Power control and timing advance are sent in the Layer 1 header of the downlink SACCH. During a call in progress, we should check (a) whether the mobile station uses the correct values for timing and level accuracy, and (b) whether it does, in fact, adjust itself to the commanded timing and level fast enough. The level accuracy for all power control levels used by the mobile station is part of the approval tests for the transmitter (output power tests, Layer 1, see Section 9.3.2). In order to perform these tests, the test equipment must be able to change the SACCH's Layer 1 header information on line during a dedicated channel link, thus allowing the test engineer to modify timing advance and power control levels at will. In addition, the test equipment

must perform *peak power* and *burst timing* measurements with sufficient accuracy to confirm compliance.

2. The mobile station's *measurement report* and Layer 1 header (containing the actual timing advance and power control level used by the mobile) has to be evaluated (decoded) and displayed on line by the test equipment. An adjustment of output power levels of the active cell and additional simulated adjacent cells (base channels) must result in corresponding changes of the RX-LEV and RX-QUAL variables. A prerequisite for taking measurements on, and the subsequent reporting on, adjacent cells is that the base channel of the serving cell transmits its neighboring channel numbers on the BCCH. The compliance with RX-QUAL reporting is tested with combinations of two signal sources (the desired signal and an interferer) at different RF power levels. Traffic channel BERs between 0.1% and 20% are acquired with combinations of channel and interferer relationships. These relationships are then used for eight tests with given BER limits.

Table 9.10 shows values and levels for RX-LEV and Table 9.11 the values and BERs for RX-QUAL with the corresponding measurement accuracy expected from the mobile station. Figure 9.8 shows the measurement report test screen of a GSM test system. The mobile station reports two cells, the active cell and one additional neighboring cell. Neighboring cells are identified by their BSIC. The BSIC is transmitted on the SCH and includes the cell identity and the number of the training sequence to be used for the detection of the BCCH.

The display reveals the distinction between reporting on the base channel of the serving cell ("FULL") and the dedicated active channel ("SUB").

9.6.2 Test of the Mobile Station's Layer 2 Protocol

The *data link protocol* used by Layer 2 makes sure that Layer 3 signaling messages are transferred in a defined format and without loss of information (acknowledged

Table 9.10
Values and Levels for RX-LEV Measurements by the Mobile Station

RX Level	*Level at MS Receiver (dBm)*
0	Less than −110
1	−110 to −109
2	−109 to −108
...	
...	
62	−49 to −48
63	above −48

Table 9.11
Values and BERs for RX-QUAL Measurements by the Mobile Station

RX Quality (Coding)	*Corresponding Bit Error Rate (%)*	*Range of Actual Bit Error Rate (%)*	*Expected MS-Reporting-Accuracy/ Probability (%)*
0	Below 0.2	Below 0.1	90
1	0.2 to 0.4	0.26 to 0.30	75
2	0.4 to 0.8	0.51 to 0.64	85
3	0.8 to 1.6	1.0 to 1.3	90
4	1.6 to 3.2	1.9 to 2.7	90
5	3.2 to 6.4	3.8 to 5.4	95
6	6.4 to 12.8	7.6 to 11.0	95
7	Above 12.8	Above 15	95

Measurement Report

BA used Off DTX used Off Measurement valid On
RX_LEV_FULL 58 RX_QUAL_FULL 0
RX_LEV_SUB 59 RX_QUAL_SUB 0
No. of identified Neighbour Cells 1
1.Neigh.Cell RX_LEV 35 BCCH 7 BSIC 14
2.Neigh.Cell RX_LEV 0 BCCH 0 BSIC 0
3.Neigh.Cell RX_LEV 0 BCCH 0 BSIC 0
4.Neigh.Cell RX_LEV 0 BCCH 0 BSIC 0
5.Neigh.Cell RX_LEV 0 BCCH 0 BSIC 0
6.Neigh.Cell RX_LEV 0 BCCH 0 BSIC 0
Call in progress
Int:0, Usr:0 MEAS REPORT Tue 12 Jan 1993 13:09:18

Figure 9.8 Test equipment display: measurement report from a GSM mobile station.

information transfer). In order to test the conformance to the technical specifications (type approval and design test), the test equipment employed for data link tests must cover a variety of functions. It must be able to:

- Transmit and receive messages in all defined frame formats;
- Establish, maintain, and terminate acknowledged data links;
- Simulate erroneous behavior by manipulating fields and frames and by sending messages out of sequence;

- Trace, interpret, and display the content of frames exchanged with the DUT.

One example of an easy-to-read Layer 2 trace has been treated in Chapter 7. The tests are, in general, performed with real Layer 3 messages contained in the Information fields, while observing the Layer 2 parameters. For Layer 1, the physical parameters are selected in a way that creates ideal radio conditions in terms of power levels and propagation. The performance of the radio logically has to be checked separately and must not influence signaling tests. As an example, the correct establishment of an acknowledged data link is discussed.

After paging a mobile station (in an idle updated state), the mobile station would request a channel (channel request) and be assigned a signaling channel (immediate assignment). The next uplink message from the mobile station would include an SABM command, coded in the Layer 2 parameters (control field). The network (here: test equipment) will answer with a UA response frame containing the same information field as the SABM command. After the transmission of the UA messages, and correct reception and processing within the mobile station, there will be no more SABM commands sent by the mobile station. It will instead only transmit fill frames (UI frames).

Additional tests verify the correct reaction on holding back the UA frame (SABM repetition) or UA frame with an information field different from the one in the SABM command.

Step by step, all possible protocol combinations can be verified with and without the simulation of errors.

9.6.3 Test of the Mobile Station's Layer 3 Protocol

When approving a GSM mobile station, the Layer 3 performance of a mobile station is tested after the verification of Layers 1 and 2. Since the network layer's tasks are manifold and the protocol implementation is a complex but structured one, the test efforts are complex and have to be structured as well. As is true for all kinds of test procedures, with Layer 3 it is not enough to check only the cases of normal use without problems. The cases in which errors occur have to be covered as well to make sure that the DUT takes appropriate action or enters clearly defined states.

The procedures in the following three sections are checked.

9.6.3.1 Radio Resources Management–Related Functions

- *Channel request procedure* with different test conditions (e.g., whether the mobile station does not transmit more requests than allowed by the network);
- *IMSI attach and detach procedures*;
- Mobile station in idle mode;

- Lower layer failure;
- Error simulation: reaction to wrong protocol discriminator, wrong transaction identifier, and wrong message type; wrong, missing, or repeated information elements in one Layer 3 message;
- *Channel assignment* (does the mobile correctly seize the assigned logical and physical channel, such as traffic channel or SDCCH?);
- Test of paging;
- Measurement report with simulated neighboring cells;
- Handover, synchronous and asynchronous (new cell is synchronous/asynchronous with old cell);
- Start of cipher mode.

9.6.3.2 Mobility Management–Related Functions

- Registration (location update procedures);
- Authentication;
- Identification: reveal IMSI and IMEI;
- Failure of registration: check that the mobile station will not repeat registration attempts after access has been denied;
- IMSI attach/detach, use of TMSI.

9.6.3.3 Connection Management–Related Functions

The call control tests are performed by checking defined call control states and transitions from one state to another. A distinction is made between the cases of an incoming call, an outgoing call, and incall functions. In addition, the following cases are tested.

- Emergency call setup: a GSM phone has to be able to perform an emergency call, even when a valid SIM card is *not* inserted.
- Call reestablishment of an interrupted connection;
- Test of supplementary services and SMS implementations;
- Test of *structured procedures*:
    - Mobile-originated call, *early assignment* of traffic channel (before a *connect* message), release is network-initiated.
    - Mobile-originated call, late assignment of traffic channel;
    - Mobile-terminated call, early assignment of traffic channel, release is network-initiated.
    - Mobile-terminated call (paging), *late assignment* of traffic channel.

9.6.4 How Can Mobile Station Layer 3 Protocol Tests Be Performed?

Here we provide an example of a practical implementation of a Layer 3 test. Although there are different approaches to GSM network tests available currently, they all share a common list of requirements.

- The test scenarios (test cases) have to be implemented as sequential tables, containing the correct message content to be transmitted to or expected from the DUT. For approval, such scenarios must be carefully verified and accessible on a high level (e.g., by naming files corresponding to sections in the test specifications).
- It must be easy to modify such signaling tables or create new ones (e.g., by using some powerful PC-based editor software). This feature is very important for the design environment. An appropriate compiler may translate such tables (written in a high-level, message-oriented language) into a preprocessed format, which is passed to the test equipment's Layer 2. Such a format may already contain an octet structure (hexadecimal representation of fields and messages).
- The test equipment must trace all the information that is exchanged with the DUT, give some interpretations (clear text), allow pass/fail verdicts, and finally store the trace information for further investigation or filling.

The example selected and shown in Figure 9.9 is a test case that verifies a *structured procedure*: mobile-originated call, with *early* assignment of a traffic channel (before the *connect* message). The messages contained in this table are to be sent and received subsequently. In the case of a received message that does not comply with the one specified in the table, the test system would issue an error message and include it in the trace. A successful execution of the test given in Figure 9.9 would result in the trace already shown in Section 7.5.

In order to read and understand the table content, the following explanations are offered. Messages expected from the mobile station are marked with *ul1* (uplink). Messages that the test equipment has to transmit to the mobile station are marked *dl1* (downlink). In addition, the logical channel type, the PD, and the Layer 2 mode (acknowledged or unacknowledged) are listed for each message. Finally, there is the message type and the content of the IEs. The letter *i* denotes a binary value, *x* a decimal, and *h* a hexadecimal value. SETT(*xxxx*) is used when a particular Layer 3 timer (e.g., 3260 for authentication) is to be set (started) by the test equipment. Upon receipt of the relevant message, the timer is reset (REST(*xxxx*)). Registers are used for the intermediate storage of information. For example, the request reference in the channel request message is stored in register number 0 (R0) and then packed into the immediate assignment message. Comments are enclosed within brackets designated by a solidus and asterisk (/* . . . */).

```
/* CHANNEL REQUEST                                   */
ul1,rach,TS(0i),pd(rr),
chan_req;


/* IMMEDIATE ASSIGNMENT                              */
dl1,agch,TS(0i),l2_unack,pd(rr),
imm_ass,
    pag_mod (=00i)          /* PM Normal paging          */
    chan_desc (=  00100i,   /*CHAN-TYPE  SDCCH/4 + SACCH/4 ,SCN=0*/
                  0x,       /* TN  Timeslot number = 0   */
                  5x,       /* TSC     same as BCC = 5  */
                  0i,       /* H  hopping off            */
                  00i,      /* FB-NO Band number=0  used if H=0  */
                  20x       /* ARFCN  = 20               */),
    req_ref  (R0),       /*read from register R0        */
    tim_adv (=000000i       /* TIM_ADV   0               */),
    mob_alloc (=0),
SETT(3101),                 /* Set Timer  T3101 on IMM_ASS     */;

/* CM SERVICE REQUEST                                */
ul1,sdcch(0i),TS(0i),pd(mm),
cm_serv_req,
    cm_serv (=0001i         /* Service-TYPE Mobile originated call */),
    ciph_key_seq (=000i/* KEY-SEQUENCE  Key not relevant        */ ),
    mob_class2 (R4          /* write into internal register R4      */ ),
    mob_id (R5),
REST(3101);                 /*  Reset Timer T3101                   */

/* AUTHENTICATION REQUEST                            */
dl1,sdcch(0i),TS(0i),l2_ack,pd(mm),
auth_req,
    ciph_key_seq (=000i/* KEY-SEQUENCE  Key not relevant        */),
    rand (=  0h             /* RAND-VAL    0              */)  , SETT(3260);
                     /* Set Timer T3260 for AUTH_RES  */

/* AUTHENTICATION RESPONSE                           */
ul1,sdcch(0i),TS(0i),pd(mm),
auth_res,
    sres(R6              /* write into internal register R6       */ ),
REST(3260);                 /*  Reset Timer T3260  on AUTH_RES  */

/* CIPHER MODE COMMAND                               */
dl1,sdcch(0i),TS(0i),l2_ack,pd(rr),
ciph_mode_cmd,
    ciph_mode_set( =0i      /* SC      No ciphering              */);

/* CIPHERING MODE COMPLETE                           */
```

Figure 9.9 Layer 3 signaling table.

```
ul1,sdcch(0i),TS(0i),pd(rr),
ciph_mode_com;

/* SETUP UPLINK                                        */
ul1,sdcch(0i),TS(0i),pd(cc),
setup,
    bear_cap (=1i,       /* EXT1      extended              */
        01i,             /* CHANRQ    full rate             */
        0i,              /* CDS       GSM standardized coding   */
        0i,              /* TRANS-MOD  circuit              */
        000i             /* TRANS-CAP  speech               */
    cd_p_bcd (R7);       /*  Dialled Number                 */

/* CALL PROCEEDING                                     */
dl1,sdcch(0i),TS(0i),l2_ack,pd(cc),
call_proc;

/* ASSIGNMENT COMMAND                                  */
dl1,sdcch(0i),TS(0i),l2_ack,pd(rr),
assign_cmd,
    chan_desc (=00001i,           /* CHAN-TYPE  Bm and ACCHs        */
        3x,              /* TN      Timeslot number = 3      */
        5x,              /* TSC     same as BCC = 5          */
        0i,              /* H       hopping off              */
        00i,             /* FBNO    Band number=0  used if H=0 */
        20x              /* ARFCN      = 20                  */
    pow_cmd (=01111i          /* POWER-LEV  13dBm              */
    chan_mod (= 00000001i)  /* CHANMOD      Speech full rate   */
SETT(3107);                   /*  Set Timer 3107                */

/* ASSIGNMENT COMPLETE                                 */
ul1,facch,TS(3x),pd(rr),
assign_com,
    rr_cause(= 00000000i          /* RR-VAL    Normal release          */
        ),
REST(3107);                       /* Reset Timer T3107                 */

/* ALERTING DOWNLINK                                   */
dl1,facch,TS(3x),l2_ack,pd(cc),
alert;

/* CONNECT DOWNLINK                                    */
dl1,facch,TS(3x),l2_ack,pd(cc),
conn,
SETT(0313);                   /* Normal stop: CONN_ACK received  */

/* CONNECT ACKNOWLEDGE UPLINK                          */
ul1,facch,TS(3x),pd(cc),
conn_ack,
REST(0313);                   /* Cause for start: CONN sent         */

/* The call connnection is established. After the connect-acknowledge-message the traffic
channel (TCH) is activated.*/
```

Figure 9.9 (continued).

CHAPTER 10

TESTING A GSM BASE STATION

It is easy to think that testing a base station may be simple; it does not seem to do much, since it merely maps data between the network and the Um interface and does not have anywhere near the tasks to perform that the mobile does. We are easily deceived. In sharp contrast to a mobile station, a base station not only supports the air interface (Um), but also the network interface (Abis) towards the BSC. The fact that there are two interfaces, two sides to the base station, has to be taken into account when testing GSM base stations. If a BTS should be completely examined, the test equipment should support both the Abis and the Um interfaces.

Throughout this book, our focus has been on the Um interface. A basic understanding of the Abis interface is necessary to perform most of the tests on base stations. In this chapter, a short introduction to the structure of a base station and the Abis interface is given. We should also mention that there are specific test solutions for interfaces quite different from the Abis, which are peculiar to certain BTS manufacturers. Because these solutions are specific to just certain BTS types and are often proprietary, they will not be investigated further. These odd BTS types run counter to the efforts of the ETSI and the network operators toward a common specification and the consequent simplification of test setups. The descriptions in this chapter focus only on the general approach for testing a GSM base station.

As we start, we soon discover that the Abis interface is not as fully specified in all its details as the Um interface is. If one compares the Um specification of some 400 pages with the Abis specification with only about 100 pages of text, the difference becomes apparent. The ETSI has specified the messages that are necessary to pass information independently of whether they are coming from the mobile station or going to it. The base station–related messages have been spared detailed descriptions. Base station messages are those that configure a BTS with the different frequencies or channel combinations, load software into the BTS, and other messages specific to the operation of an active base station. These are pieces of information that are important for setting up a base station and testing it, but it has been left to the manufacturers to implement these messages in the *operation and maintenance* (O&M) protocol, which then (sadly) is different for each manufacturer. Moreover, the behavior and tolerances of the base stations are different from each other. It is no problem to disconnect some BTSs from the BSC; it is a small problem for others, causing some error messages, and still others simply die when they are disconnected from their BSC; that is, they need the connection to the BSC (via the O&M messages) to perform even the simplest functions. The ETSI is aware of the need to clean up this awkward situation, at least to allow for common test procedures, and is working on a solution.

10.1 STRUCTURE OF A BASE STATION

The general structure of a base station consists of a *base station control function* (BCF) and between one and sixteen transceivers (TRX). The control function takes care of the software administration and the necessary control via the appropriate O&M functions within a base station. A single transceiver is the object of all this control effort, and provides one single frequency with eight time slots on the air interface. A block diagram of this general configuration can be seen in Figure 10.1.

Since most of the tests are performed on a single transceiver, the different functional blocks of such a transceiver should be explained in further detail. The different entities are identified in Figure 10.2 and are described below.

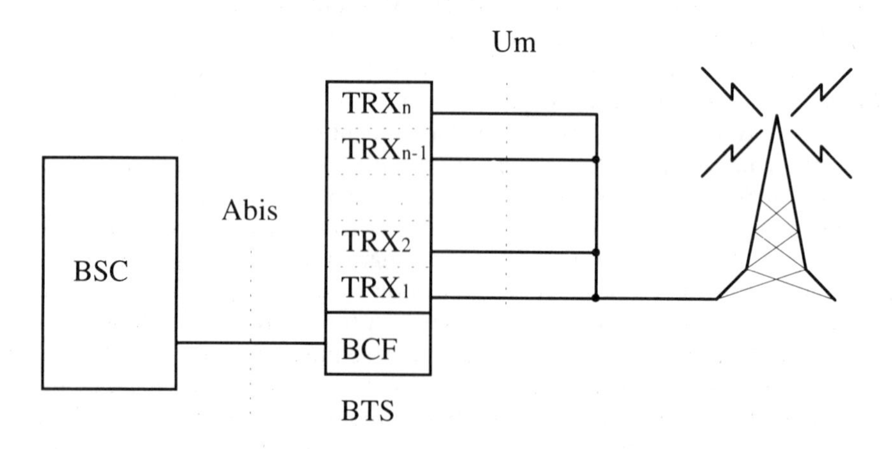

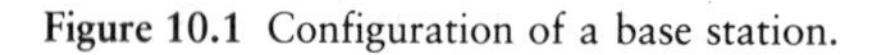
Figure 10.1 Configuration of a base station.

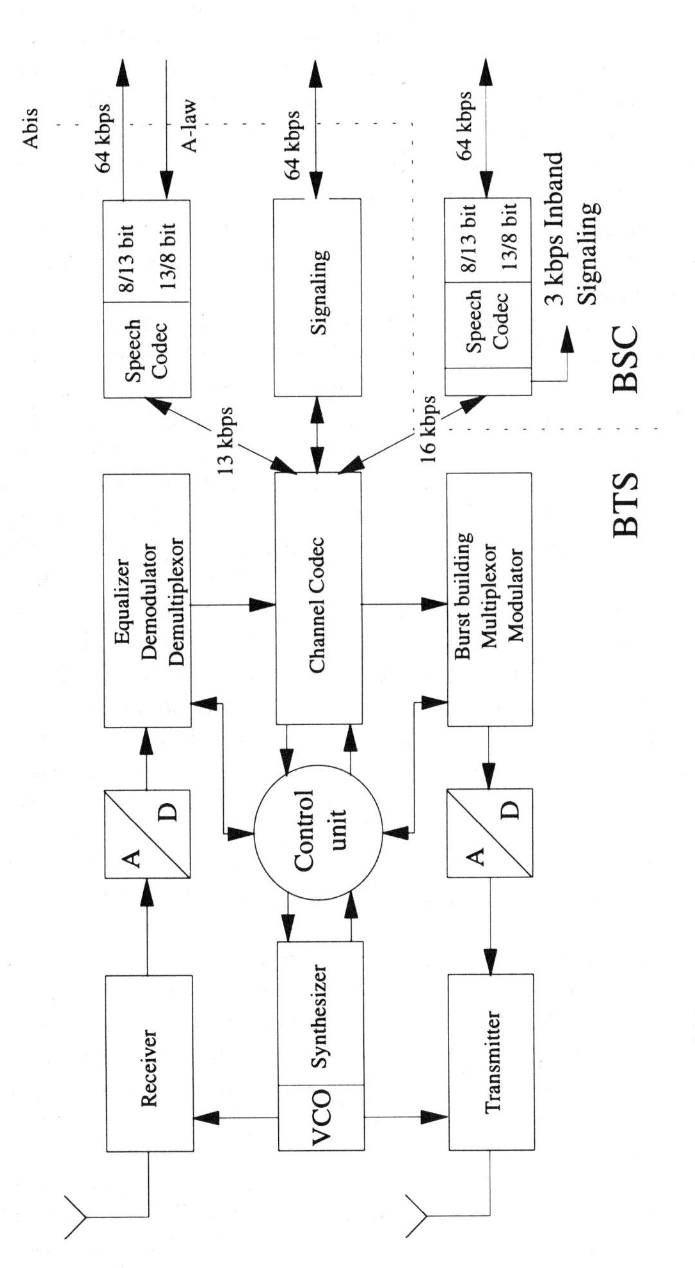

Figure 10.2 Block diagram of a base station.

- The receiver contains the receiving filter which blocks frequencies other than the desired receiving band from the base station. Behind the filtering section, the signals are down-converted to an IF or directly to the baseband frequency, where the signals are sampled and quantized with an ADC.
- *Multipath propagation* applies to base stations as well as to mobile stations. The *equalizer* compensates for the influences coming from the mobiles in the outside world. The *demodulator* extracts the bit stream from the equalized signal and passes it to the demultiplexer. This entity assigns parts of the received data flow to the different time slots and logical channels; the allocation is sorted from among different mobile stations.
- The *channel codec* decodes the bit streams from the different logical channels. It decides whether the recovered data are a signaling channel spread over four bursts, which it passes to the signaling entity, or whether the data are a traffic channel spread over eight bursts, which it passes to the *speech codec*. The channel codec is also able to detect errors that have been introduced into the RF path and correct them up to a point. If the errors are unrecoverable, then the codec simply discards the damaged frame. For the speech codec, two possibilities need to be considered: (1) If the speech codec is located within the base station, the speech data are directly fed into it at a rate of 13 kbps. If it is located in the BSC, then inband signaling is added before the data can be transmitted to the BSC at a rate of 16 kbps on the Abis interface (see Section 10.2.3.).
- The *speech codec* (which is not necessarily incorporated in the BTS) transforms the 13-kbps speech data rate into the original data stream of 104 kbps before it is reduced again to 64 kbps in the uplink direction toward the network. In the downlink direction toward the mobile, the 64-kbps data rate is transformed into 13 kbps, which is passed to the channel codec, where it is error-protected before it is transmitted on the air. The implementation of the speech codec is independent of its position, be it located in the base station or in the BSC. There are advantages to locating it in the BSC; these are discussed in further detail in Section 10.2.3.
- The *signaling* unit is the logical interface for the control messages between the network and the mobile stations. Many of these signaling messages are transparent to the base station and are simply passed on via the channel codec to the mobile station; the BTS takes no action on these messages besides mapping the data properly for the Um interface and then transmitting it properly. The relatively few messages and functions that require an action from the BTS are passed to the control unit. These include such functions as ciphering and frequency hopping. The O&M messages are only passed to the control function because they have nothing to do with normal mobile station operation.
- The *control unit* performs all the internal control tasks of the base station that has access to the O&M messages from the BSC, all of which are delivered via the Abis signaling path.

- The *burst-building* function adds the training sequence and the tail bits to the subblocks of coded data coming from the *channel codec*. After this has been done, the multiplexer maps the single bursts onto the single time slots bound for the individual mobile stations. The *modulator*, as the name indicates, modulates the digital signals onto the radio frequency carrier, which is an analog process requiring conversion by a DAC.
- The *transmitter* contains the output filters to band-limit the signals that should not disturb other radio services. It also controls the output level depending on the base station's power class, and if power control is implemented in the BTS, it is possible to set different power levels for each individual time slot.
- The *synthesizer* provides the necessary frequencies for the different entities in the BTS. The synthesizer is usually synchronized with the Abis clock which comes from the BSC. The advantage here is that, in this case, it is possible to synchronize all the base stations and have a synchronous network, since there is one central clock reference. Alternatively, it is possible to have a local clock reference in each base station.

10.2 THE ABIS INTERFACE

All the data, both signaling and user data, move between the base station (the BTS part) and the BSC on the Abis interface. The Abis is implemented when the BTS and BSC are located at different sites. If both are positioned at the same location, even in the same cabinet or rack, different solutions are possible, depending on the manufacturer. The original intent for the introduction and specification of this interface in the GSM system was to promote competition among the infrastructure manufacturers. If competition was successful, then it would, in theory, be possible to select the BSC and BTS from different manufacturers. In the initial phase of the GSM networks, this goal has not been achieved, because of different interpretations of how the O&M protocol should be implemented among the operators as well as among the manufacturers. The operators wanted to be sure their networks were running properly with the least number of problems, and one prerequisite seemed to be to use BSC and BTS equipment from the same manufacturer. There are still attempts to work toward a common standard so that it will be possible to use equipment from different manufacturers. It might even be possible that one manufacturer adopts the standard of another just to be able to supply equipment to network operators.

10.2.1 Physical Parameters of the Abis Interface

In a manner similar to the air interface, the Abis interface also uses a layered structure, Layers 1, 2, and 3. Though the three layers in the Abis have identical functions to those on the Um interface, their details are somewhat different. We will not explore the details of the Abis in as much detail as we did for the Um interface, but will confine our explanations just enough to make it possible to get a good understanding of the requirements for testing and understanding a base station. Additional information may be obtained from [1–4].

Layer 1 on the Abis is also the physical layer on which we find the digital data (speech and signaling) moving between the base station and the BSC at a rate of 2,048 kbps. It makes use of a TDMA structure using 32 time slots, each at a rate of 64 kbps. It is also possible to transmit these data on a single physical 64-kbps link (without a TDMA structure). This latter possibility is not further explained for two reasons: first, it is not a very practical solution because of its low capacity and the effort necessary for implementation, and, second, it is similar to the transmission on a single 64-kbps time slot on a 2-Mbps link.

Due to its structure and *speech coding*, the 2-Mbps link is also referred to as a *PCM30* link. PCM stands for the type of modulation used on Layer 1, *pulse code modulation*, and the number 30 indicates that out of the 32 time slots 30 are used for user data communication between the base station and its controller. The other two time slots, indicated by the gray squares in Figure 10.3, are dedicated to synchronization tasks (on TS 0) and the signaling required between the base station and the BSC simply to maintain Layer 2 of the Abis link (on TS 16).

In addition to the allocation of time slots on the 2-Mbps frame, the specifications allow a further variation. A 64-kbps channel may be subdivided into four *subslots* of 16 kbps each. Such a subslot is not only addressed by its time slot number (in the Abis sense), but also by its subslot number. The subslot may be used for signaling purposes or traffic channel assignments (see Figure 10.3). The purpose of this scheme is to assign a Um traffic channel directly to one subslot, which has approximately the same data rate as the speech frames. The advantage will become more obvious in a later section when we consider the physical placement of the speech codec within the network. Today, the 16-kbps traffic channels are supported

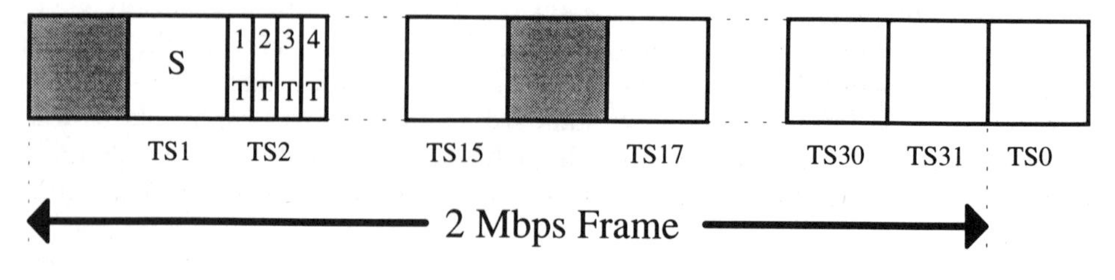

T traffic channel in time slot 2, sub slots 1, 2, 3, 4, each 16 kbps

S signaling channel in time slot 1 at 64 kbps

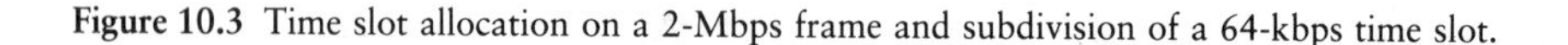

Figure 10.3 Time slot allocation on a 2-Mbps frame and subdivision of a 64-kbps time slot.

by all manufacturers, but the network's signaling channels on the 16-kbps subslots are not very common. Table 10.1 shows all possibilities for transmitting data on the Abis interface.

10.2.2 Configurations of the Abis Interface

With the resources provided by the Abis interface, different possibilities exist to connect base stations to their controllers (see Figure 10.4). If a BTS supports only one transceiver, it is enough to support a maximum of only eight time slots on the Um interface. Even if a base station supports several transceivers (BTS3), one 2-Mbps Abis link is sufficient. If, instead, a large number of transceivers are used within one cell in one base station, then there is a need for more Abis links (BTS2). A different philosophy is to control each entity within one base station via its own Abis link (BTS4). This is a very costly scheme and is not, therefore, a common solution. A structure not presented here is a circular arrangement. For this constellation, several base stations, with between one and two transceivers in each, are connected via an Abis line in a ring. This allows us to control several BTSs with only one Abis link, where each Abis is addressed via a different time slot.

10.2.3 Position of the Speech Codec

By definition, the *speech codec*, as it was introduced in Chapter 5, on the network side, is part of the base station. In spite of this fact, two different implementations to position the speech codec are possible. It works within the base station itself or it can be placed within the BSC.

Table 10.1
Possible Transmission Rates on the Abis Interface

<table>
<tr><th colspan="4">2.048-Mbps Connection (PCM30)
32 Time Slots Each at 64 kbps</th></tr>
<tr><td colspan="2">Speech codec in the BTS =>
speech transmission via single channels, each 64 kbps</td><td colspan="2">Speech codec in the BSC =>
speech transmission via single subslots, each 16-kbps, multiplexed on 64-kbps channels</td></tr>
<tr><td colspan="2">Signaling via</td><td colspan="2">Signaling via</td></tr>
<tr><td>64-kbps channels</td><td>16-kbps channels, multiplexed on 64-kbps channels</td><td>64-kbps channels</td><td>16-kbps channels, multiplexed on 64-kbps channels</td></tr>
</table>

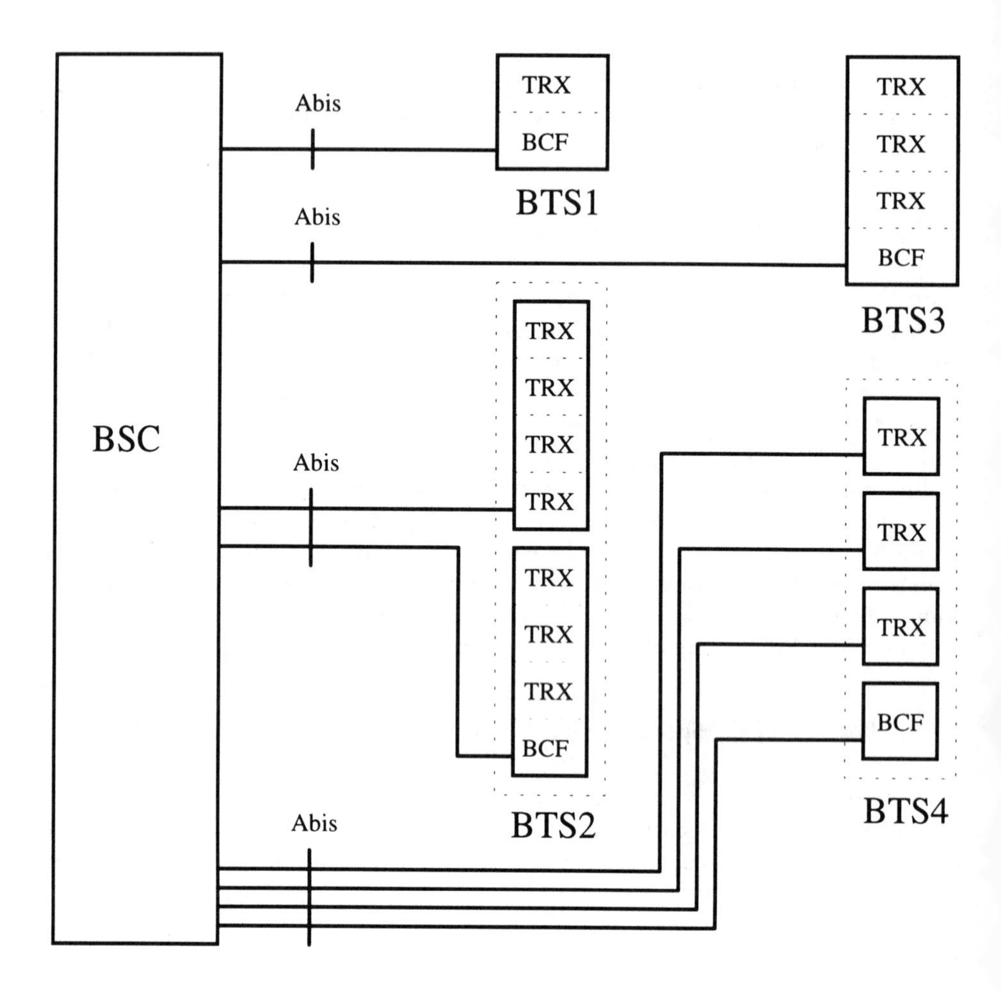

Figure 10.4 Possibilities for connections between BTS and BSC.

When the codec is located within the base station (BTS), the speech data must be converted from the net data rate of 13 kbps of the Um interface to the ISDN-specific data rate of 64 kbps. In order to increase the data rate, the speech blocks are decoded from the class Ia, Ib, and II bits into a rate of 104 kbps. From this high rate, it is again coded according to the *A-law* method at a data rate of 64 kbps, which appears as PCM in order to compress the data and reduce the effect of quantizing errors by giving lower (speech) signal levels more significance. (This technique, the A-law method, is described in [5].) This means that one traffic channel uses one 64-kbps channel just like a signaling channel.

If the speech codec is instead operated outside of the base station (e.g., within the BSC), it retains the net data rate of 13 kbps after channel decoding, which is already the rate on the Um interface not accounting for the coding. Only some *inband signaling* with a data rate of 3 kbps is added, which handles the correct timing of the speech blocks between the air interface and the speech codec. With this mechanism it is clear that the data coming from the codec arrive in the base station exactly at the time when they have to be transmitted on the Um interface (time alignment). The resulting 16-kbps data rate fits exactly onto a single subslot on the Abis interface. The advantage of this implementation lies in the more economic use of the physical resources of the Abis interface (cables). It is possible, with this second option, to pack up to four traffic channels from the air interface onto the Abis link instead of only one traffic channel on a single 64-kbps link, to which the first option confines us when we install the speech codec in the base station.

10.2.4 Structure of the Abis Interface

Similar to the radio interface, there is a structure to the Abis interface for the transmission of user data (speech or data bits) and the signaling data (between the mobile station and the network). Depending on the location of the speech codec, each transceiver uses between two (codec in the BSC) and eight (codec in the BTS) time slots on the Abis interface for user communication. In addition, each transceiver needs a signaling time slot dedicated to the message transfer between a mobile station and the network. Since there are also signaling messages exchanged on the FACCH for handover or longer signaling messages during a call in progress, this signaling time slot is needed even if a transceiver is only equipped with traffic channels. One signaling channel for a complete transceiver is sufficient, since it uses a very high data rate of 64 kbps. If we recall the rather slow data rates used on the radio channel, it becomes apparent that a SDCCH (782 bps) along with its SACCH (391 bps, which is just an approximate value, since the BTS adds its own measurements for the BSC) uses only 1,173 bps. Each SACCH associated with a traffic channel uses 382 bps. This means that for a signaling channel on the Abis with a configuration of four SDCCHs plus a SACCH and seven traffic channels (which do not use any signaling in the first instance) with their associated SACCHs use approximately 7,366 bps. Even if some signaling is necessary on a FACCH at a data rate of 9,200 bps, it is far from using up all the signaling capacity of the Abis's 64-kbps channels.

Figure 10.5 is a simple example of how a general configuration of three Abis time slots for one transceiver can appear. We can see that each Abis subslot carries traffic data assigned to an individual time slot on the air interface. The Abis time slot reserved for signaling (S) has access to each time slot on the Um interface.

10.2.5 Addressing on the Abis Interface

It has already been shown that one time slot on the Abis interface carries the signaling information for multiple time slots on the Um interface. Since the different links

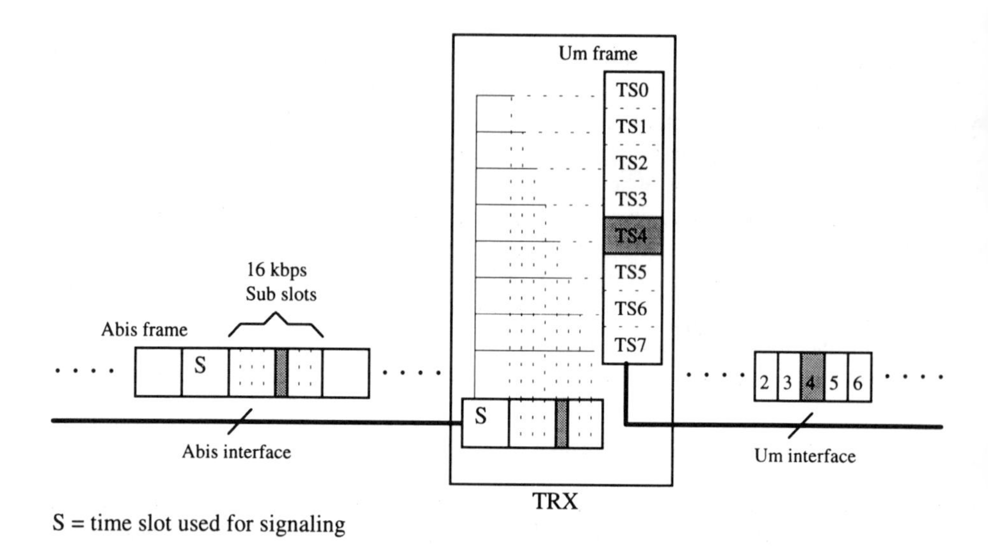

Figure 10.5 Mapping of the time slots between the Abis and the Um.

have to be distinguished from each other, the time slots of one transceiver are addressed via the header of the Abis message indicating the channel (traffic, signaling, or common control) and the applicable time slot. It is also possible that one 64-kbps Abis time slot carries the signaling information of several transceivers configured only with traffic channels, which have a limited need for signaling. In this case, a further distinction has to be made for the individual transceivers. The *terminal equipment identifier* (TEI) is used for this purpose. TEIs might be numbered from 0 to 127 within each Abis signaling time slot.

Figure 10.6 illustrates the addressing of the different entities in a base station. The following SAPIs are distinguished (see also Section 6.3.1):

- SAPI = 0 for the *radio signaling link*;
- SAPI = 3 for the *SMSs*;
- SAPI = 62 for the *O&M* messages.

The SAPI for the *Layer 2 management link* (SAPI = 63) is missing from this list, but since this is implemented in the Abis L2, there is no need to mention it further. Since all messages for the different units can be mapped onto one single Abis time slot, they have to be distinguished from each other by individual "signs." The first sign we see is the SAPI. The Abis L2 routes messages toward destinations indicated by the SAPI. Within a single SAPI, the TEI distinguishes different entities. Examples of TEIs within a SAPI are an individual transceiver or different parts of the O&M. Within a transceiver, the messages are marked by the individual time

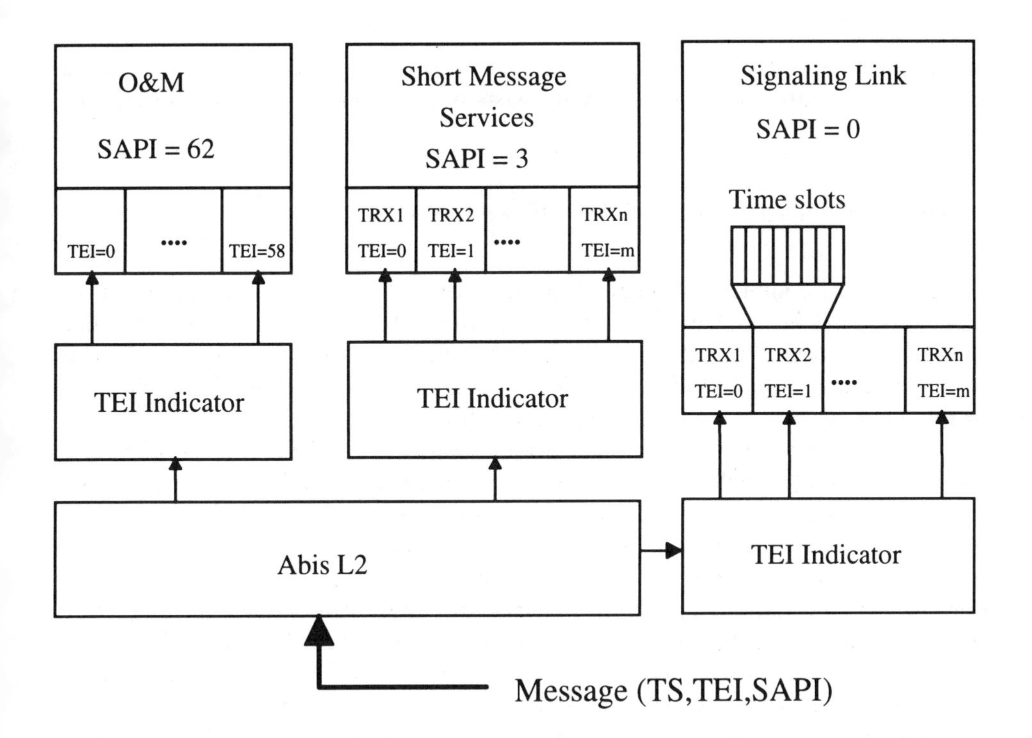

Figure 10.6 Mapping of the different Abis signaling links onto the different BTS entities.

slot numbers on the Um interface. This three-point addressing mechanism shows the capabilities of the Abis interface, which can appear in several forms, depending on the manufacturer. Some manufacturers might have all signaling information on one time slot, and others allot an individual Abis time slot for each transceiver. All these factors amplify the advantage of a remote-positioned speech codec in the BSC rather than in the BTS. In the case described here, the requirement for one transceiver is three time slots on the Abis interface (one 64 kbps for signaling plus two time slots with four 16-kbps subslots in each for speech channels). See Figure 10.5. Placing the speech codec within the base station would require nine time slots (one for signaling and eight for each of the traffic channels). For this latter, expensive implementation, the speech transmission is no better in quality.

Another consideration enters into the positioning of the speech codec. There are implementations of the TRAU, which converts the speech data from 13 to 64 kbps close to the MSC. By definition, the TRAU is remotely connected to the BSC with a special internal interface, with the A-interface only a few meters' distance away. This configuration is even more economic for the network operator. One simply has to remember that it does not make any difference where the speech codec is located because there is no significant interference incorporated on the network interfaces; they are all fixed wires with twists or shielding to prevent interference.

10.2.6 Signaling Connection Between BSC and BTS

So far, only the communication between the network and the mobile station via the base station has been mentioned. This connection is handled with something called the *radio signaling link* (RSL), which includes messages that are specified in great detail and that are common for each BTS. Further information about the significance of the messages and coding may be found in [6].

It should be obvious now that messages between the BSC and the BTS have to be exchanged (1) to download the system software onto the BTS, (2) to configure the BTS, (3) to control the BTS, and (4) to control the Abis link. For these purposes, different types of connections are selected. The *operation and maintenance link* (OML) is intended for the use of *network management*. It is sufficient to provide one link per base station. Since these messages are dependent on the implementation by the manufacturer, they will not be explained in any detail. The *Layer 2 management link* (L2ML) controls the Layer 2 connection of the Abis link, which uses the dedicated time slot number 16. This link is always present and no specific Layer 2 establishment has to be performed. This link has the same needs as the OML; one link is sufficient for each BTS. This link is responsible for the organization of Layer 2 on the Abis interface; it assigns different TEIs to different time slots and manages these links.

The RSL includes two different types of messages for the base stations: the *transparent* and *nontransparent messages*. Transparent messages are passed on by the base station without any interpretation. These are messages coming from either the BSC or the MSC and are intended for direct communication between these two entities and the mobile phone. These are, in general, messages of the MM and of the call control functions. A few of the RR messages coming from the BSC are also part of these messages. A good understanding can be gained from Figure 7.5, which indicates the general organization of the signaling architecture in a GSM network. For example, an authentication request by the MSC does not need any kind of interpretation by the base station. The nontransparent messages are interpreted by the base station and usually require that some kind of action be performed by the base station, or some kind of response is assumed. Most of the RR messages fall in this category. Some examples are (1) a channel activation that the BTS has to be aware of, and (2) the start of ciphering. For testing a base station, this means that the effort expended on the signaling messages is much smaller than that for a mobile station, which also has the MM and call control functions built in. On the other hand, some network-specific tests on a base station have to be added.

10.3 PREPARATIONS FOR THE MEASUREMENTS AT A BTS

In principle, there are two different constellations for testing a GSM base station: (1) the base station is connected to the BSC, and (2) the base station is operated in a standalone configuration.

10.3.1 Base Station Within a Network

When a base station is operated within a network, the BSC provides it with the necessary software, the configuration data, and the O&M messages. The measurements in this configurations are restricted to the air interface and, therefore, are limited to the transmitter tests.

With some additional effort, it is also possible to perform the receiver and signaling measurements in this configuration. This additional effort includes a *drop and insert switching matrix*, which gives the test system access to individual time slots on the Abis interface, which are used for different tests. This switching matrix can be implemented in the test equipment, and when in the idle mode (no test is running), is transparent to all the normal Abis messages; that is, time slots pass freely between the BTS and BSC (see Figure 10.7). When a test is started, the test system drops one predefined time slot of the Abis link coming from the BSC and

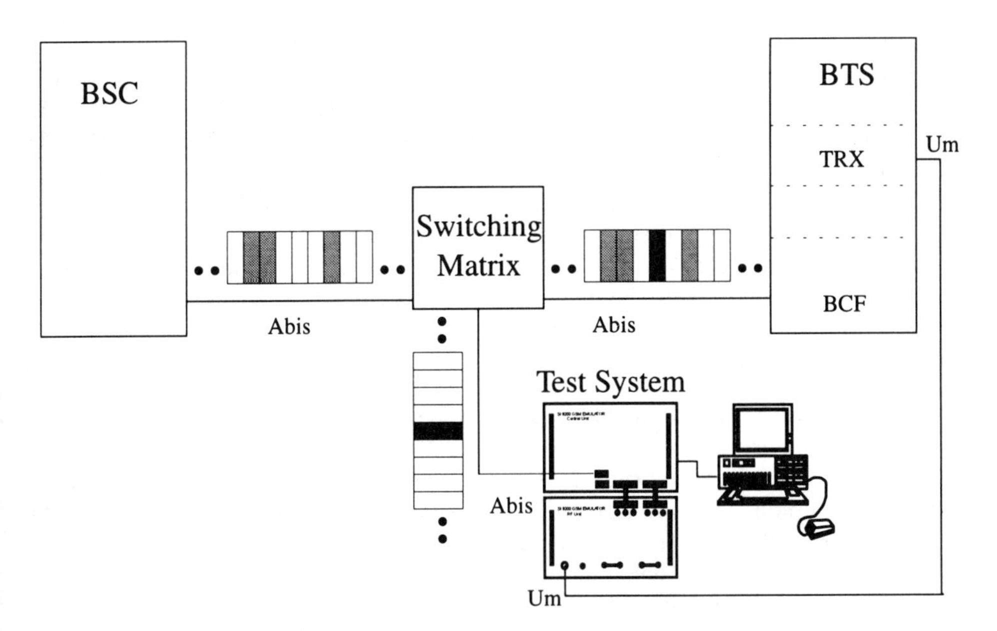

Figure 10.7 Test configuration of a BTS within a network with a digital switching matrix.

inserts its own messages needed for testing. So it is possible to communicate on only one time slot from the test set to the BTS, and on all other time slots between the BSC and the BTS.

10.3.2 Measurements on a Standalone Base Station

Of course, it is more flexible to perform the measurements on a *standalone base station*, without the link to the BSC. In this configuration, the test system controls the BTS. Due to the fact that a base station is heavily dependent on certain messages on the Abis interface, these tests must be prepared very carefully and require a good knowledge of the particular type of base station. Since even the configuration of the different time slots on the Abis interface is peculiar to a certain BTS type, this must be well known in order to set it up properly. Usually, the system software—necessary for the basic procedures—is already loaded into the BTS before a test session starts. If this is not the case, then the software must be loaded (1) by the test system, (2) by the BSC using a drop and insert switching matrix, or (3) with the use of manufacturer-specific tools using a *local maintenance terminal* (LMT) interface.

With the system software loaded in the BTS, the configuration (time slot and frequency allocation) is setup by using O&M messages. This is usually incorporated in the software download process. After the BTS is provided with the software and is configured properly, the system information units have to be sent to the transceiver that is transmitting the BCCH. These information units consist of system information 1 to 4, which are mapped onto the BCCH, and of system information 5 and 6, which are mapped onto the SACCH for each individual channel. For the SACCH, of course, the L1 header, which contains the downlink power control and timing advance parameters, is different for each mobile station. The SACCH contains the most important system information items in case some parameters change during a call. This is to ensure that the mobile will get notice of the changes immediately, since it has to listen to this channel anyway.

When a single base station has to be tested, the following important parameters have to be set up and checked:

- Time slot allocation on the Abis interface for the signaling and traffic data transmissions;
- Electrical impedance of the Abis link (75Ω asymmetrical or 120Ω symmetrical);
- Transmission of the speech data on a 16- or 64-kbps link along with the position of the speech codec (remote or built into the BTS);
- Transmission of the signaling data on a 16- or 64-kbps link, where the latter is the more commonly used link;
- Necessary commands via the Abis link, which gets the base station to transmit the BCCH or other channels;

- Necessary Layer 2 management link messages (necessity of the TEI assignment procedure).

When all the parameters are set properly, the actual testing can start. A short functional test may be the initial configuration of the base station with the system information units, and then getting it to transmit them with a subsequent verification that the data are sent correctly on the Um interface; that is, the test set can synchronize properly. Figure 10.8 shows a trace from such an initial test at a base station, where

frame number	Um downlink (MS ← BTS)	Abis uplink (BTS → BSC)	Abis downlink (BTS ← BSC)	**BSC** = test system (Abis) **MS** = test system (Um)
FN 00284 00:24:29	*Establishment of the Layer 2 by sending of a SABME, which is equal to the SABM on Um*		Abisdl30 Address: SAPI = 0, Command TEI = 001 Control:SABME: N(R)=-,P/F = 1, N(S)=- Length = 0	
FN 00285 00:25:30		Abisul30 Address: SAPI = 0, Response TEI = 001 Control:UA : N(R)=-, P/F = 1, N(S)=- Length = 0		*Acknowledgment with a UA frame*
FN 00289 00:03:34	*Transmission of the L2 Information (which will not be shown in the following)*		Abisdl30 Address: SAPI = 0, Command TEI = 001 Control: I : N(R)=000, P/F = 0, N(S)=000 Length = 30 Info = 0C 11 01 80 1E 01 0B 00 15 06 19 00 00 00 00 00 00 00 00 20 00 00 00 00 00 00 01 51 00 00	
FN:00289 00:03:34 00:03:34	*The translated L2 Information corresponds with this L3 Information (BCCH 1, System Information 1)* *Embedded Um message, which is another word for transparent, i.e. the message will appear like this on the Um interface*		AbisdlE,l2_ack,md(COM), bcch_info, ch_no(= 10000i,0h), sys_info_typ(= 1h), L3_info(= /* Embedded UM message */ TI(0),pd(rr), sys_info_1, cchan_desc(= 0h, 00000000000000020000000000000001h), rach_par(= 1h,4h,0i,1i,0000h));	
FN 00291 00:05:36		Abisul30 Address: SAPI = 0, Response TEI = 001 Control:RR : N(R)=001, P/F = 0, N(S)=- Length = 0		*Acknowledgment of the Layer 2 about receiving this message (which will not be displayed anymore in the following)*
FN:00299 00:13:44	*Transmission of System Information No. 2 on the BCCH*		AbisdlE,l2_ack,md(COM), bcch_info, ch_no(= 10000i,0h), sys_info_typ(= 2h),	

Figure 10.8 Trace of the synchronization with a base station.

```
                                         L3_info(= /* Embedded UM message */
                                         TI(0),pd(rr),
00:13:44                                 sys_info_2,
                                         neig_cell(= 0h,0i,
                                         00000000000000000000000000000
                                         0h),
                                         plmn_perm(= 01h),
                                         rach_par(= 1h,4h,0i,1i,0000h)
                                         );
=============================================
The System Information Nos. 3-6 have been erased from the trace!
=============================================

Here the trace record of the Um interface starts. The Abis part is finished here in this test!

Um part of the tester synchronizes onto the FCCH:
              Channel Type = FCCH,  TN = 0,
              FN = 9374 ( FN mod 51 = 41 ),  CV = 4788

From the SCH BSIC, T1, T2 and T3 are read, with this information the FN Number is updated:
              FN = 451 ( FN mod 51 = 43 ), Channel = SCH, BSIC = 0,
              T1 = 929,  T2 = 21,  T3' = 0,  FN-Diff. = 1231710

FN 1232213   Umdl1 BCCH                          Again first the L2 data are read
0929:21:02   Length = 16, M = 0, EL = 1          (these are taken  ut in the following)
             Info = 06 1A 00 00 00 00 00 00 00
             00 00 00 00 00 00 00 00 01 51
             00 00

FN:1232213   Umdl1,BCCH, TS(0),TI(0),pd(rr),     The previously recorded L2 data is
0929:21:02   sys_info_2,                         translated which correspond with
             neig_cell(= 0h,0i,                  the System Information 2
             00000000000000000000000000000h),
             plmn_perm(= 01h),
             rach_par(= 1h,4h,0i,1i,0000h);

FN:1232366   Umdl1 BCCH, TS(0),TI(0),pd(rr),     Reading of System Information 1
0929:18:02   sys_info_1,
             cchan_desc(= 0h,
             0000000000000020000000000000001h),
             rach_par(= 1h,4h,0i,1i,0000h)
```

Figure 10.8 (continued).

the specific manufacturer-related data has been erased. At the beginning, the system information items are transmitted (numbers 3 to 6 are not displayed here) from the test system to the BTS via the Abis. After the BTS starts to transmit the BCCH (on the Um), the test system synchronizes with the FCCH, the SCH, and then the BCCH. Comments are added in italics.

Frame numbers do not exist for any of the network links—whether these are the Abis, A, or any other interface—in the sense they are known for the Um interface. For a better comparison of the data flow through the base station, the tester used for this example adds the same numbers as those that are current for the Um interface. The trace also shows the significance of transparent messages; the system information, just as it is transmitted on the Um interface, is transmitted in the same way from the BSC to the base station.

10.4 MEASUREMENTS ON GSM BASE STATIONS

Independently of the test setup used to make measurements on a base station in the following chapters, practical tests on base stations are described. Moreover, it is assumed that the BTS is operational and only single channels are activated. The test limits are also given, and if additional test equipment is necessary, then the complete setup of each individual test is described. Exactly like the case with mobile stations, RF tests (split into transmitter and receiver tests) and signaling tests are listed.

10.4.1 Transmitter Tests

The *transmitter tests* are usually performed in the frequency-hopping mode over three frequencies selected from all available channels of the GSM band. If a base station is only equipped with one transceiver, then three different frequencies are tested consecutively. They are separated into a bottom (B, ARFCN = 1), a middle (M, ARFCN = 62), and a top (T, ARFCN = 123) frequency of the accessible GSM frequency range.

10.4.1.1 Static Layer 1 Measurement

Before starting with the actual tests, we should make sure that the information transfer from the BSC side (Abis interface) to the mobile side (Um interface) is correct. It is for this reason that a BER measurement is performed for both signaling and traffic channels. This test checks the encoder and the multiple-access mechanisms. To execute the test, a known PRBS is transmitted from the test set to the base station via the Abis interface. Within the BTS, these data are encoded (error-protected), the single bits are reordered and restructured, interleaved onto several bursts, and then, finally, mapped onto a time slot of the Um interface. The test system receives the signal from the base station, decodes it again, and then compares the received bit stream with the one previously transmitted via the Abis interface. This test is performed for all frequencies and channel combinations. During this test, no bit should be altered by the base station.

On the Abis interface, signaling, data, or traffic information is transmitted without any *error protection*. Error protection is performed within the base station, and this functionality is verified with this test.

10.4.1.2 Phase and Frequency Error Measurement

A PRBS is sent to the base station via the Abis interface, which maps this data onto the air interface. The PRBS is used to produce all possible modulation transitions.

The test system evaluates the bursts on the Um interface for the *phase error* and the *frequency error*, as was described in Section 8.3.2. The limits for the phase error are 5 deg for the rms (average) value and 20 deg for the maximum value. The frequency error must not exceed a maximum of 0.05 ppm (parts per million), which at 900 MHz equals a frequency error of 45 Hz.

Since most of the base stations get their frequency reference from the Abis clock, care has to be taken in the standalone test setup that the test system provides a time base accurate enough to measure against these requirements. Otherwise the test system measures the accuracy of its own clock and not the ability of the BTS to synchronize itself with the Abis clock.

10.4.1.3 Power Measurements

A base station transmits continuously on the base channel; every time slot is used. The *power measurement* is performed on the base channel and at first concentrates on the *peak power*. The levels that have to be reached for the different power classes are given in Table 10.2. A transmitter must not only be able to send on these power levels, but also at least on six more power steps, each of these steps being 2 dB further down. The base station has to meet these steps with an accuracy of ±0.5 dB.

Optional *transmission power control* exists for base station transmitters, which is similar to the mobile station's mandatory power control feature. Its purpose is to always provide the mobile stations with the appropriate power level and not much more. When this option is implemented, the base station must be able to transmit nine additional power steps of 2 dB each. The power of step number 0 equals the power level of the respective power class, and the power of step number 15 equals the smallest power level. The individual steps must be accurate within ±1.5 dB. An

Table 10.2
Power Levels of Base Stations

Power Class	*Maximum Peak Power/ (dBm)*	*Tolerance (dB)*
1	320W (55)	–0, +3
2	160W (52)	–0, +3
3	80W (49)	–0, +3
4	40W (46)	–0, +3
5	20W (43)	–0, +3
6	10W (40)	–0, +3
7	5W (37)	–0, +3
8	2.5W (34)	–0, +3

additional requirement is that the power levels must build on an increasing sequence with a step size of 2 dB (±1.5 dB) up to the highest power level. This is only a relative requirement. The absolute accuracy for each step must be within ±3 dB. Figure 10.9 shows an example of the evaluation of a BTS with power control of power class 1.

Besides measuring peak power, we also evaluate the *power-time template*. The power-time template requirements for mobile stations also apply to base stations. The only difference is that, in the base station case, no short burst (random access) exists. To measure the template, the base station has to be operated in its *bursting mode*. This test cannot be performed on the base channel, since a base station has to transmit continuously in all time slots. In most cases, an additional channel besides the base channel has to be activated or the transmitter under test is not considered to be configured as a base channel transmitter. The average peak power from the previous part of the test is used as a reference. The time reference is provided by the training sequence in the middle of the burst. The template has to meet the requirements of Figure 10.10. Again, the noise level of –70 dBc is dependent on the

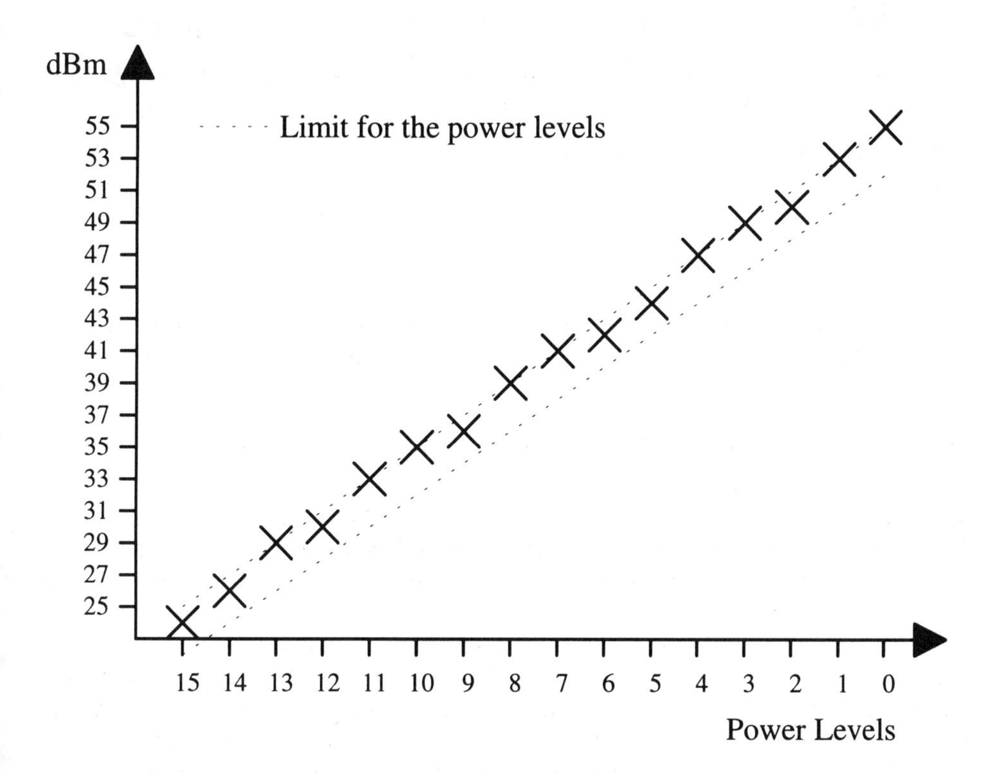

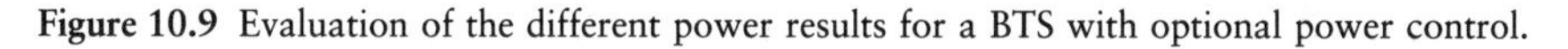

Figure 10.9 Evaluation of the different power results for a BTS with optional power control.

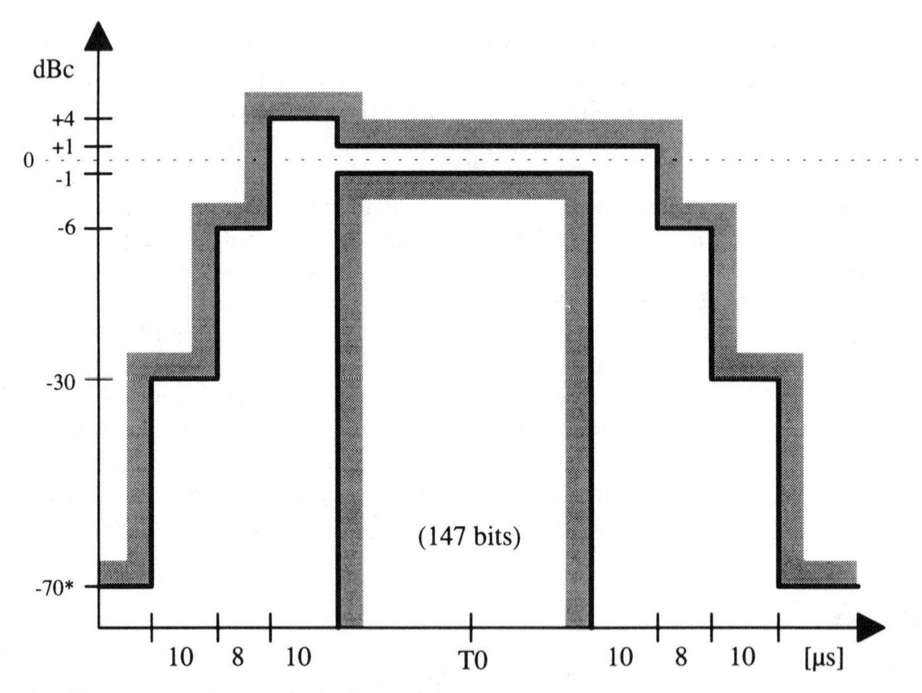

Figure 10.10 Power-time template.

actual transmitted power level. If this level is below −36 dBm, the absolute level applies. The template is usually evaluated along with the peak power level.

10.4.1.4 Adjacent Channel Power Measurement

The *adjacent channel power measurement* quantizes the power falling into a neighboring channel from the transmitted signal. Two different parameters or processes are investigated: (1) the *spectrum due to modulation*, and (2) the *spectrum due to switching transients.*

To find the modulation spectrum (caused by the modulation), the base station must transmit on only one time slot in the burst mode. A PRBS is sent from the test equipment via the Abis interface. This random sequence is mapped onto a single time slot on the Um side by the base station. The measurement is only performed over the coded data that are a random sequence and that exhibit a random characteristic in order to measure all possible modulation products. Since the remainder of the time slot (the training sequence and tail bits) is static, the modulation spectrum would not change for these and is, therefore, not measured.

A spectrum analyzer is required and is set to a resolution bandwidth of 30 kHz, a video bandwidth of 30 kHz, and a frequency span of zero. The reference power level is the average peak power signal of the burst. The power is measured for the following offsets from the carrier: 100, 200, 250, and 400 kHz, and between 600 and 1,800 kHz in steps of 200 kHz. We should note that the resolution bandwidth of 30 kHz is (1) small enough not to measure components outside a 200-kHz band (e.g., when measuring at only a 200-kHz offset from the carrier, no spectral component stemming from the carrier will be measured, which would be the case for a resolution bandwidth of 300 kHz or 1 MHz), and (2) is still great enough to see the ramping nature of the signal. The requirements for power levels below the carrier at these offsets from the actual transmitted channel are given in Table 10.3. If the measured values are below the absolute level of –36 dBm, the absolute level is used as the limit. The measurement is performed consecutively for the three specified frequencies.

As we described in Section 8.3.4.1, a spectrum analyzer with a trigger input has to be used to allow a *gated sweep*. The trigger is provided either by the test system, internally from the spectrum analyzer itself, or from the base station. Only the gated sweep ensures that only the spectrum of the modulated coded data is observed; spectral influences caused by the modulation have to be distinguished from those caused by the switching transients. For the measurement of the switching-transients spectrum, different switching characteristics from the base station are classified and measured.

- Even if all time slots are allocated and the base station is transmitting continuously, it may ramp down in between the time slots; this makes it easier for the manufacturer to use the same hardware configuration for both the base channel

Table 10.3
Test Limits for the Spectrum Due to Modulation

Power Level (dBm)	*Maximum Relative Level (dBc) at the Distance From the Carrier (kHz)*					
	0	100	200	250	400	600–1,800
≥43	0	+0.5	–30	–33	–60	–70
41	0	+0.5	–30	–33	–60	–68
39	0	+0.5	–30	–33	–60	–66
37	0	+0.5	–30	–33	–60	–64
35	0	+0.5	–30	–33	–60	–62
≤33	0	+0.5	–30	–33	–60	–60

and all the other channels. The power can be reduced by approximately 30 to 35 dB between the individual bursts.

- If the optional *power control* feature is available, the transmitter can set different power levels for each time slot, which may be, in the extreme case, 15 · 2 dB or 30 dB. On the other hand, it is possible to switch the power completely off if an individual time slot is not used. To test this influence, the base station must use the switching characteristic specified in Figure 10.11, which represents the worst case and, therefore, also has the greatest impact of the switching transients on the adjacent channels.
- If the optional power control feature is not available, then the base station can only switch the power on or off without assigning individual power levels for each time slot. The transmitter sends, then, the set power level in one of the six steps. The worst case scenario represented by this case is given in Figure 10.12.

Depending on the option implemented in the base station, the switching transients spectrum is evaluated for the three different frequencies. The spectrum analyzer is set to a resolution bandwidth of 30 kHz, a video bandwidth of 100 kHz, a zero frequency span setting, and a peak hold display setting. This measurement does not require a trigger because the impact of the different switching transients is recorded with the peak hold function. The offset from the transmitting frequency and the required minimum level distance from the carrier power are given in Table 10.4.

10.4.1.5 Spurious Emissions From the Transmitters

The influence of the transmitter on its own GSM frequency band is evaluated with the adjacent channel power measurements. The measurements of the *spurious emissions*

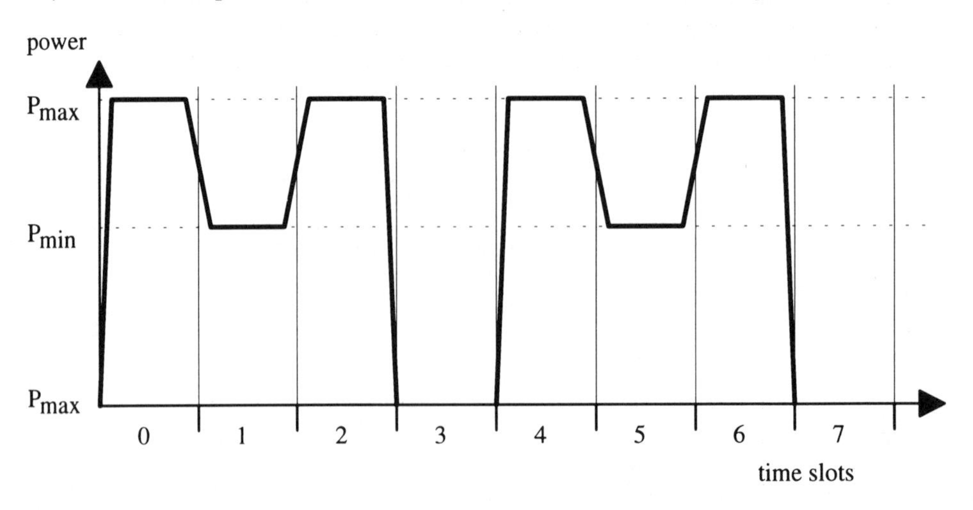

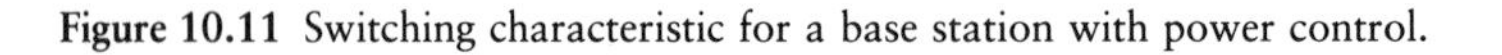
Figure 10.11 Switching characteristic for a base station with power control.

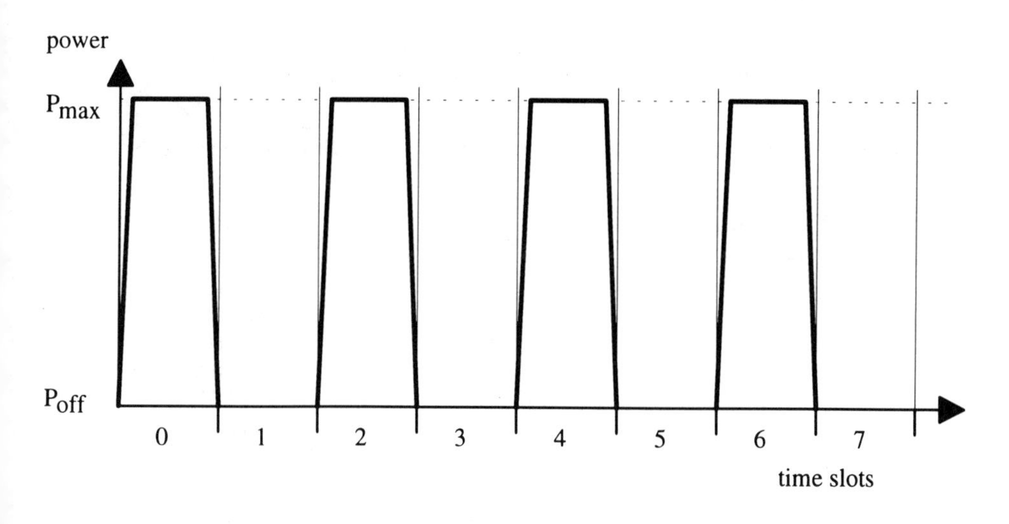

Figure 10.12 Switching characteristic for a base station without power control.

Table 10.4
Test Limits for the Spectrum Due to Switching Transients

Distance From the Carrier Frequency (kHz)	*Minimum Distance From the Carrier Power (dBc)*
400	–60
600	–69
1,200	–75
1,800	–79

check the purity of the transmitted signal (i.e., whether other undesired frequencies are transmitted in addition to the desired emissions, which might disturb other services). The base station is, again, operated in the frequency-hopping mode over the three frequencies B, M, and T, just as was done in previous tests. The evaluation is performed separately for each frequency. Spurious emissions may be radiated from three different places into the spectrum:

- The power transmitted into a 50Ω load at the antenna output;
- The effective radiated power from the cabinet;
- The RF power from the power leads.

To measure the cabinet radiation, a specially shielded RF chamber with special receiving antennas is used.

Table 10.5 shows the spectrum analyzer settings for the spurious-emission measurements in different frequency bands. Since the measurements are carried out

Table 10.5
Analyzer Setup for the Measurement of Spurious Emissions

Frequency Band (MHz)	*Frequency Offset From the Carrier*	*Resolution Bandwidth*
9 kHz–100 kHz		1 kHz
100 kHz–10 MHz		10 kHz
10–890	≥2 MHz	30 kHz
	≥5 MHz	100 kHz
	≥10 MHz	300 kHz
	≥20 MHz	1 MHz
	≥30 MHz	3 MHz
890–915		30 kHz
915–935	≥2 MHz	30 kHz
	≥5 MHz	100 kHz
	≥10 MHz	300 kHz
	≥20 MHz	1 MHz
	≥30 MHz	3 MHz
935–960	≥600 kHz	10 kHz
	≥1.8 MHz	30 kHz
	≥6.0 MHz	100 kHz
960–12,750	≥2 MHz	30 kHz
	≥5 MHz	100 kHz
	≥10 MHz	300 kHz
	≥20 MHz	1 MHz
	≥30 MHz	3 MHz

during the operation of the base station, the frequency offsets from the transmitting frequency are variable.

Table 10.6 shows the requirements for passing the test. The measured levels will not exceed the limits. Special care must be taken in the receiving band of the base station.

10.4.1.6 Intermodulation Attenuation

The *intermodulation attenuation* is another indication of the transmitter's quality, indicating the linearity of the nonlinear parts within the output stage, primarily the amplifiers. This test gives evidence of how an interfering signal at a certain offset from the desired or assigned signal would alter the output spectrum through intermodulation (see Section 8.3.4.4).

For this measurement, an *interfering signal* 30 dBc below the actual carrier power is transmitted via a directional coupler into the base station's transmitter. In

Table 10.6
Test Limits for the Spurious Emissions From the Transmitters

Measuring Position	*Frequency Band*	*Maximum Level*
Power leads	9 kHz–10 MHz	40 dBμV(pd)
	10 MHz–30 MHz	60 dBμV(pd)
Cabinet radiation	30 MHz–1,000 MHz	–36 dBm
	1 GHz–12.75 GHz	–30 dBm
Antenna	9 kHz–1,000 MHz	–36 dBm
	1 GHz–12.75 GHz	–30 dBm

older versions of the technical specifications, a circulator was suggested instead of the directional coupler, but since the ferrite within a circulator can create intermodulation products itself, one could never be really sure where the measured intermodulation originated. These systematic problems led to the suggestion of the directional coupler. In Figure 10.13, the general test setup is given. Care must be taken when setting up the power level of the interfering signal and the base station. Since the power levels may attenuate each other, they have to be measured separately before they are switched on together at the coupling device. The interfering signal is generated at different offsets from the carrier frequency (1) to see the general influence of the intermodulation products, and (2) to position the intermodulation products of the

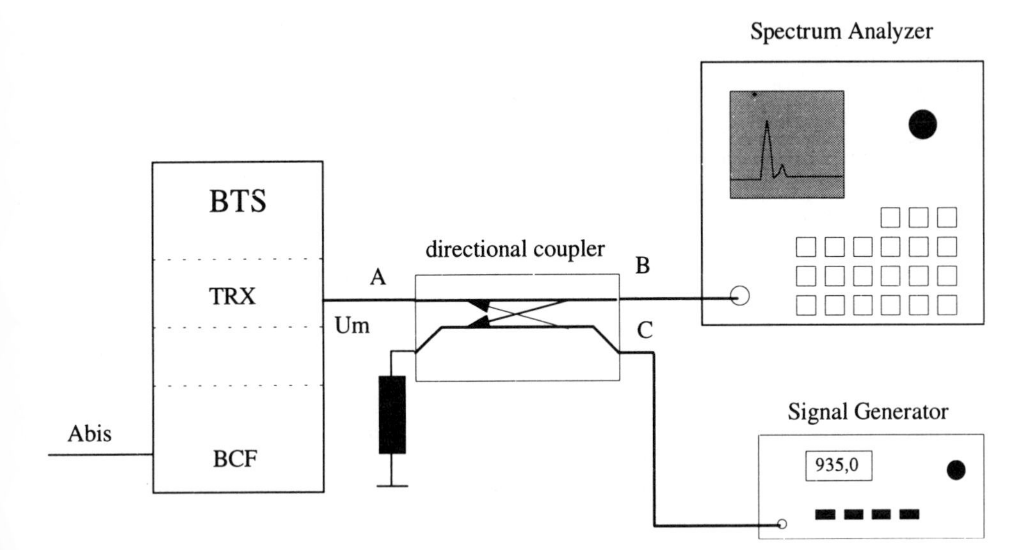

Figure 10.13 Test setup for the intermodulation attenuation measurement.

third, fifth, and seventh order in such a way that they are exactly in the middle of the receiving band. This is the requirement of the specifications, but in practice it does not make much sense to look for intermodulation products of the seventh order. The highest impact comes from the intermodulation product of the third order, and if the attenuation is greater than the allowed limit, then the base station is probably okay for this test. The exact requirements are given in Table 10.7. In the first case, where the intermodulation products are evaluated for the three different frequencies, the impact on the receiving band is only evaluated for one middle frequency.

10.4.1.7 Intrabase Station Intermodulation Attenuation

For a single transmitter in a base station in which we find other transmitters, it makes no difference whether the interfering signal is any kind of transmitter operated close to the base station's antenna, or whether it is a second transmitter within the same base station; the two situations are equivalent. For this reason, the impact of several transceivers within one station must also be investigated for their intermodulation attenuation. All transmitters are set to their maximum power levels for this test and to the minimum frequency offset specified for the base station. Two adjacent channels are usually not occupied by the same base station within a network, but, instead, at a certain frequency/known offset that has to be used for this measurement. The test has to be started with channel number 3. When, for example, a base station provides a channel spacing of 2.2 MHz (11 channels), then the channels numbered 3, 14, 25, and so on are used for this test. This sequence is continued in accordance with the number of transceivers available. In addition, the frequency offset is set to such a value that the intermodulation products of the third, fifth, and seventh orders fall exactly into the receiving band, just as was described in the previous section. The test limits are given in Table 10.8.

Table 10.7
Test Limits for the Intermodulation Attenuation Measurement

Transmit Frequency	*Offset of the Interferer From the Transmit Frequency*	*Limit*	*Frequency Band*	*Measurement Bandwidth*
B, M, T*	800 kHz	−70 dBc	100 kHz–12.75 GHz	300 kHz
M	11.25 MHz 15 MHz 22.5 MHz	−103 dBm	890 MHz–915 MHz	30 kHz

*If measured at a frequency offset 1.8–6 MHz from the carrier, the measurement bandwidth shall be at 30 kHz; below 1.8 MHz, the level from the spectrum due to modulation applies.

Table 10.8
Test Limits for the Intra-BTS Intermodulation Attenuation Measurement

Frequency Band	*Limit*	*Measuring Bandwidth*
890–915 MHz	−103 dBm	30 kHz
935–960 MHz*	−36 dBm resp. −70 dBc	300 kHz

*If measured at a frequency offset 1.8–6 MHz from the carrier, the measurement bandwidth shall be at 30 kHz; below 1.8 MHz, the level from the spectrum due to modulation applies.

10.4.2 Receiver Measurements

Due to the technique applied to *receiver measurements* (BER measurements), it is important that the test equipment used for these types of tests supplies both the Um and the Abis interfaces to the base station. A base station does not "loop" any recovered bits back from the receiving path to the transmitting path. Instead, the received data appear on the Abis interface. Besides the RF performance reflected in BER measurements, the correct decoding of signaling and user data is tested. Different test setups are possible; some use a standalone BTS that is completely disconnected from the BSC, and others only need a connection to the test system (Figure 10.14) or a setup similar to what is depicted in Figure 10.7, with the test system between the BSC and the BTS. Receiver tests, like the transmitter tests, are performed using frequency hopping on three frequencies selected from the bottom, middle, and top ranges of the GSM band. If the base station does not support frequency hopping,

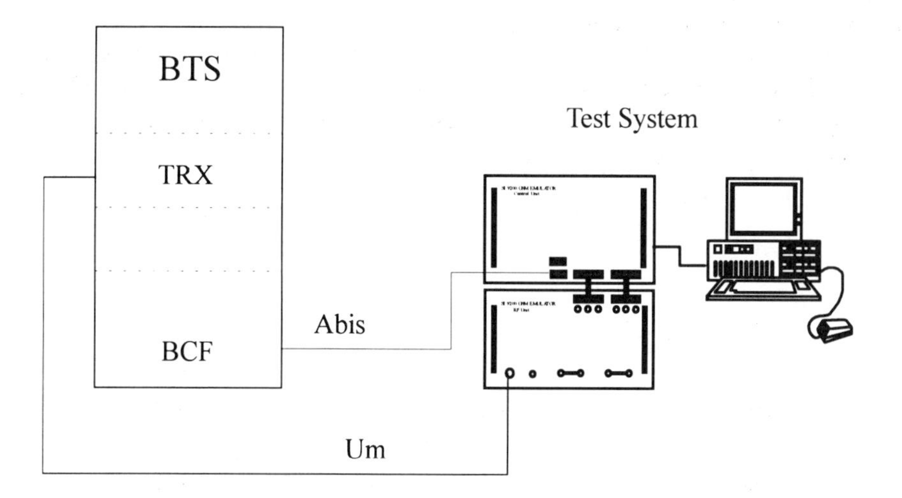

Figure 10.14 Test configuration for a standalone base station.

then the three frequencies are tested separately. This applies for all tests if not specified otherwise.

10.4.2.1 Static Layer 1 Measurement

Before getting started with the actual receiver tests, the demodulation and decoding process (Layer 1) has to be verified. This test is similar to the static Layer 1 test for the transmitter side, but only the traffic channel is evaluated for the receiver, and then only the unprotected bits of class II. Two different propagation conditions are tested: (1) static and (2) equalizer test at a simulated speed of 50 km/h (EQ50). The level of the test signal varies between the values specified in Table 10.9. This table shows, furthermore, the maximum error ratio allowed for these power levels.

10.4.2.2 Error Detection Mechanism

The *error detection mechanism* allows the receiver to detect unrecoverable errors which could not be corrected during the decoding process. If such an error occurs, a traffic frame is discarded and a signaling frame is not acknowledged, which, in the signaling case, means that it has to be repeated by the mobile station. For traffic data (speech), the bits of class Ia are evaluated, and if these do not fit the parity bits of the block code, it is marked as a bad frame, thus setting the *bad frame indication* (BFI) bit. For the signaling data case, the parity bits of the block code of all the bits are evaluated, and, if necessary, the *frame erasure indication* (FEI) is set. It may be said, in general, that in case the codec finds an error in a data block within the CRC bits, or the parity bits, this block will not be decoded further.

The reliability of this important feature has to be checked. To do so, a PRBS is transmitted to the receiver of the base station at a power level of –85 dBm. The BTS is configured to expect the following channel combinations:

- TCH/FS + FACCH + SACCH;
- SDCCH + SACCH;
- RACH.

Table 10.9
Level of Test Signal and Maximum BER for the Statistical Layer 1 Test

Multipath Characteristic	*Level of the Test Signal (dBm)*	*BER (%)*
Statistic	–85 up to –10	0.01
EQ50	–85 up to –40	0.3

Since a PRBS carries no information, the received frames do not fit any channel combination and the BTS simply regards every received frame (speech or signaling) as a bad one. It may be possible that there are certain combinations of bits that the decoder regards as a valid frame. Taking this remote possibility into consideration, the following requirements apply for the above channel combinations:

- TCH/FS: less than one undetected bad frame (BFI = 0) shall occur on average in a period of 10 seconds;
- SDCCH, FACCH, and SACCH: less than 0.002% of the frames shall be detected as error-free (FEI = 0);
- RACH: less than 0.02% of the frames shall be regarded as error-free (FEI = 0).

A second step evaluates the influence of the noise floor of the receiver. No signal is injected into the receiver, which still expects one of the above-mentioned channel combinations. Again, the requirement for the PRBS signal must be met to test the noise floor impact.

10.4.2.3 Sensitivity Measurement

The *sensitivity measurement* evaluates the reference sensitivity level of the receiver; a very-low-level signal is injected into the base station's receiver. The signal level used for testing is set to –104 dBm, at which power level the decoder may decode only a certain number of wrong bits. The maximum error rate is given in Table 10.10. The sensitivity measurement is divided into two different types of tests with

Table 10.10
Maximum Bit Error Rates for the Sensitivity Measurement

Channel Type	*BER Type*	*Multipath Propagation Characteristic*			
		Statistical (%)	*TU50(%)*	*RA250(%)*	*HT100(%)*
SDCCH	FER	0.10	13.0	8.0	12.0
RACH	FER	0.50	13.0	12.0	13.0
TCH/F9.6	BER	0.001	0.50	0.10	0.70
TCH/F4.8	BER		0.01	0.01	0.01
TCH/F2.4	BER		0.02	0.001	0.001
TCH/H4.8	BER	0.001	0.50	0.10	0.70
TCH/H2.4	BER		0.02	0.01	0.01
TCH/FS	FER	$0.10 \cdot \alpha$	$6.0 \cdot \alpha$	$2.0 \cdot \alpha$	$7.0 \cdot \alpha$
Class Ib	RBER	$0.40/\alpha$	$0.40/\alpha$	$0.20/\alpha$	$0.50/\alpha$
Class II	RBER	2.0	8.0	7.0	9.0
TCH/HS	t.b.d.	t.b.d.	t.b.d.	t.b.d.	t.b.d.

different propagation conditions: (1) *static propagation* and (2) *multipath propagation.*

The static reference sensitivity level specifies the behavior of the receiver with the active time slot at a power level of –104 dBm and the two adjacent time slots at a 30-dB higher power level. This test setup is similar to the *adjacent time slot rejection measurement* for a mobile station in the previous chapter, with the difference that the requirements here are more stringent.

The multipath reference sensitivity level specifies the behavior of the receiver under multipath propagation conditions. Different speed and environmental conditions are evaluated and entered into the fading simulator. The applicable profiles and error rates are indicated in Table 10.10. The only control channel listed in this table is the SDCCH. The performance for the other control channels such as the BCCH, the AGCH, and the PCH is the same as for the SDCCH. The performance of the FACCH and the SACCH is expected to be better than that of the SDCCH. Therefore, the same requirements apply for these signaling channels as for the SDCCH. The α-value used in the table shall be in the range of 1 to 1.6, but shall be the same value within one propagation characteristic. The α-value is derived from the same definition as that explained in Section 9.4.1.

10.4.2.4 Reference Interference Level

In a real GSM environment, a base station does not only have to cope with weak signals but also with stronger interfering signals at certain frequency offsets from the carrier. The capability of the receiver to maintain communication with a mobile station even if such an interfering signal is present is expressed by the reference interference level. The signal level of the desired signal remains at –85 dBm for the whole test. Depending on the offset from the frequency of the desired signal, two different principal measurements are performed.

- The *cochannel rejection* measures the impact of an interferer at the same frequency but at a power level 9 dB below the effective signal.
- The *adjacent channel rejection* measures the impact of an interfering signal on one of the adjacent (frequency) channels. The power level used depends on the offset. A one-channel offset (200 kHz) means the interfering signal is transmitted at a level 9 dB above that of the desired on-channel signal, and a two-channel offset (400 kHz) means that the signal is transmitted at a level 41 dB above the desired signal.

The frequency-hopping option is switched off during this test. One exception is the test using the typical urban propagation characteristic at 3 km/h (TU3). The interfering signal uses a PRBS-modulated signal transmitted along with the on-channel signal. The maximum error rates that may occur are indicated in Table

10.11. The α-value used in the table shall again be in the range of 1 to 1.6, but shall be the same value within one propagation characteristic.

10.4.2.5 Blocking and Spurious Response Rejection

One test setup here covers two aspects at the same time: *blocking* and *spurious response rejection*, which are measured one after another. For both measurements, there is only a difference in the test requirements, not in the procedure itself. Since the spurious response rejection is a more relaxed requirement, this test is performed first.

Blocking and spurious response rejection is a quality indication of the base station's receiver that, even in the presence of a very strong interfering signal, it is still able to maintain the quality of the radio link to the mobile station. A practical justification for this measurement is that GSM antennas may also be located close to the antennas of TV or radio broadcast stations, which transmit at much higher power levels than the typical GSM BTS does.

The two signals applied to the receiver are the normal desired GMSK-modulated signal and the interfering signal, which is frequency-modulated with a 2-kHz tone using a deviation of 100 kHz. During the execution of this test, frequency hopping is switched off. Preferably, only one frequency located in the center of the receiving band should be used. For performance evaluation, the residual BER of class II (RBER II) on the traffic channel is measured.

Table 10.11
Maximum Bit Error Rates for the Reference Interference Level Measurement

Channel Type	*BER Type*	*Multipath Propagation Characteristic*			
		TU3 Without SFH (%)	*TU3 With SFH (%)*	*TU50(%)*	*RA250(%)*
SDCCH	FER	22.0	9.0	13.0	8.0
RACH	FER	15.0	15.0	16.0	13.0
TCH/F9.6	BER	8.0	0.30	0.80	0.20
TCH/F4.8	BER	3.0	0.01	0.01	0.01
TCH/F2.4	BER	3.0	0.001	0.001	0.001
TCH/H4.8	BER	8.0	0.3	0.8	0.2
TCH/H2.4	BER	4.0	0.01	0.01	0.01
TCH/FS	FER	$21 \cdot \alpha$	$3.0 \cdot \alpha$	$6.0 \cdot \alpha$	$3.0 \cdot \alpha$
Class Ib	RBER	$2.0/\alpha$	$0.20/\alpha$	$0.40/\alpha$	$0.20/\alpha$
Class II	RBER	4.0	8.0	8.0	8.0
TCH/HS	t.b.d.	t.b.d.	t.b.d.	t.b.d.	t.b.d.

10.4.2.6 Intermodulation Attenuation

The frequency hopping is switched off for this measurement. The three frequencies in the bottom, middle, and top ranges of the frequency band are investigated separately and the desired signal uses these frequencies one after another. The two interfering signals are a GMSK-modulated signal carrying PRBS information and an unmodulated carrier. The modulated signal is operated at a frequency offset of eight channels (1,600 kHz) and the unmodulated signal four channels (800 kHz) above the desired signal. The level of the desired signal shall be at 12 dBμV (–101 dBm) and that of the interfering signals at 70 dBμV (–43 dBm). The interfering signals shall be located below the desired signal with the same offset in the second test run. As a criterion for fulfilling the test requirements, the BER values for the statistic propagation in Table 10.10 apply.

When a test system provides two signals, one as the desired on-channel signal and the other as the interfering signal, then an additional third signal from a signal generator has to be added to the RF path. This can be done through a *coupler* or a *combining network*. Figure 10.15 shows the principle for this kind of test setup.

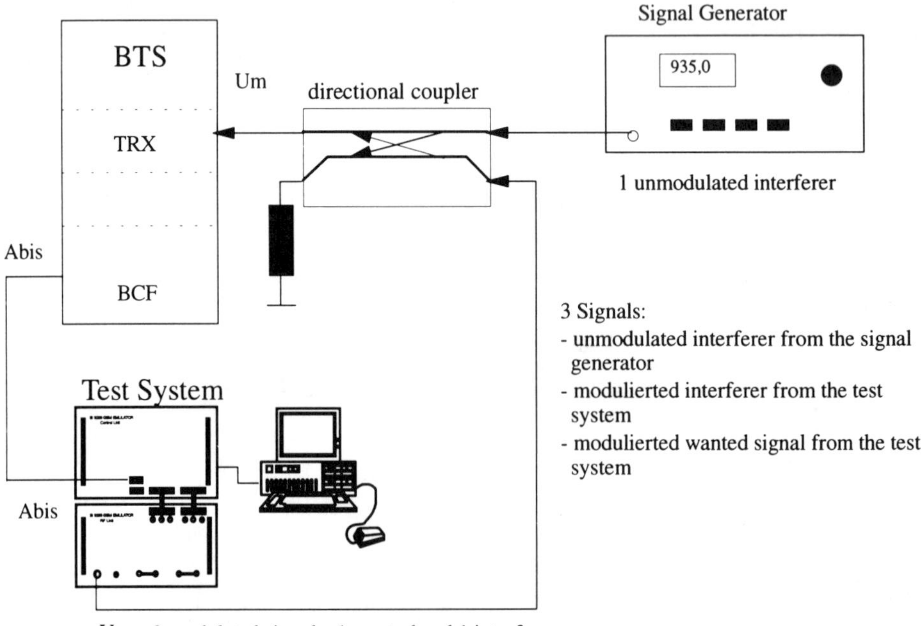

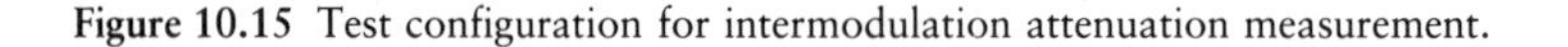

Figure 10.15 Test configuration for intermodulation attenuation measurement.

10.4.2.7 Spurious Emissions From the Receivers

After the spurious emissions have been measured for the transmitter, the test is repeated for the receiver. We must make sure that no signal originating from internal frequency sources is radiated via the receiving antennas to the outside world. The base station shall be operated in the frequency-hopping mode, but the frequency of the base channel (BCCH) shall be switched off. No signal shall be applied to the receiver.

The radiated power is again measured at the three locations already specified for the spurious emissions from the transmitters (i.e., antenna output, cabinet radiation, and the power leads). The setup of the spectrum analyzer is similar to the one specified in Table 10.5. The requirements valid for this measurement are indicated in Table 10.12.

10.4.3 Radio Link Management Tests

As was the case with the mobile stations, the base station must meet specific requirements to maintain a usable radio link with the mobile station. This control functionality in the BTS is the task of the *radio link management* entity, which is located in the control unit.

10.4.3.1 Timing Tolerance

The *timing tolerance* test measures the timing relationships among different transmitters within one base station. There should be no timing offset among the transmitters, since this would cause problems when assigning a channel from one transceiver to another. The timing reference is the BCCH's frequency.

To verify the timing tolerance, the test system sets up a traffic channel on an arbitrary frequency in time slot 0. Then the relative timing of the training sequence

Table 10.12
Maximum Level of the Spurious Emissions From the Receivers

Measuring Position	*Frequency Band*	*Maximum Level*
Power leads	9 kHz–10 MHz	72 dBμV(pd)
	10 MHz–30 MHz	72 dBμV(pd)
Cabinet radiation	30 MHz–1,000 MHz	–57 dBm
	1 GHz–12.75 GHz	–47 dBm
Antenna	9 kHz–1,000 MHz	–57 dBm
	1 GHz–12.75 GHz	–47 dBm

between the traffic channel and the other channels on the base frequency is measured. Since the base channel also includes an SDCCH, SACCH, or CCCH, any of these may be used for evaluating the reference. Frames with either an FCCH or an SCH are not used for this measurement because they do not contain the training sequence.

After this first test, the traffic channel is established consecutively on the three channels on the bottom, middle, and top frequencies of the GSM band. The frequency of the base channel is not used in the evaluation. The complete measurement shall be repeated with the base channel as one of the three frequencies (B, M, T). This sounds like a huge task, but as many frequency combinations as possible should be used to make sure that there is really no frequency offset (and thus timing offset) between any two carriers within one base station. The maximum allowed timing tolerance is one-quarter bit.

10.4.3.2 BCCH Multiframe

The first thing a mobile station looks for within a network is the FCCH; then it looks for the SCH, and finally the BCCH with the system information. We must make sure, therefore, that these channels are located within their correct frames (within the signaling (51) multiframe). The test system acts like a mobile when it first looks for the FCCH, which must be located in the following frames: 0, 10, 20, 30, and 40 of the 51-multiframe structure.

The SCH must be located on the subsequent frames with the numbers 1, 11, 21, 31, and 41. This is verified with the test equipment. From the SCH, the reduced frame number, indicating the current position (timing) within the 51-multiframe, and the training sequence code can be extracted. With this valuable information, it is possible for a mobile station, as well as a test set, to decode the system information in the BCCH.

10.4.3.3 Radio Link Measurements

Just as the mobile station performs measurements on the received signal (the active channel and adjacent base stations), the base station also has to perform such measurements on the radio link. The base station continuously receives measurement reports from the mobile station, which are sent to the BSC along with measurements performed by the base station itself. From the radio link measurements carried out by the mobile station and the base station, the BSC decides, among other things, when is the best time for a handover and to which neighboring cell the handover should be be carried out. These measurements are characterized by two criteria: (1) *signal strength* and (2) *signal quality*.

The signal strength is referred to in the measurement report as RX-LEV and indicates the level of the incoming signal. The base station has to measure this level with a certain accuracy. There are 64 different values defined for the signal strength variable indicating power levels from −110 dBm to −48 dBm, each of which has to be verified. The test system transmits the RF signal at specific levels in 1-dB steps to the receiver, which evaluates this level with a RX-LEV value.

The measured values over a 20-dB range, somewhere between −110 and −48 dBm, must not deviate more than ±1 dB from the regression line over the same range. This is an indication of the relative accuracy; the absolute accuracy shall be better than ±4 dB over the whole range. The relationship between RX-LEV and the absolute power levels is shown in Table 10.13.

The signal quality is indicated in the *measurement report* as RX-QUAL. The quality of a link is calculated from an internal BER measurement within the decoder on the active logical channel (either a traffic channel or an SDCCH). This internal BER measurement should not be confused with the receiver measurement, since it is calculated purely from criterion within the decoder itself. The value indicates the quality of the radio link, since the decoder would have to correct more bits on an inferior link. The RX-QUAL value is divided into eight steps, each of which corresponds to a certain BER. Table 10.14 shows the relationship between RX-QUAL and the BER values.

10.4.3.4 Continuous Frame Adaptation

A base station continuously measures the reception time of a burst coming from a mobile station. It does this to make sure that the burst timing does not cross the time slot limits due to movement of the mobile station. The base station has to continuously monitor the timing and eventually command the mobile station to use different timing advance values, which appear in the SACCH. This procedure is referred to as *adaptive frame alignment*. The base station's capability to handle this

Table 10.13
Relationship Between RX-LEV and the Absolute Power Levels

RX-LEV	*Absolute Power Level (dBm)*
0	Smaller than −110
1	−110 up to −109
2	−109 up to −108
3	−108 up to −107
.	.
.	.
62	−49 up to −48
63	Greater than −48

Table 10.14
Relationship Between RX-QUAL and the BER Ranges

RX-QUAL	*BER Range (%)*
0	<0.2
1	0.2 up to 0.4
2	0.4 up to 0.8
3	0.8 up to 1.6
4	1.6 up to 3.2
5	3.2 up to 6.4
6	6.4 up to 12.8
7	> 12.8

should be evaluated with this test. Two moving mobile stations, which are moving in opposite directions, have to be simulated for this test. The first mobile simulates driving toward the base station and the second simulates driving away from the base station. The base station has to measure the timing continuously and assign new timing advance values to the mobile stations. The values that have been assigned and the values used by the simulated mobile stations are registered. The difference between both values (i.e., the timing error on the Um interface) shall not exceed ±1 bit.

10.4.4 Layer 2 Signaling Tests (LAPDm)

The *Layer 2 signaling* tests verify the correct information transfer over the air (Um) interface. In addition to this test, erroneous cases are performed which offend the LAPDm protocol on purpose and force the base station to recover from these situations. Tests that are performed for the Layer 2 signaling were named in Section 8.5. In Figure 10.16, a trace can be seen that shows such a test, as seen on the test equipment's display. The evaluation of such a test is usually done within the test equipment itself. The sequences are, furthermore, usually predefined. The user simply has to start these tests from some kind of menu. The example shown here simulates the loss of a BTS-originated UA frame. This means that the test equipment repeats the SABM frame once more after the expiration of the Layer 2 timer T200. The base station responds once more with a UA frame, and the link can then be established correctly. The verification of the correct behavior of the base station is completed with the repetition of the UA frame. The test stops after this UA frame. Comments are added in italics to the trace.

frame number	Um uplink (test system → BTS)	Um downlink (test system ← BTS)
FN 0000825 0000:19:09	Umul1 rach Info = E1	*The test system requests a channel*
FN 0000838 0000:06:22	*The base station assigns a channel*	Umdl1 agch Length = 0C, M = 0, EL = 1 Info = 06 3F 00 43 00 01 E1 01 33 00 01 00 2B 2B 2B 2B 2B 2B 2B 2B 2B
FN 0000882 0000:24:15	Umul1 sdcch Address: SAPI = 0, Command Control: SABM: N(R)= -, P/F= 1, N(S)= - Length = 10, M = 0, EL = 1 Info = 05 24 01 03 00 00 00 08 19 32 54 76 98 10 32 54 FF FF FF FF	*The test system wants to establish the layer 2 link (acknowledged mode)*
FN 0000918 0000:08:00	*The base station acknowledges the establishment of the L2 connection with a UA frame*	dl1 sdcch Adress: SAPI = 0, Response Control: UA : N(R) = -, P/F = 1, N(S) = - Length = 10, M = 0, EL = 1 Info = 05 24 01 03 00 00 00 08 19 32 54 76 98 10 32 54 2B 2B 2B 2B
FN 0000933 0000:23:15	Umul1 sdcch Address: SAPI = 0, Command Control: SABM: N(R)= -, P/F= 1, N(S)= - Length = 10, M = 0, EL = 1 Info = 05 24 01 03 00 00 00 08 19 32 54 76 98 10 32 54 FF FF FF FF	*The UA frame seems not to be received by the test system which therefore transmits the SABM once more*
FN 0000969 0000:07:00	*The BTS transmits once more a UA frame which had to be checked* Q.E.D.	Umdl1 sdcch Address: SAPI = 0, Response Control: UA : N(R) = -, P/F = 1, N(S) = - Length = 10, M = 0, EL = 1 Info = 05 24 01 03 00 00 00 08 19 32 54 76 98 10 32 54 2B 2B 2B 2B

Figure 10.16 Trace of a Layer 2 signaling test.

10.4.5 Network Function Tests

Only certain aspects of the air (Um) interface have been considered in the previous sections. But obviously a base station must also perform other tasks in order for the network to function. Such network tasks are evaluated by the *network function tests*. These tests have to be performed on both the Um and the Abis interface to verify the correct realization of the commands. These tasks include the following:

- *Radio link layer management* controls the establishment and clearing of a radio link, the reception and transmission of transparent messages in the acknowledged and unacknowledged modes, and the detection of errors on the air interface. In the transparent mode, the base station has to simply pass the messages from the Um to the Abis interface (and the other way around) without any kind of interpretation.
- *Dedicated channel management* controls channel activation and modification, the start of ciphering, passing on measurement reports, deactivation of the SACCH, the release of a channel, and the power control of a mobile station.
- *Common channel management* takes over the detection of channel requests from the mobile station, the paging of a mobile station, the channel assignment, the changing of the system information, and the transmission of the CBCH.
- *Transceiver management* controls the reporting of the frequency resource (RF interference level) to the BSC, the altering of the fill information on the SACCH, the flow control to indicate overload in the BSC, and error reports on erroneous messages from the BSC.

These tests are very much like the simpler signaling tests we saw for the mobile station. Unlike the mobile situation, however, they have to be performed on the Um and Abis interface, which is to say that messages have to be verified and exchanged over both interfaces. The tests themselves will not be discussed further except to offer the example in Figure 10.17, which shows how such a test appears on the tester. In this test, a transparent message is passed from the Abis interface to the Um interface (and further on to the mobile station).

The test system simulates both the BSC and the mobile station, thus "clamping" around the BTS with both ports (Um and Abis) in the tester's "jaws." This trace shows very clearly how the mobile station (in this case the test system) transmits the RACH and how this results in a channel-required message from the base station. Then follows the channel activation procedure and the immediate assignment.

10.4.6 Additional Tests on Base Stations

The most important tests on base stations have been described in this chapter. There are still additional functions related to a base station which have to be verified. These are mentioned here, but not treated in detail.

1. The Abis interface provides an internal interface to identify different internal units (TEIs). The TEIs allow us to address different entities on the same Abis time slot within a BTS (e.g., different transceivers).
2. For the speech codec, the different directions (to and from the mobile station) and the *rate adaptation* from 13 to 64 kbps (when the codec is located within the BTS) or from 16 to 64 kbps (when located in the BSC) have to be verified and tested.

frame	Umul	Umdl	Abisul	Abisdl
number	(test system)	(base station)		(test system)
FN:2651595: 1999:11:03	Umul1,rach, TS(0),TI(0),pd(rr), chan_req, chan_req(= 7h,01h);		*channel request from the* *mobile test system*	
FN:2651599: 1999:15:07	*Notification of the* *reception of a RACH* *to the BSC*		Abisul1,md(COM), ch_rqd, ch_no(= 10001i,0h), req_ref(= E1h,0Fh,03h,0Bh), acc_del(= 03h);	
FN:2651606: 1999:22:14	*Channel activation* *from the BSC*			Abisdl1,l2_ack,md(DED), ch_act, ch_no(= 01000i,1h), act_typ(= 000i), ch_mode(= 0i,0i,3h,01h,00h), ch_ident(= 08h,1h,0h,0i,0h,64h ms_pow(= 06h), tim_adv(= 00h);
FN:2651612: 1999:02:20	*Acknowledgment of* *the channel activation* *from the BTS*		Abisul1,md(DED), ch_act_ack, ch_no(= 01000i,1h), frm_numb(= 0Fh,10h,03h);	
FN:2651621: 1999:11:29 1999:11:29	*Only now follows the channel* *assignment from the BSC to* *the BTS and later on to the MS* *Insertion of the transparent* *imm_ass message*			Abisdl1,l2_ack,md(COM), imm_ass_cmd, ch_no(= 10010i,0h), imm_ass_info(= /*Embedded UM: */ TI(0),pd(rr), imm_ass, pag_mod(= 3h), chan_desc(= 08h,1h,0h,0i,0h,64h), req_ref(= FFh,1Fh,3Fh,1Fh), tim_adv(= 00h), mob_alloc(=)
FN:2651649: 1999:13:06		Umdl1,agch, TS(0),TI(0),pd(rr), imm_ass, pag_mod(= 3h), chan_desc(= 08h,1h,0h,0i,0h,64h), req_ref(= FFh,1Fh,3Fh,1Fh), tim_adv(= 00h), mob_alloc(=)		*Passing on of the imm_ass* *message to the mobile station*

Figure 10.17 Network function test: transfer of a transparent message.

3. The speech delay through the base station is another interesting aspect worth testing. Each entity within the GSM infrastructure contributes to the *speech delay*. For this reason, the contribution to the delay of each individual entity has to be limited to a specific value. If the total amount of delay passes beyond 200 ms, it annoys the user. An example of such an excessive delay is a long-distance phone call from one continent to another through a satellite, which already requires some discipline from the callers not to interrupt each other.

FN:2651709: Umul1,sdcch(0), TS(1),TI(0),pd(mm),
1999:21:15 cm_serv_req,
cm_serv(= 1h),
ciph_key_seq(= 6h),
mob_class2(= 0h,0h,0h,0i,0h,00h),
mob_id(= 4h,0i,00000000h);

mobile station notifies assigned channel of its wish

FN:2651720: *The BTS passes this wish on to the BSC (embedded)*
1999:06:26

Abisul1,md(RLL),
est_ind, ch_no(= 01000i,1h),
link_id(= 0h,0h,0h),
L3_info(= /* Embedded UM message */
TI(0),pd(mm),
1999:06:26
cm_serv_req,
cm_serv(= 1h),
ciph_key_seq(= 6h),
mob_class2(= 0h,0h,0h,0i,0h,00h),
mob_id(= 4h,0i,00000000h)

FN:2651728: *Following this the BSC transmits an embedded message of the MMlayer*
1999:14:34

Abisdl1,l2_ack,md(RLL),
data_req, ch_no(= 01000i,1h),
link_id(= 0h,0h,0h),
L3_info(= /* Embedded UM msg */
1999:14:34
TI(0),pd(mm), auth_req,
ciph_key_seq(= 6h),
rand(=
00000000000000000000000000h)
);

The BTS passes exactly this message on which is the requirement for passing this test:

FN:2651796:
1999:04:00

Umdl1,sdcch(0), TS(1),
TI(0),pd(mm), auth_req,
ciph_key_seq(= 6h),
rand(=
00000000000000000000000000h);

Figure 10.17 (continued).

REFERENCES

[1] GSM TS 8.54, "BSC-TRX Layer 1: Structure of Physical Circuits."
[2] GSM TS 8.56, "BSC-BTS Layer 2 Specification."
[3] CCITT Rec. G703.
[4] CEPT Rec. T/S 46-20.
[5] CCITT Rec. G711.
[6] GSM TS 8.58, "BSC-BTS Layer 3 Specification."

CHAPTER 11

APPLICATIONS FOR GSM TEST EQUIPMENT

The manufacturers, network operators, service operations, and regulatory bodies making up the whole GSM industry have very different and specific requirements for GSM test equipment. The most important aim is to satisfy the cellular customer with the best quality of service achievable. The quality of service cannot be guaranteed without the use of appropriate tools that help the industry to develop, maintain, and improve its products and quality standards.

With the introduction of the GSM services and the accompanying new technologies, application-tailored test equipment has been brought to the market. First-generation products for network and radio tests sometimes did not advance beyond the prototype stages. Sometimes conventional laboratory, and thus already commercially available equipment, could be used and often was used with different degrees of success.

The initial need was for test equipment in the research and design laboratories. Then type approval equipment, especially for mobile phones, was in demand, and finally high-volume manufacturing was a major issue. Eventually, with the start of commercial service in some countries, the industry also asked for tools applicable for service and maintenance functions.

The products offered by the radio test equipment industry are targeted toward five major applications:

1. Research and development, and design of mobile stations and BTSs;
2. Type approval and acceptance testing;

3. Manufacturing and quality assurance;
4. Service and maintenance;
5. Network planning.

The following chapters will deal with the specific applications and the technical requirements imposed on test equipment by its various applications.

11.1 DESIGN LABORATORY

When designing mobile terminals or base stations, design engineers work in two fields: hardware design and software design. Both fields require specific test gear that allows users to verify their designs and make them compliant with the specifications.

11.1.1 Design of Hardware and Analog (Radio) Components

During the development of hardware components such as RF sections (e.g., modulators, demodulators, power stages, receivers, filters), digital and analog modules, as well as application-specific semiconductor solutions (ASICs), test equipment is used to verify single functions against physical processes, the proper interface among different components, and support of the integration effort. Particularly in the case of hand-portable phones, many functional entities (speech codec, equalizer, channel codec, modulator) are realized with very-large-scale integrated circuits (VLSI), where all the functions are often squeezed onto a single chip.

The test and measurement equipment that can be found in design laboratories includes signal generators, spectrum analyzers, measuring receivers, digital oscilloscopes, logic analyzers, audio analyzers, audio generators, modulation analyzers, and fading simulators. The instruments are usually very general in their use and often lack any specific functions peculiar to specific communications system. Commercial test instruments specific to a protocol or system are usually not found in labs, but things are changing with digital radio and complicated systems.

GSM-specific radio test equipment finally has its place on the designer's test bench. In order to cope with new signal forms, modulation schemes, bandwidth constraints, dynamic range specifications, pulsed transmission, and the coding of data, GSM test systems for the radio engineer differ significantly from the conventional equipment seen in labs before GSM.

11.1.2 Design of Software

The enormous impact of the GSM signaling environment on the amount of software employed in any GSM entity caused design engineers to change the meaning of the three letters in GSM as a remedy for relentless stress. For some time, GSM stood for "great software monster" as a testament to the revolutionary efforts that had to be expended in software design. Software design engineers need test systems with which they can work comfortably and reliably, and which they can learn how to operate quickly. Especially for the implementation and verification of Layer 2 and Layer 3 protocols, there have to be tools that easily simulate any signaling environment or sequence. Powerful editors and trace functions, preferably with software and hardware trigger functions, have to be ready to support the engineering work.

In case there is no working Layer 1 (complete coding and radio link) available, the existence of digital interfaces for baseband protocol development and verification is useful. Another key need is the verification of hardware control software and inherent functions.

11.1.3 GSM Test Equipment for the Design Laboratory

The design application requires a multitude of functions with flexible access to all the relevant key parameters found in the *GSM Technical Specifications*. Testing power and flexibility tend to be mutually exclusive. The more power a tester and the more complete its simulation of a system, the less flexible it tends to be unless a great deal of money is added to its cost to include all the hooks and flexibility the R&D functions will need. For the development of GSM mobile stations, a test system should be capable of at least simulating two GSM cells (base stations) and perform all relevant Layer 2 and Layer 3 signaling procedures with an easy-to-use editor and trace functions. In addition, it must comply with GSM-specific radio transmissions and be able to test all the radio parameters. The parallel use of additional equipment, such as spectrum analyzers and fading simulators, should be possible without creating a whole new project within the design staff. Additional test features, which are not defined in the *GSM Technical Specifications*, have proven to be useful for design engineers: trace of Layer 1 (burst content and other relevant information like coding errors and timing), baseband signaling tests, display of measurement report data, and RF tests without signaling.

GSM BTSs are transparent to most of the Layer 3 signaling messages, and thus the signaling tests are not as extensive when compared to those on mobile phones. Of course, the transparency (the correct mapping of signaling between the network side and the radio) has to be verified one direction at a time. Still, the layer that transports the Layer 3 data, the data link layer, needs some attention. Thus, a GSM base station tester needs signaling test capabilities to verify the LAPDm functions, along with the standard radio test features. In addition, an integrated Abis interface tester within the GSM radio test system allows intensive design verification. Figure 11.1 shows a GSM test system for the design of mobile phones. Such a test system typically uses a number of microprocessors and DSPs to cope with the tasks of transmitting and receiving signals and performing measurements in an accurate way.

Figure 11.1 GSM test system for the design laboratory.

11.2 TYPE APPROVAL AND ACCEPTANCE TEST

Type approval and acceptance testing is an obscure function often ignored by many system engineers, since it is often buried in tedium and vast documentation exercises.

11.2.1 Type Approval for GSM Mobile Stations

The concept of approval of mobile terminals is based on its acceptance by all the MoU signatories. It was agreed that a type of mobile station that has passed the approval procedures in one MoU country is automatically approved for all the other MoU countries. This only works if the procedures are very precise and clear, and are not open to broad interpretations.

For the type approval of GSM mobile stations, a test system has been specified by the regulatory bodies, the so-called *system simulator* [1]. The system simulator is capable of performing all tests and measurements as described in the "GSM Mobile Station Conformity Specification 11.10" [2]. The tests include intensive RF measurements, signaling tests, audio tests, and speech codec tests.

Because of the complexity of the subject and the incomplete nature of the test specifications, there was, and still is, a continuing process of maintenance and definition of the test procedures. Due to this flow of specifications, and also because the enormous efforts that were needed to meet the challenges were underestimated, the completion of the whole system simulator and the test scenarios (test cases necessary to run the tests) was constantly delayed. It soon became obvious that a full type approval (FTA) would not be possible in time for the start of network services, for which approved phones were necessary. So the MoU participants decided to perform an interim type approval (ITA) using a subset of tests with less stringent requirements for the test equipment. The networks "opened" with ITA-approved terminals. Now that FTA capabilities are available, manufacturers have to achieve this approval for their products, so a number of terminal types must be retested or removed from the market.

11.2.2 Type Approval for GSM Base Transceiver Stations

Things are a bit different in the area of GSM base stations. The approval or admission of base stations was kept a national matter from the start. Still, ETSI developed and issued a test specification for GSM BTSs [3]. This test standard is the general basis for the commissioning and acceptance of base station equipment among the manufacturers, operators, and regulatory bodies. In a similar way as was done for the specification of mobile tests, the BTS procedures describe measurements and procedures, as well as measurement limits and requirements. Besides the verification of radio performance and radio-related parameters, Layer 2 tests and network function tests are laid down in this standard.

In order to support the recommended tests, the test system used for GSM base station verification has to incorporate six functions: (1) a GSM radio test set including an Abis simulator/tester for the GSM-specific radio parameters, (2) two GSM signal sources, (3) an additional signal generator (interferer), (4) a spectrum analyzer, (5) a filter network and combining network, and (6) a fading simulator. Figure 11.2 shows a block diagram of such a test system and Figure 11.3 the corresponding test equipment setup.

11.3 PRODUCTION TEST

The manufacturing process for both mobile phones and base stations is always accompanied by test and measurement procedures. Test equipment is used to verify the functions of the products or the parts of the products at different stages in their manufacture. Faulty components have to be detected at an early stage in the process, and measurements are also necessary for adjustments of the product. After the product is finally assembled, a kind of end test is often performed in order to check the basic operations, the operations that include as many individual functions as possible.

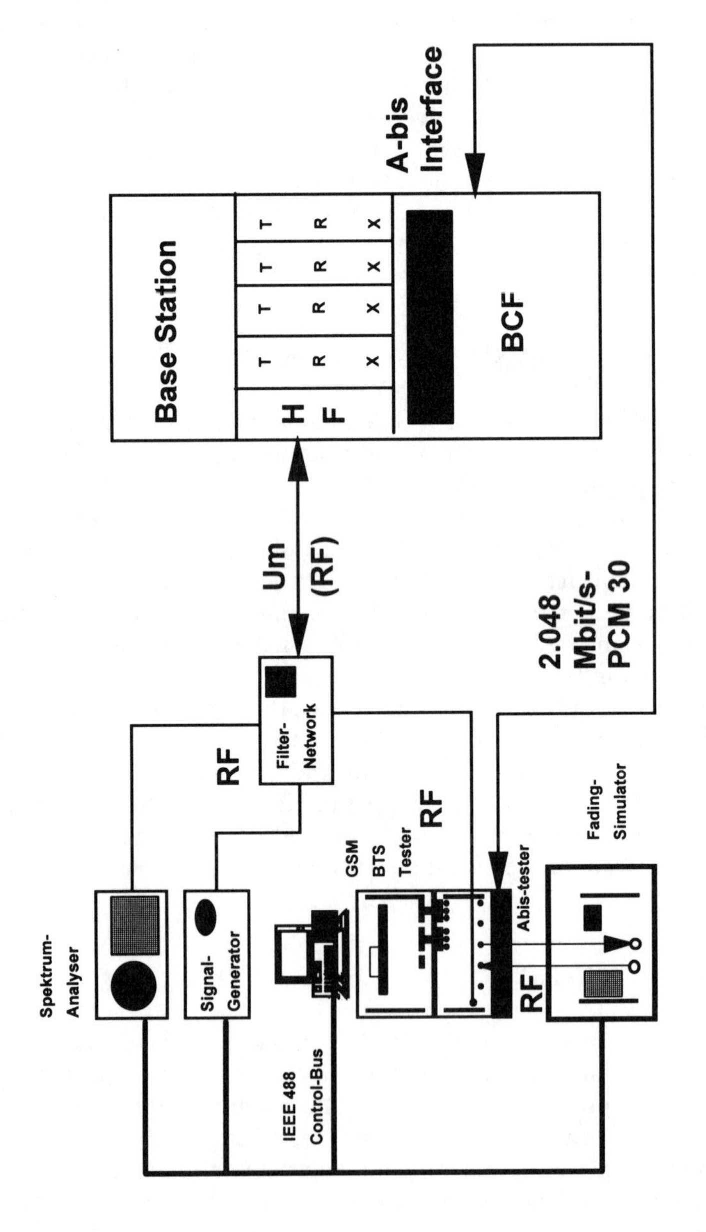

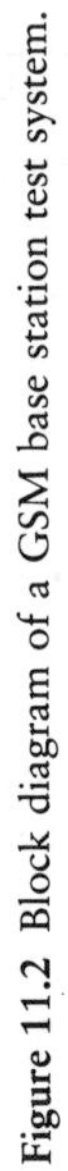
Figure 11.2 Block diagram of a GSM base station test system.

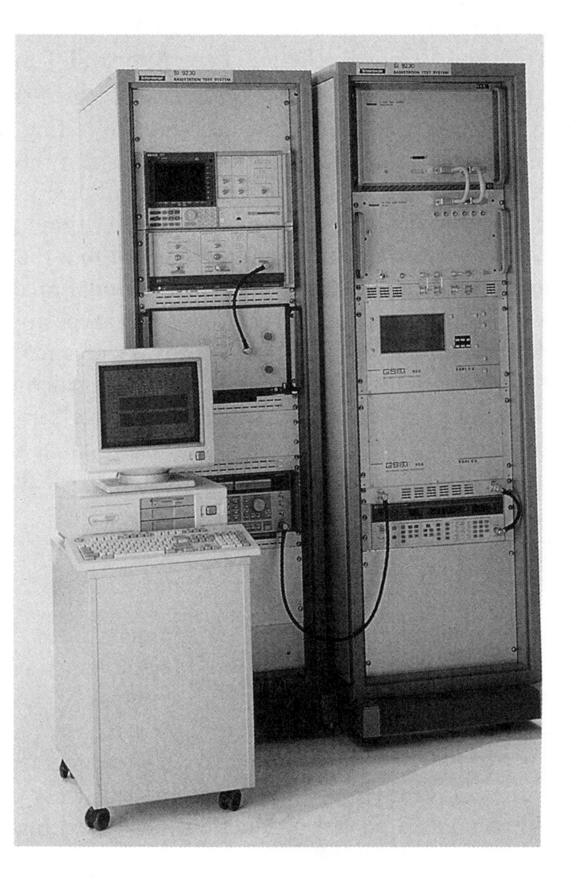

Figure 11.3 GSM test system for base station approval.

Besides the testing of components or modules, where board testers and incircuit testers are used, a production line for GSM radios cannot exist without appropriate test equipment. The production environment has its own demands. The following eight requirements for the testers apply: (1) transmission of analog and digital modulated signals and measurements of all relevant radio parameters, (2) high measurement rate, (3) only *sufficient* measurement accuracy (too much accuracy increases test time and costs money), (4) basic signaling capabilities (e.g., set up a call and change radio parameters during the call), (5) remote control of all functions (integration in an automated test line with a network of test equipment, controllers, and peripherals, (6) fast delivery of measurement results to the host or controller, (7) flexible software structure that allows frequent modifications, and (8) high reliability to ensure low downtime.

11.3.1 Production Testing Versus R&D and Type Approval Testing

We can construct a hierarchy of testing where production testing stands at the top of a list if we confine our consideration to testing speed and the precision of test procedures (not necessarily the precision of the tests themselves). At the other end of the spectrum, at the bottom of the list, we see the R&D applications, which call for almost the same tests performed in production, but without the strict procedures and great speed requirements; flexibility and some additional precision in the measurements are more important. It is interesting to note that type approval is a hybrid of production and R&D applications. Type approval demands almost as much flexibility as the R&D functions need, but they also need the structure and simulation details production testing requires (lack of flexibility). The service and network planning applications are in a world all their own, with peculiar requirements that seem to have nothing to do with the applications treated so far in this chapter.

11.4 SERVICE AND MAINTENANCE

Design, production, and maintenance of GSM equipment, both the terminals and the infrastructure, are based on the most advanced technologies available. State-of-the-art manufacturing processes now put out high volumes of mobile terminals of extraordinarily high quality. Still, there are equipment failures and breakdowns. Malfunctions in, for example, a mobile phone are due to undetected problems in the manufacturing process, dead components, or physical stress (shock and heat). Such breakdowns often occur after the product has been used for a period of time to the great dissatisfaction of the user (in many cases, right after expiration of the warranty!). Certainly, these kinds of systematic problems in the manufacturing processes that cause failures in the customer's hands are steadily tracked down in order to approach zero defect quality as the problems are resolved.

The service technician who has to take care of failures and complaints after the product has been sold and used by a customer has to discover the symptoms and causes of failures quickly and at a very low cost. The device has to be checked, repaired, and finally its functions have to be guaranteed again.

The fact that an excellent level of service can be provided even if problems occur is considered a benefit justifying the right choice of technology, product, and service provided to the cellular customer. In a universe of similar phones, when a purchase is considered, the prudent customer often looks to the manufacturer who can provide high-quality service. This advantage, this added value, can only be guaranteed if manufacturers, service technicians, radio retail shops, network operators, and service providers have the proper test equipment in their service shops.

The GSM test equipment must cater to all the measurements and tests necessary to track down defects and to calibrate and qualify the device. The following section gives a detailed view of practical service testing of GSM mobile stations. Because this subject can be described in a similar way for the majority of manufacturers and their products, it is treated with more detail than the other subjects in this chapter. Advanced after-market service is a real challenge and will be one of the new arenas of competition in digital cellular. Though base station service does not have the competitive consequences terminal repair does, a paragraph on the service and maintenance of GSM base stations is added to complete the presentation. The universe of parameters that should be tested in service applications is much smaller than in the other applications explored earlier in this chapter. The service industry has to constantly check to make sure it is not testing too many software functions and design constraints (they should seldom be tested) at the expense of making sure radios are returned to the best level of service at the least cost.

11.4.1 Service of GSM Mobile Terminals

Testing GSM terminals employs methods that differ from conventional analog radio testing. Digital voice processing means that there are no analog audio signals (tones, 1-kHz sine waves, SINAD tests) which were so useful for stimulating and measuring the radio part in analog phones (Figure 11.4).

There are a number of tests that check the RF performance of a mobile station. New test methods have been introduced to cope with the new digital modulation

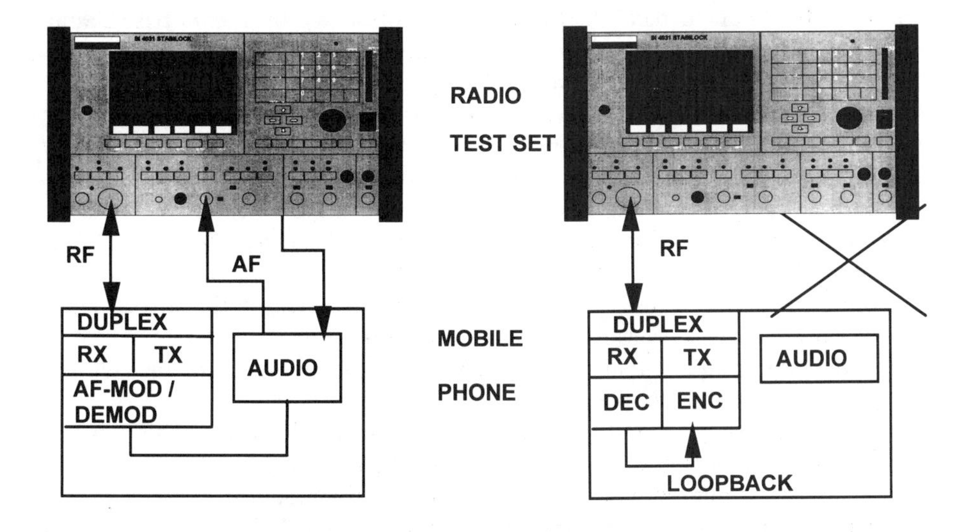

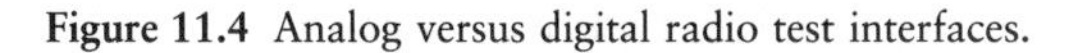

Figure 11.4 Analog versus digital radio test interfaces.

methods and pulsed signal transmission. Some of the checks do not particularly require measurements, but merely need some kind of environment in which to simulate a live network in most of its details. Manufacturers and design engineers also provide *test modes* and service interfaces for their particular models to ease in troubleshooting and performing the necessary adjustments (electronic calibration and tuning). Real and more traditional radio measurements must be performed to check the mobile phone's modulator and output power and to support the necessary adjustments and alignments.

11.4.1.1 Checking a Broken Mobile Station

What can be checked? When a mobile station is out of order or not performing well, a number of sources for the failures have to be considered. The following tentative list indicates the checks that can be carried out, starting with basic checks that do not, at first, require measurements. This checklist does not, however, claim to be complete and may have to be extended with different models of GSM terminals with different approaches for troubleshooting. The list is, rather, a comprehensive starting point for specific problem resolution, a way to think about broken phones.

1. Does the phone accept a SIM and PIN? (-> "INSERT SIM" and "ENTER PIN")
2. Does the mobile station start to search for a network? (-> "SEARCHING FOR NETWORK," etc.)
3. Does the phone detect a GSM network (represented by a *valid* base channel at an acceptable RF level (in this case, the test equipment simulates a base channel)? Does it indicate "service" at a reasonable RF level (e.g., –80 dBm) for the base channel? (Note: "service" is expected to be indicated by the phone when location parameters stored on the SIM card in use are *deliberately* made equal to those transmitted on the base channel.)
4. Is there a response to paging (the mobile station has to transmit access bursts)? (Note: The test equipment must page the mobile station with identification parameters stored on the inserted test SIM.)
5. Does the phone adjust its output power correctly (within the correct limits) as power control levels are entered into the test set?
6. Is the pulse shape (bursts, e.g., in response to paging) in compliance with the GSM power-time template?
7. Are the phase error and frequency error of the modulation within limits?
8. Can the mobile station set up a call with the test equipment?
9. Is the dialed number correctly transmitted to the network (test set)?
10. Does the mobile station correctly report the received signal level (measurement report)? Is the BER at a certain signal level within limits?

11. Does the audio entity work (e.g., speech loopback test)?
12. Can the mobile station correctly clear the call?

11.4.1.2 Receiver Test

When, for example, step 3 in the previous section is answered with a "no," this would indicate that there may be a problem in the receiver section of the phone under test. The applicable entity can be the analog RF/IF section or the digital part of the mobile station that is positioned further back in the system closer to the user and further from the air interface.

The receiver quality in analog radios is usually referred to as the SNR or the SINAD (see Chapter 8). The GSM recommendation specifies receiver sensitivity as an equivalent to a BER. This BER result is measured at a certain low RF level of the simulated BTS. A special test mode, traffic channel loop, has to be invoked via GSM signaling in the mobile station before starting the BER test. In this mode, the mobile station must receive, decode, encode, and retransmit (loopback, see Figure 11.4) the bit patterns sent from the test equipment. The test equipment compares the received and sent bits and calculates the bit error ratio.

For service applications, this method is not very practical. At relatively high signal levels (greater than –100 dBm), it is possible that no bit errors will be detected. The transition from good to bad is a rather sharp one, because the decoder employed in the GSM system provides an error correction mechanism that covers, or compensates, for disturbances and thus also for bad radio receiver performance to a considerable degree. Two other drawbacks of this method are the measurement's long duration (the higher the number of bits, the more accurate the results will be) and the fact that a call has to be established in order to test the BER on the traffic channel; the phone has to work rather well just to perform the test. The service technician who is familiar with testing an analog receiver looks for the RF level at which a certain SNR is detected or a certain SINAD value applies. A similar explicit search of the digital receiver's sensitivity limit (at what level the connection fails or the voice quality is unacceptable) has not been defined in the GSM standards. There are *alternative* approaches to test the receiver. The following suggestions describe three straightforward methods that can be applied to test the receiver.

Quick Test

A very simple approach is to check the mobile station's service indication in the so-called *idle updated* state. The test set transmits a base channel with synchronization and system information. Using a special test SIM card inserted in the phone with a location area identity (LAI) field which is equal to the LAI transmitted on the base channel from the test set, the operator can discover at which level the service indication is suspended. For this simple check, the operator has only to modify (decrease) the RF output power level of the test set and check the mobile station's display. This test can be performed without establishing a call.

Page and Listen

The RF level at which the mobile station is no longer able to detect the base channel (at its sensitivity level) can be searched for automatically. This process is described in the following procedure.

1. The test set transmits the base channel at a user-defined START LEVEL; the test set pages the mobile station.
2. When the mobile station synchronizes and reads the paging data, it will request a channel by transmitting access bursts.
3. The test set ignores the channel requests from the mobile and lowers the output power by a user-defined STEP size; it continues to page the mobile station.
4. After receiving no response from the base station, the mobile station will restart synchronization (cell reselection), and then, when it is able to read the paging data again, repeat sending channel requests.
5. Steps 3 and 4 are subsequently carried out automatically over and over until the mobile station does not respond to the paging at all. The last RF output level at which the mobile station responded can be displayed on the test set's monitor screen.

Note: This test is also carried out without establishing a call. The duration and accuracy of this test depends on the selection of the START level and the STEP size.

Measurement Report

Once a call is in progress, the mobile station has to constantly report the received signal level and quality back to the network. This feature can be regarded as a receiver self-diagnosis of the mobile station. Again, by varying the RF output power level on the test set, the operator is able to check the receiver's compliance with its reported signal levels. The test equipment has to decode and display the measurement report data.

In the case where there is a service indication, but the mobile station does not transmit access bursts at all or does not transmit them at the expected power levels (see step 4 above), this can be an indication that there is something wrong with the transmitter section within the mobile station.

11.4.1.3 Transmitter Test

The transmission path has to be checked for correct power levels (maximum output power, power control levels and steps), correct burst shaping, and modulation accuracy (phase error and frequency error). Measurements carried out on the mobile station's transmitter are as follows:

- Peak (burst) power;
- Burst template/length/flatness of burst's active part;
- RMS phase error;
- Peak phase error;
- Frequency error;
- Spectrum due to modulation and spectrum of test signal.

11.4.1.4 Call Procedures

When the basic RF performance has been checked and is within the expected limits, it should be possible to set up a call with the mobile station. The case where a call procedure fails is an indication that there is probably something wrong in the digital logic (call processing or lower layer failure). Once the mobile station is assigned (call setup) to a traffic channel, the following checks will be useful.

- Checking the audio entity and speech processing: The test set has to provide a voice loopback facility (AF LOOP) so that the performance of speech coding, channel coding, and analog audio components (cables, wires, loudspeaker, and microphone) can be verified. The voice data transmitted from the mobile are acquired, delayed, and then looped back (on the RF path) by the GSM service test set. An operator merely speaks into the mobile's microphone and listens for his or her voice returned out of the phone's loudspeaker after a short delay. A digital storage of voice (to be repetitively transmitted on the traffic channel by the test set) in order to distinguish between, for example, microphone *or* loudspeaker failure is very useful.
- Checks on RF performance during a call in process: It is possible (a GSM feature) to control the mobile station's output power level over up to 16 levels and measure its compliance (amplifier level linearity) with the levels. Furthermore, phase error and frequency error can be checked, as well as the burst's shaping (power-time template).
- Measurement report during call;
- Mobile station information: When a call is set up with a mobile station, some useful information is transferred (upon request) to the base station (test set), which can be acquired and displayed by the test set: dialed number, power class of the mobile station, capabilities, IMEI (serial number), and IMSI (on SIM card).

11.4.1.5 *Service Test Modes*

The tests mentioned above can be performed with a mobile station in normal operation with a test SIM; it is operated and tested as if it were using a real network. Additionally, manufacturers of GSM terminals have implemented service modes in order to better support the service technician's efforts. These modes may include diagnostic routines and transmit/receive functions, as well as some means for calibration. These are all product (manufacturer) dependent. In many cases, an additional test adapter is required. These service modes enhance diagnostic methods, since they add capabilities for the service technician to track down malfunctions and to adjust radio parameters (electronic calibration). They also add value to the mobile phone. The mobile station's display may be used as a monitor, and the keypad is a means to control the phone in its test modes. Test mode features show the full range among manufacturers; some have very sophisticated modes, and others have almost none at all.

Features can include:

- Selection of a transmit and receive channel pair;
- Generation of a test signal (ramping or CW);
- Control of output power level;
- Traffic channel loopback;
- Internal voice loopback;
- Display of received signal level;
- Tuning of radio parameters.

11.4.1.6 *Repair*

Once a failure has been tracked down to a component or module, the technician will replace the defective part. High integration and surface-mounted technology (SMT) makes repair to the component level technically difficult or impossible in some special cases (particularly on layered "sandwich" boards) and often unprofitable even if the attending technician has the special training and equipment SMT work requires. A module exchange is the more reasonable and practicable solution (most GSM mobile stations are built from only a few modules). Some exchanged RF components on the modules require electronic or even manual tuning or calibration on a voltage-controlled synthesizer or frequency standard, or on a quadrature modulator or demodulator.

Adjustments and checks in the transmitter section are performed, for instance, on some or all power levels used by the mobile station. The test set has to provide a means for synchronizing the mobile station (by supplying a base channel) and must continuously measure the mobile station's output power so that the operator can make adjustments. This can be done by using a service test mode or with (or sometimes without) setting up a call as described above. The mobile station power can be controlled by the simulated base station (test set). The compliance of phase error and frequency error with the standards is tested. The requirements are phase error rms <5 deg, peak phase error <20 deg, and frequency error <90 Hz.

The receiver section often requires adjustments in the automatic gain control (AGC) at different RF levels. For such adjustments, the test set has to supply RF signals at various levels (modulated and unmodulated).

11.4.1.7 Test Automation

The calibration routines mentioned above can be automated so that the operator does not need to set up the test set again and again, thus opening the whole procedure up to operating errors. Each RF level setting and the selection of different test channels are examples of operations ripe for automation, since they are subject to operating errors. An interactive test sequence saves time and makes sure that no steps are left out. The sequences will, of course, be different for different brands and models of phones. After repair and calibration, a final check, including call setup and call clearing, makes sure that the phone works well again. This final check can be performed with the help of an automated test program, and the program may produce a printed test report at the end of the test session. The test report can serve as a certificate for the customer and serves as a reference for later service calls. A printed report also states the competence of the service laboratory and restores the customer's confidence in the phone.

11.4.1.8 Test Equipment

In order to provide all the necessary functions for the service of GSM terminals, including test, repair, and calibration, the test equipment must comply with the minimum requirements listed below. It should be noted that in the service shop, no type approval or design tests are performed. Testing is confined to those functions likely to fail for some real reason; the service center should carefully avoid testing software and the phone's design integrity. GSM service test equipment features include:

- Man-machine interface (MMI): Easy to use with quick access to parameters and results.
- Analog signal generator to provide analog test signals for terminals with service test modes.
- Analog measuring receiver: Measurement of analog signals from the terminal in the service test mode (e.g., frequency counter, power meter, spectrum analyzer).

- GSM signal generator: Simulation of a GSM base station on selectable channels and power levels. It is also a requirement to generate unmodulated carrier signals which are used as test signals for receiver tuning.
- GSM measuring receiver for transmitter testing: Transmitter tests must be possible without setting up a call or during a call in progress. The measurements to be performed are power, pulse shape (burst template), phase error, frequency error, spectrum of modulation, ramping, and other output signals (CW test signal). The test receiver must not be restricted to valid pulse shapes and data content.
- Voltmeter and ammeter to check battery and power consumption of the phone in standby and during a call.
- Basic receiver test by reduction of the RF generator level (base channel).
- Basic call procedures to establish and clear calls (mobile-originated/terminated call setup, mobile-originated/terminated call clearing).
- BER test on a traffic channel.
- Audio echo function (AF loop) for a traffic channel during call in progress.
- Continuous interpretation and display of measurement report data from the mobile station during a call in progress.
- Display of mobile station parameters (e.g., power class, IMEI, A5 algorithm implementation, SMS capability, extended GSM frequency band capability).
- Automated test sequencing to ease the calibration procedures, such as stepping through certain power levels, and to provide an automated and interactive final check protocol for printouts.

In addition, the following tools may be present on the test bench, again depending on the type of phone to be serviced:

- Test adapters;
- PC with monitor program;
- Programmable power supply.

Figure 11.5 shows the basic MMI of a GSM service test set for mobile phones. Parameters can be entered on the left-hand side, and results appear on the right-hand side. Call procedures and additional functions are available via softkeys.

There are, today, clear demands from the radio service market (retailers), that test sets used in the analog service should be usable on GSM phones, too. In other words, existing test equipment, already in place, should accept an upgrade for GSM work. Lower retail margins, aggressive price policies, and the cost of first- and second-generation GSM test equipment often do not justify investments in GSM-only test gear. This is especially true for specialized radio retail shops, which still deal with all kinds of analog radios, but also have a major share of GSM phone sales. On the other hand, manufacturers who enable their retailers to service the phones at the first customer contact point must make sure that the service shop is equipped with sufficient test capabilities. To cope with these requirements, test

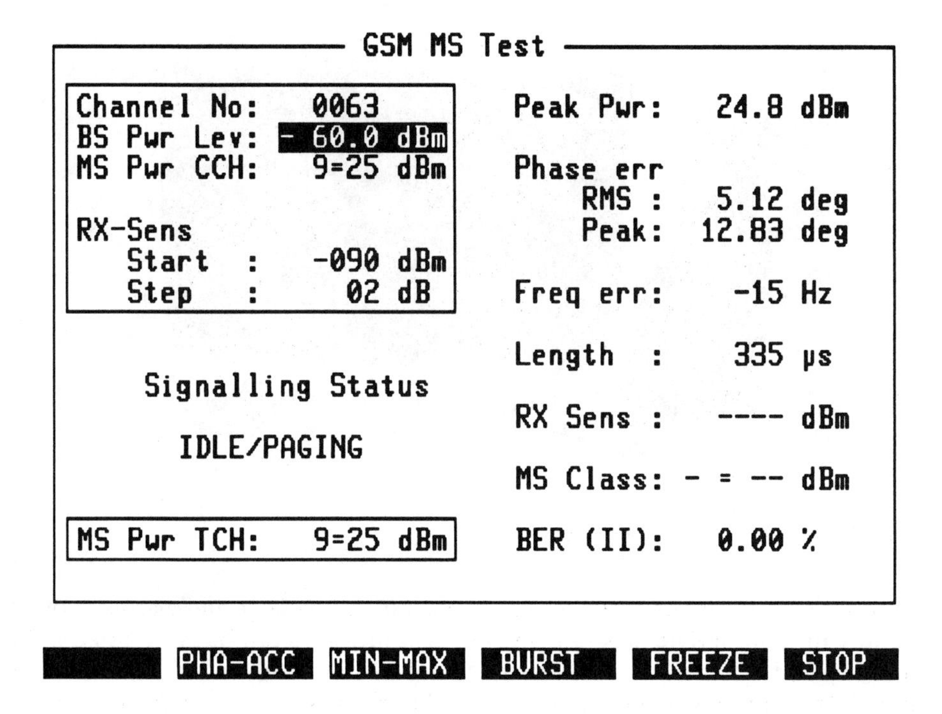

Figure 11.5 GSM mobile station test, man-machine interface.

equipment manufacturers are challenged to put appropriate test sets in place. Single-mode (GSM-only) test equipment has been designed and is sold at price levels comparable to analog test equipment. Dual-mode test equipment (GSM plus analog) covers the requirements of service technicians who still have to face test applications in both worlds—analog and digital. Among the requirements for GSM service test equipment, analog functions (signal generator/measuring receiver) are still used and are required. Another good example is the D-AMPS system (analog and digital traffic channels) in the United States (NADC, see Chapter 12). Dual-mode phones require dual-mode test equipment.

A feasible approach for dual-mode test equipment is an open system architecture in an analog test set, where additional digital test capability can be supplied with plug-in modules and appropriate software. Figure 11.6 shows an example of an upgradable analog-digital test set for the service and manufacture of radio terminals and base stations.

11.4.2 Commissioning and Service of GSM Base Stations

GSM base stations are connected to OMCs and to base station controllers. These components in a GSM network carry facilities for monitoring the functions of base

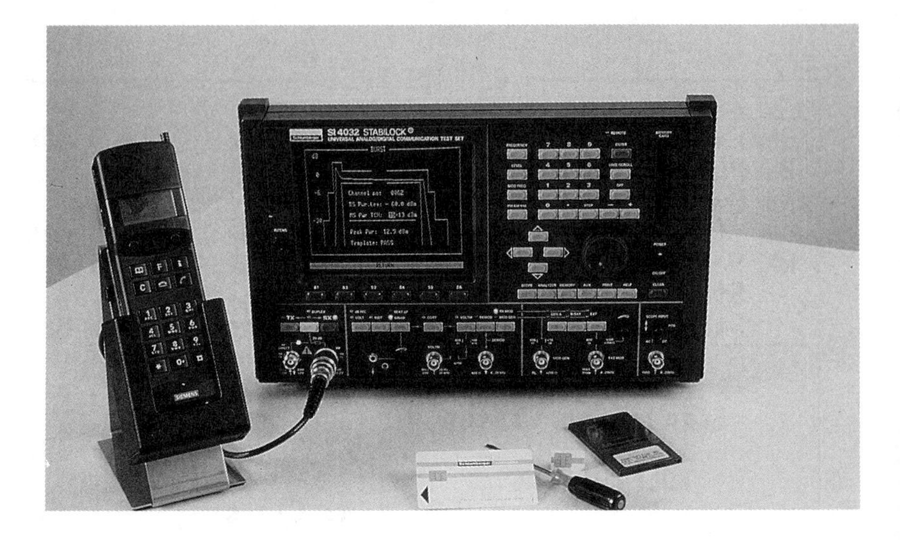

Figure 11.6 Analog-digital test set for the service.

stations. The base stations themselves use diagnostic functions and alarms in order to notify the appropriate network entity so that measures can be taken to keep the station up and maintain reliable service. An appropriate reaction to a failure can be the activation of redundant components within the base station rack (if supplied) or the reconfiguration of the base station. Still, technicians will have to cope with on-site service, track down defects, exchange faulty modules, and measure different radio parameters. Exchanged modules may then be returned to the factory, looped into the manufacturing flow, or repaired separately. The test equipment necessary for service and for installation (commissioning) has to have certain qualities. Since strategies are different among the various manufacturers and operators, time and experience with running networks will show the most efficient ways to approach this challenge.

Service testing of GSM base stations can be supported by manufacturer-specific test interfaces, special test mobile stations (with monitor functions), and local maintenance terminals (e.g., a laptop computer with diagnostics and control software), which is the typical case for manufacturing and commissioning procedures. The features of additional radio test equipment will have to cover the following requirements:

- Analog and digital measuring receiver for power and modulation measurements;
- Spectrum analyzer function for modulation and ramping spectrum according to the test specifications (see Chapter 10);
- Synchronous (to base channel) GSM signal generator for receiver tests;
- Optional Abis interface tester or monitor (in case it can be connected physically and logically);

- Portable (small size and weight);
- Easy to use;
- Recording of measurement data.

11.5 NETWORK PLANNING

Network planning is a critical issue before and during the installation of a GSM network. Every network operator will expend considerable effort on achieving an optimized network structure. The network operator has to make sure that service is available anywhere (coverage) and the blocked call rate (service temporarily not available) is kept low (capacity). These two important cellular network parameters have to be aligned with another important variable: cost. Of course, coverage and capacity can be achieved by setting up cell sites with a high geographical density and with a high number of channels. The trick (or call it the art) is to provide *sufficient* capacity with a maximum of coverage aligned with the terrain and propagation, and demographic factors (population, vehicular traffic, employment, income, housing data). In other words, the operator will not install a high number of channels—at high cost—in a deserted area with low expected traffic. An operator will also not save on infrastructure equipment in a densely populated area with high average income levels, in which cellular phone use is likely to be very high.

Another important topic is *frequency planning*. In order to avoid interference problems, radio channels must be selected among operators and then for single cells or clusters in a way that the reuse of channels does not cause interference with other cells nearby.

Power planning is another issue. The provision of service to different power classes of terminals requires certain considerations of radio propagation. In the case of having to provide service to higher classes of terminals (power class 5 hand-portables with 2W maximum output power), the base stations must be set up with a higher density. A 2W hand-portable phone may be able to read a 150W base station at a 20-km distance, but the base station may not be able to read the phone. In such an unbalanced case, service will not be available for the subscriber. Battery life is also an important factor for service quality. A mobile station that has to transmit at full power will use up more energy, and thus the talk time between battery charges will be shorter. A higher density of base stations in a network designed for low-power mobile stations also means that the base stations can transmit at lower power levels.

Tools that are used by the network designers during the planning and installation of infrastructure equipment are:

- Propagation models, including terrain and sector parameters (positioning, photogrammetry), power and frequency schemes (coverage), and expected handover boundaries;

- Demographic data: population, density, expected subscriber rate, expected traffic loading (capacity);
- Data on available cell sites and fixed lines (leased-line usage);
- Traffic simulation tools for expansion planning and cell splitting (directional antennas).

Hardware used by the network designers consists of signal generators for cell site simulation (a base station simulator with a power amplifier), measuring receivers for signal level and quality measurements, positioning systems (satellite-based) for geographical coordinates, and test mobile stations (for monitoring signal levels and signal quality). All the data are collected and processed by powerful workstations loaded with specific and sophisticated software tools such as stereo or three-dimensional terrain images, demographic/traffic data, and propagation models. Many test drives are carried out in order to collect real-time data or to verify the predicted propagation parameters and handover boundaries. As a result of such planning, a plot of a map with proposed cell sites, propagation parameters (received signal levels from base stations, power control levels from mobile stations, signal quality), handover boundaries/areas, and a host of interpretations can be generated.

Network employees, and particularly the cell site technicians are given phones for their private and official use. Since the cell site technicians are very sensitive to the network's quality and can quickly suggest adjustments in the system, this practice is an efficient way to fine-tune the network. Network planning and optimization is a constant task that requires careful consideration of how to use the restricted radio spectrum resource as efficiently as possible and with a guaranteed level of service.

REFERENCES

[1] *GSM Technical Specifications*, ETSI, Sophia Antipolis, series 11.40.
[2] *GSM Technical Specifications*, ETSI, Sophia Antipolis, series 11.10.
[3] *GSM Technical Specifications*, ETSI, Sophia Antipolis, series 11.20.

PART IV

COMPETING CELLULAR STANDARDS

CHAPTER 12

North American Digital Cellular and Personal Digital Cellular

The United States has the distinction of running the world's largest cellular system (the highly successful AMPS system), and the Japanese hold the honor of installing the world's first commercial system (Nippon Telephone and Telegraph Company's (NTT) MCS system in December 1979). Both of these innovators have proposed and are installing their own versions of TDMA digital cellular: the NADC system in the United States, and the personal digital cellular (PDC) or JDC system in Japan. The two systems have as much in common with each other as they differ from GSM. The differences between them serve to shed additional light on GSM and clarify its prominent place in the world's cellular markets.

The NADC and PDC systems share an interesting type of modulation ($\pi/4$-DQPSK), which we will explain early in this chapter. The time slot structures, frame structures, and certain details of their air interface definitions are also similar in both systems, though the PDC versions are slightly more reminiscent of GSM. The two systems, however, diverge in all other matters. The PDC system is a purely digital TDMA system just as GSM is, and some of its operating modes, organization, and deployment schemes are similar to GSM. The NADC system graphs digital traffic resources onto the existing AMPS and expanded AMPS (EAMPS) system in a fascinating overlay scheme. It holds almost no resemblance to GSM in this regard.

Since this is a book on GSM, we will not investigate all the details of the NADC and PDC systems. We will reserve such a detailed analysis for the next chapter, Chapter 13, when we have a close look at CDMA. It is only the differences in certain aspects that distinguish GSM from its TDMA cousins, NADC and PDC. CDMA has nothing in common with GSM. We have to take a broader and closer look at CDMA in order to judge the relative benefits of each system and increase our understanding of GSM in the same way as a selective look at NADC and PDC does.

12.1 TDMA IN THE UNITED STATES

The United States has a long history of and a noted ability to extend the life of various communications systems by devising methods to maintain compatibility between older users of a system with traditional technologies and new users employing enhanced capabilities and technologies. The reasons for this are primarily economic ones. The United States is a large and wealthy country. American communications systems tend to become very large by world standards, so that ripping out one system and replacing it with another is usually too expensive to even consider. This was the case with the American National Television Standard Committee (NTSC) color television system (fully compatible with the black and white receivers of the 1940s and 1950s), and remains true with the NADC cellular system today.

12.1.1 Frequency Bands

Figure 12.1 lays out the frequency bands and channel scheme for the EAMPS and NADC systems. The reader will discern two frequency bands of 25 MHz each. The mobile transmitter's uplink is in the lower 824- to 849-MHz band, and the base station transmitter's downlink is offset 45 MHz up to the 869- to 894-MHz band.

Each of the uplink and downlink segments is divided into 832 channels of 30-kHz width each. The numbering of the channels is a bit strange. The enumeration starts at number 1 near the bottom of the bands, continues up to channel number 799 at the top, then folds back on itself to channel number 991 at the bottom edge of the bands, and finally ends at number 1,023 just below the starting point. The awkward numbering comes from an expansion of the original 666 channels of the AMPS system in the mid-1980s, which had to be coordinated with some existing land mobile services in the 860-MHz band to yield the EAMPS system.

12.1.2 Competing Operators

The two 25-MHz bands in the EAMPS system are further divided into equal-sized A and B channel groups. The A group is indicated by some dark bands in Figure 12.1 (e.g., between channel numbers 666 and 716) to distinguish it from the unmarked B

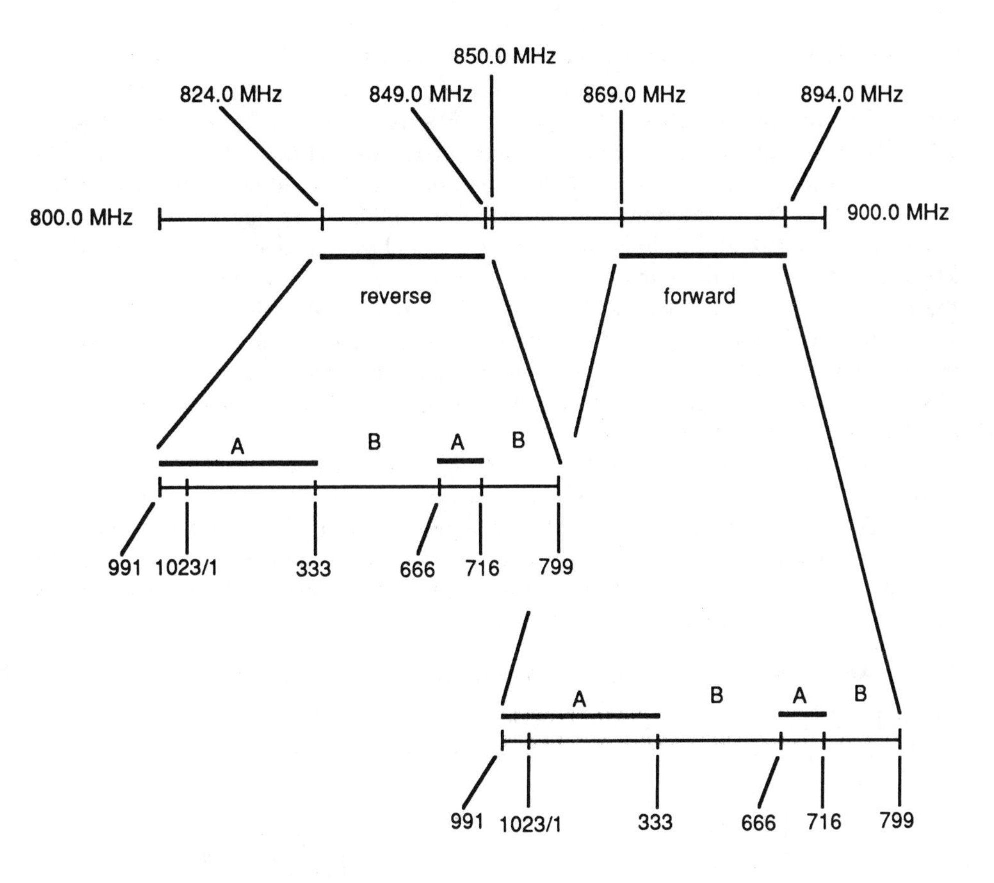

Figure 12.1 U.S. cellular frequencies and channels.

group. Each operating area in the U.S. EAMPS markets has two competing service providers. One is the local fixed-network service provider, the B system, sometimes called the *wire-line operator*, and the other, the A system, someone else called the *non-wire-line operator*. The number of channels allotted to each is identical. A few dozen channels in the region near channel number 333 are reserved for use by each operator for control channels in their respective systems.

12.2 DUAL-MODE ACCESS

The most important feature of all the NADC enhancements is that they are dual-mode systems. This means that all systems, regardless of their enhancements, can provide an appropriate traffic channel for even the oldest analog AMPS terminals that may seek cellular phone service.

The NADC system adds digital traffic channels (DTC) to an EAMPS system by simply replacing existing analog channels. The frequency bands, therefore, are identical in the two systems. The digital additions are overlayed onto the existing EAMPS system, and the two are linked with a common set of control channels. The control channel resources carry broadcast, paging, and access messages to the mobiles, where the three functions are usually combined on one physical channel. Both the downlink and uplink channels are Manchester-coded FSK running at 10 kbps, where the uplink channel is called the *reverse control channel* (RECC), and the downlink channel is called the *forward control channel* (FOCC). The RECC and FOCC are block-coded and repeated a fixed number of times. The FOCC, moreover, uses the state of some "busy-idle" bits inserted roughly every ten message bits to indicate the current status of the accompanying RECC, thus offering a first echelon of access control for the mobiles.

The large box in the center of Figure 12.2 represents the FOCC and RECC resources in an EAMPS system. The casual observer performing a signaling trace on the FOCC would have considerable difficulty deciding if the FOCC were from an EAMPS system or an NADC system, for there are only a few bits among the myriad originally defined in the EAMPS protocol that identify the host system as one capable of enhanced digital operation.

12.2.1 Analog Single-Mode Operation Within an NADC System

Let us look at the operating modes. We will start with the trivial case in which a single-mode analog phone, depicted at the bottom of Figure 12.2, seeks a traffic

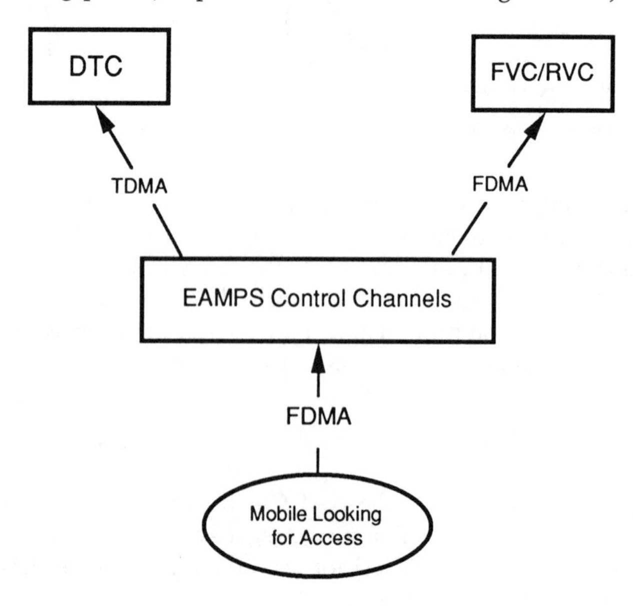

Figure 12.2 NADC overlayed dual-mode access.

channel in an NADC system. When switched on, the mobile scans the assigned frequency band (Figure 12.1) in search of an FOCC. Mobiles are attracted to an FOCC as flies are to ice cream, or as insects are to light. Upon registering with the system, which it does with some uplink RECC transmissions on a control resource, the mobile identifies itself with a mobile identity number (MIN) and an electronic serial number (ESN). After some verification checks, the network assigns a forward voice channel (FVC)/reverse voice channel (RVC), which is a pair of frequencies separated by the normal 45-MHz duplex offset. The completion of the connection, be it cell initiated or mobile-initiated, is performed on the assigned analog voice channel with some simple signaling tones. Handoffs and mobile RF power adjustments are made on the voice channel with some blank and burst signaling of exactly the same format as is used on the FOCC and RECC. The mobile's initial access to the control channel is accomplished with FDMA radio techniques, and the subsequent assignment of a voice channel (the right-hand path above the control channel box in Figure 12.2) continues in an FDMA mode.

12.2.2 Dual-Mode Operation Within an NADC System

Now, consider a dual-mode mobile's access to the same NADC system, the left-hand path in Figure 12.2. The initial access to a control channel is exactly the same. The dual-mode mobile's RECC messages, however, contain a few extra bits (not defined in the original EAMPS protocol) to tell the host system the mobile is, in fact, a dual-mode mobile. The identifying messages are the station class mark (SCM) and the enhanced protocol indicator (EPI). The SCM was defined in the original EAMPS protocol to indicate the mobile's maximum RF power capability, whether the mobile could assume a DTX operating mode, and whether it could operate in the extended channels numbered beyond channel number 666.

A few undefined bits in the mobile's RECC data steam are mustered into service to expand the SCM and add the EPI. A dual-mode mobile has some additional (lower) power levels it can be commanded to assume, and it should tell the network at registration time what kind of enhanced operation it is capable of. There are currently three enhanced modes:

1. NAMPS (narrow amps—a clever analog enhancement championed by Motorola, which adds another layer of FDMA to the voice channel assignment. There are three analog subchannels in each 30-kHz NAMPS voice channel.);
2. CDMA (see Chapter 13);
3. NADC.

Upon recognizing the mobile's enhanced digital capabilities, the network will assign a DTC, if available at the time, to the mobile. The phone and the network employ some familiar TDMA techniques as they move to the DTC. If a DTC is not available, then an analog FVC/RVC channel is assigned instead. The DTC is defined by a channel number, a time slot number, a timing advance value, a cell identifying code called coded digital verification color code (CDVCC), and a mobile power setting. The channel is maintained with some more familiar TDMA techniques using logical channels (SACCH and FACCH), and proceeds until disconnect time in a manner vaguely reminiscent of GSM. The particular procedures, signaling, and supporting frame structures are *much* simpler than we saw in GSM.

12.2.3 A Dual-Mode Mobile Within an EAMPS System

What happens if an NADC dual-mode mobile enters an older EAMPS system? Does the mobile or the system explode? The answer is quite simple. There were a few precious undefined (reserved) bits in some of the message fields in both directions, both in the control (FOCC and RECC) and voice (FVC and RVC—blank and burst messages) channels. A properly designed phone (or network) should carefully ignore any undefined bits. An NADC mobile, then, looks like any other AMPS or EAMPS phone to an older network, and the NADC phone is smart enough to recognize the default settings of the network's undefined (reserved) bits as such.

12.3 MODULATION

Both the NADC and PDC systems use $\pi/4$-DQPSK for modulation in both the uplinks and downlinks. The NADC system, however, confines this type of modulation to the DTC; the control channel retains the EAMPS control channel's FSK modulation.

12.3.1 $\pi/4$-DQPSK

Refer back to Figure 5.41 to see a simple example of high-level modulation. The carrier in this figure can assume four different phases, where each phase can represent four different messages (00, 01, 10, and 11). A data stream can be carried, two bits at a time, with QPSK. Note, particularly, the circumstances in Figure 5.41 when the message content does not change from bit pair to bit pair; the phase of the carrier also does not change. Note that a phase transition can pass through the origin, thus putting considerable strain on RF amplifier design.

Now take a look at Figure 12.3 and $\pi/4$-DQPSK. The lines represent all the possible phase changes allowed in the carrier. Take careful note of the chart at the bottom of Figure 12.3 to see that there is *always* a phase change (45 deg or 135 deg) with each pair of bits to be encoded. There is a phase change even if the message content does not change. The chart at the bottom of Figure 12.3 indicates, just as was the case with QPSK, that $\pi/4$-DQPSK carries two bits per symbol, which seems to mean that there are four possible carrier phases. But Figure 12.3 shows eight

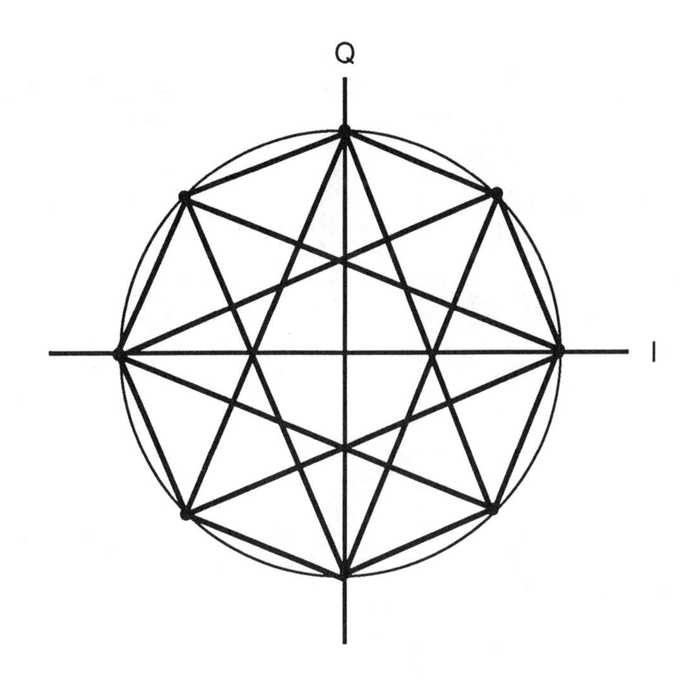

Data	Phase Change
0 0	45°
0 1	135°
1 0	-45°
1 1	-135°

Figure 12.3 $\pi/4$-DQPSK.

possible carrier phase changes separated by 45 deg from each other. Has four suddenly become equal to eight at the end of the twentieth century? What is going on here? This is often a source of consternation for readers new to digital radio, and Figure 12.4 should clear up the confusion.

12.3.2 Phase Transitions

Say the carrier's phase at some time t can be represented in the top inphase/quadrature-vector component (I/Q) diagram of Figure 12.4 as a point at 0 deg, and that

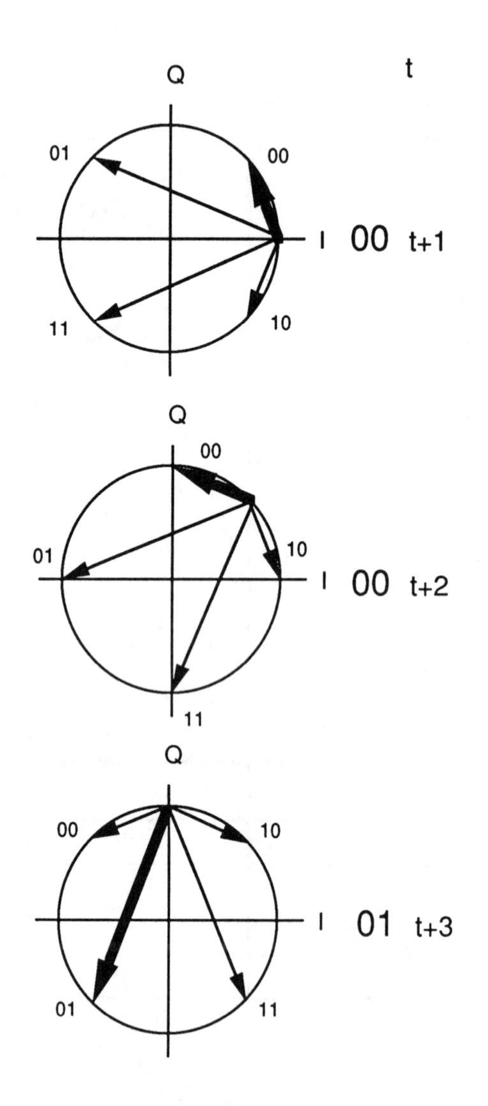

Figure 12.4 Phase changes.

at some later time $t + 1$, the message content is 00. We do not care what the message (two bits) was at time t. The modulation scheme allows four possible new phases, which are also plotted in the top diagram of Figure 12.4. The chart in Figure 12.3 says we have to change the carrier's phase +45 deg when the data are 00, and this particular change is shown in Figure 12.4 with the heavy arrow in the top diagram.

Now, at still a later time $t + 2$, the next two bits to be transmitted are, again, 00. We have the same four phase changes to select from that we did at time $t + 1$, but our chart (Figure 12.3) says we are still obliged to change the phase +45 deg. The second plot in Figure 12.4 shows the correct phase change with a dark arrow. At time $t + 3$, the data change to 01, and the chart says to change the phase +135 deg.

If the data we must transmit have a random character, and if we watch the phase changes long enough, we will eventually see the star plot in Figure 12.3. DQPSK stands for *differential quadrature phase-shift keying*, and the $\pi/4$ term means the I/Q diagram looks like it is shifted 45 deg with each encoded symbol (45 deg = $\pi/4$ rad). The reader should note some important characteristics of this type of digital modulation.

1. There are two bits in each symbol, which is twice as efficient as GSM's GMSK.
2. The phase transitions avoid the center of the diagram, which removes some design constraints on amplifiers. QPSK (Figure 5.41) is more difficult to amplify efficiently than $\pi/4$-DQPSK is, but GSM's GMSK (constant envelope) is much easier to handle than either QPSK or $\pi/4$-DQPSK.
3. Since there is *always* a phase change with each symbol, the modulation has a considerable self-clocking characteristic.

12.4 NADC FRAME STRUCTURE

In contrast to the rather complicated, though efficient, modulation in NADC, the frame structures are so simple as to be trivial when compared to GSM. Figure 12.5 depicts the only frame structure of note in NADC [1].

Transmissions on the DTC (not on the control channel) use this structure in both directions. The base station's emissions are allotted, by time slot, to different mobiles, and the mobiles burst their transmissions according to the time slots assigned to them. As with GSM, two traffic rates are defined, full rate and half rate, where the half-rate mode has yet to be implemented. In the current full-rate mode, time slots are assigned in pairs. If the two dark time slots (1 and 4) in Figure 12.5 are assigned to a user, then a second user would be assigned time slots 2 and 5, and a third would be assigned slots 3 and 6. The time slots are identified with a characteristic synchronization pattern. Half-rate voice data can be accommodated in only one time slot in each frame, and the number of users that can be multiplexed onto a single 30-kHz NADC channel will increase from three to six when the half-rate mode is implemented.

12.5 NADC TIME SLOTS

The NADC time slot structure (see Figure 12.6) is more complex than GSM's, but there are only two versions in NADC, one for the uplink and one for the downlink. There is a third structure called the *shortened burst*, which the mobile uses for its

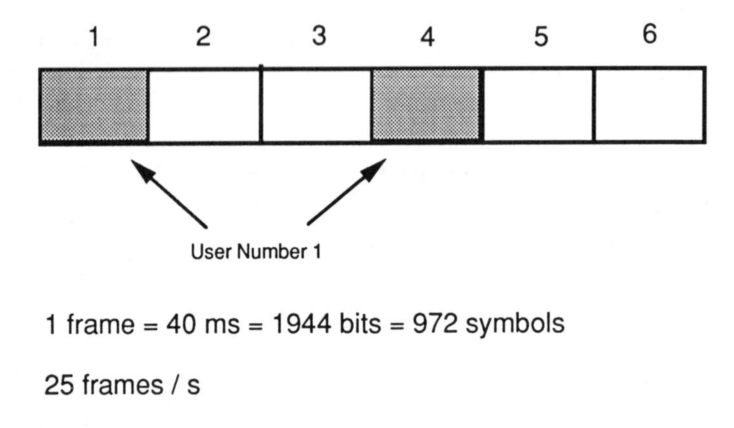

Figure 12.5 NADC frame structure.

initial access attempts in a cell, but we do not illustrate it here. The simplicity comes from the fact that NADC's digital operation is confined entirely to the traffic channels.

12.5.1 Uplink Time Slots

Speech data are encoded according to a type of code excited linear predictive coding (CELP) algorithm. The process is similar to that used in GSM, but the details are different. The payload data (260 bits) appear in three places in each uplink time slot. The SYNCH channel holds a fixed pattern peculiar to a time slot. In the full-rate mode, there are only three possible patterns. Six additional patterns appear in the half-rate mode. The SACCH channel performs exactly the same function as it does in GSM, and usually contains the mobile's measurement reports destined for the base station. The measurement reports support NADC's mobile-assisted handoff (MAHO) feature, which is similar to the corresponding GSM function. The CDVCC channel can take on any one of 256 values (12 bits – 4 parity bits = 8 bits, and 2^8 = 256). All transmitters, both mobile and base stations, within a cell have the same CDVCC value in both the uplink and downlink time slots. The 00h value is not used, and many network operators equate the cell site number (from a terrestrial map) to the CDVCC value.

A guard time (G) of three symbols (six bits) is attached to the start of each uplink time slot. This serves exactly the same function that it serves in GSM uplink

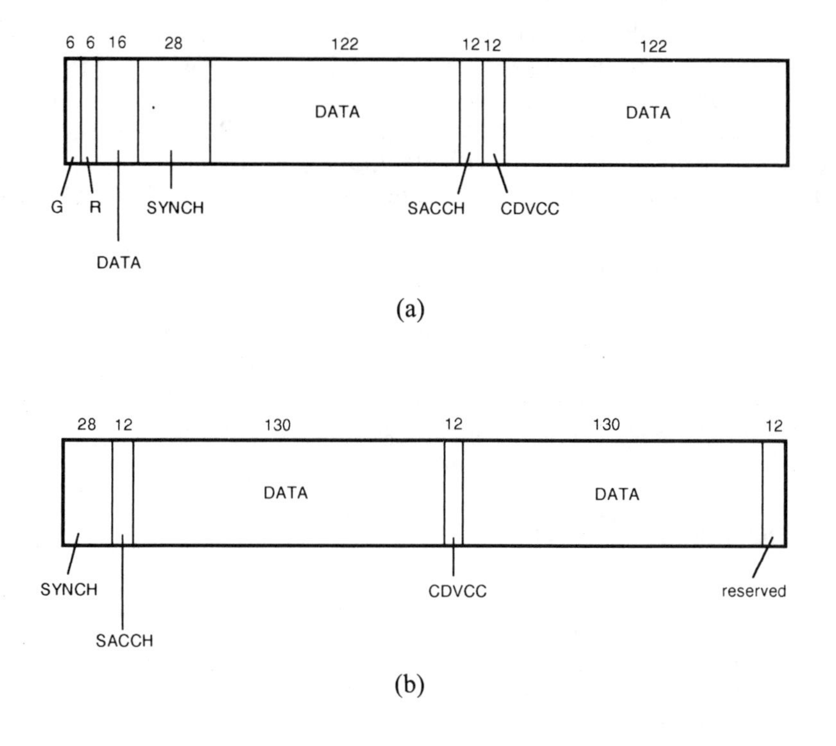

Figure 12.6 NADC time slots: (a) uplink, (b) downlink.

bursts; it eliminates the need for time alignment adjustments when the mobile is close to the cell site. The guard time and the subsequent time alignment adjustments prevent the slot collisions that occur when bursts from different mobiles overlap at the base station's receiver. Another six bits of ramp time (R) are reserved at the start of each mobile's burst during which time no important data are transmitted. If the mobile's power ramps up too fast, then too much energy will appear outside the allotted 30-kHz bandwidth assigned to the mobile. The slang term for this effect is *AM splash* (explained in Section 5.15).

Channel coding is almost identical to GSM. The speech data bits are divided into three classes just as they are in GSM. The important classes of bits are block-coded before all the bits are convolutionally coded in a half-rate coder over six bits at a time. Speech blocks are diagonally interleaved over two time slots. Signaling data is block-coded and then convolutionally coded. SACCH messages are 132 bits long and are interleaved over 22 time slots after doubling their number in a half-rate channel coder. A logical FACCH channel can appear (usually at handoff time), and when it does, it replaces the voice data bits. Since the FACCH channel is more likely to occur during the worst channel conditions, it is channel-coded in a quarter-rate channel coder. The stealing flags we saw in conjunction with GSM's FACCH channel are missing in the NADC system. Instead, the receiver's decoder merely assumes it is decoding speech in the data fields. When a CRC error is encountered during data bit recovery, the decoder shifts gears and tries decoding the received data bits again, this time using the algorithm appropriate for the FACCH channel.

12.5.2 Downlink Time Slots

Downlink time slots are similar to the uplink ones, except that the ramp and guard times are missing and there are six symbols of undefined data at the end of each time slot. The mobile imitates the base station's SYNCH and CDVCC values, and the SACCH carries signaling designed to maintain channel quality such as power adjustment and timing advance commands.

12.6 NADC IMPLEMENTATION

Increased capacity was the primary concern for the designers of the NADC system. In accordance with this emphasis, an efficient modulation scheme was selected with a minimum of signaling overhead and simple structures. The voice quality merely had to be as good as the analog experience. The supporters of GSM were more concerned with improved quality and with a greater number of features than they were with increased capacity. A relatively efficient TDMA scheme was selected for GSM primarily to allow the improvements to fit within the assigned GSM frequency band, with due regard for the increased popularity of cellular radio that GSM was sure to bring.

GSM was designed to completely *replace* all the national systems in Europe, so optimum techniques and technologies were selected. The United States, however, had no choice but to maintain compatibility with existing systems and mobile phones. The design of the NADC system was tempered almost entirely by this constraint, together with the desire to increase capacity. Installation is relatively simple. An operator merely pulls a 30-kHz analog channel out of service and replaces it with a digital radio, which at least triples the capacity on the same 30-kHz allocation. Some cellular infrastructure vendors have come up with all kinds of clever schemes to expedite a "plug-and-play" philosophy.

12.7 Personal Digital Cellular

Even though the PDC system has so much in common with the U.S. NADC system (the $\pi/4$-DQPSK modulation is identical), its implementation is more like GSM. A close examination of Figure 12.7 tells the tale.

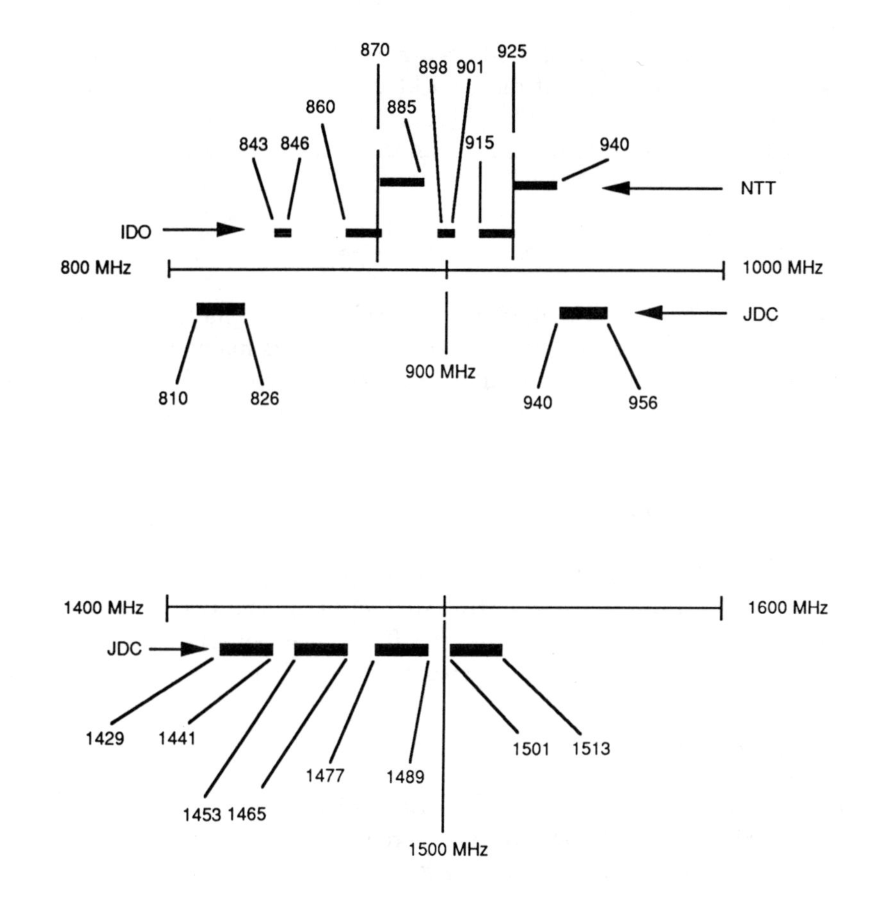

Figure 12.7 Cellular frequencies in Japan.

12.7.1 Japan's Frequency Bands

The PDC system completely replaces a collection of incompatible analog systems in Japan. Its designers, therefore, were not constrained by the need to keep a relatively limited control channel resource as the NADC system does. PDC, then, is a fully digital system just as GSM is.

12.7.1.1 NTT

NTT is a nationwide operator of an analog system called *MCS-L2* with the frequency allocations shown in Figure 12.7. The figure refers to frequencies rather than channel numbers. Its predecessor, MCS-L1, was the world's first cellular system. MCS-L2 has a 55-MHz duplex offset and 2,400 channels of 6.25-kHz interleaved spacing. The modulation is analog PM of only 2.5-kHz maximum deviation. The control resource uses a rather slow 2,400-bps biphase modulation, and the voice channel is maintained with an associated logical channel of 100 bps. Space diversity is used on both ends of the mobile channel.

12.7.1.2 IDO

IDO is an RCC in Tokyo, as well as some other big cities, and has subsidiaries all over the rest of the Japanese countryside. These operators maintain some MCS-L2 systems together with some JTACS and NTACS analog systems. NTACS is another analog system which uses very narrow channels similar to NAMPS (see Section 12.2.2).

12.7.1.3 Personal Digital Cellular

PDC operates on three sets of bands shown in Figure 12.7. One set is in the 800- and 900-MHz bands with a duplex offset of 130 MHz, and two other sets are in the 1,500-MHz band with 48-MHz duplex offsets. The new PDC system excludes the traditional analog systems from its channel assignments.

12.7.2 PDC Time Slots

The similarity of PDC time slots with some of GSM's time slot structures is evident. In its partnership with the NADC system, there is nowhere near the variety of time slot structures we find in GSM. Figure 12.8 illustrates most of the examples.

12.7.2.1 Traffic Channels

Note, first, the appearance of GSM's stealing flag, which is absent in the NADC time slots. We also see the familiar guard and ramp times, which are labeled *G* and *R*, respectively. The other fields are typical of ones we find in digital radio, but some of their names have changed. *P* is a preamble, *SW* is a synchronization channel, and *CC* is a color code indicator similar in function to NADC's CDVCC. We always find an associated control channel (SACCH) with the voice data (TCH), and the voice data can be replaced by an FACCH.

12.7.2.2 Control Channels

The control channels have their own uplink and downlink time slot structures, in which PCH, SCCH, and BCCH data are included in the CAC bits. *EC* is assigned

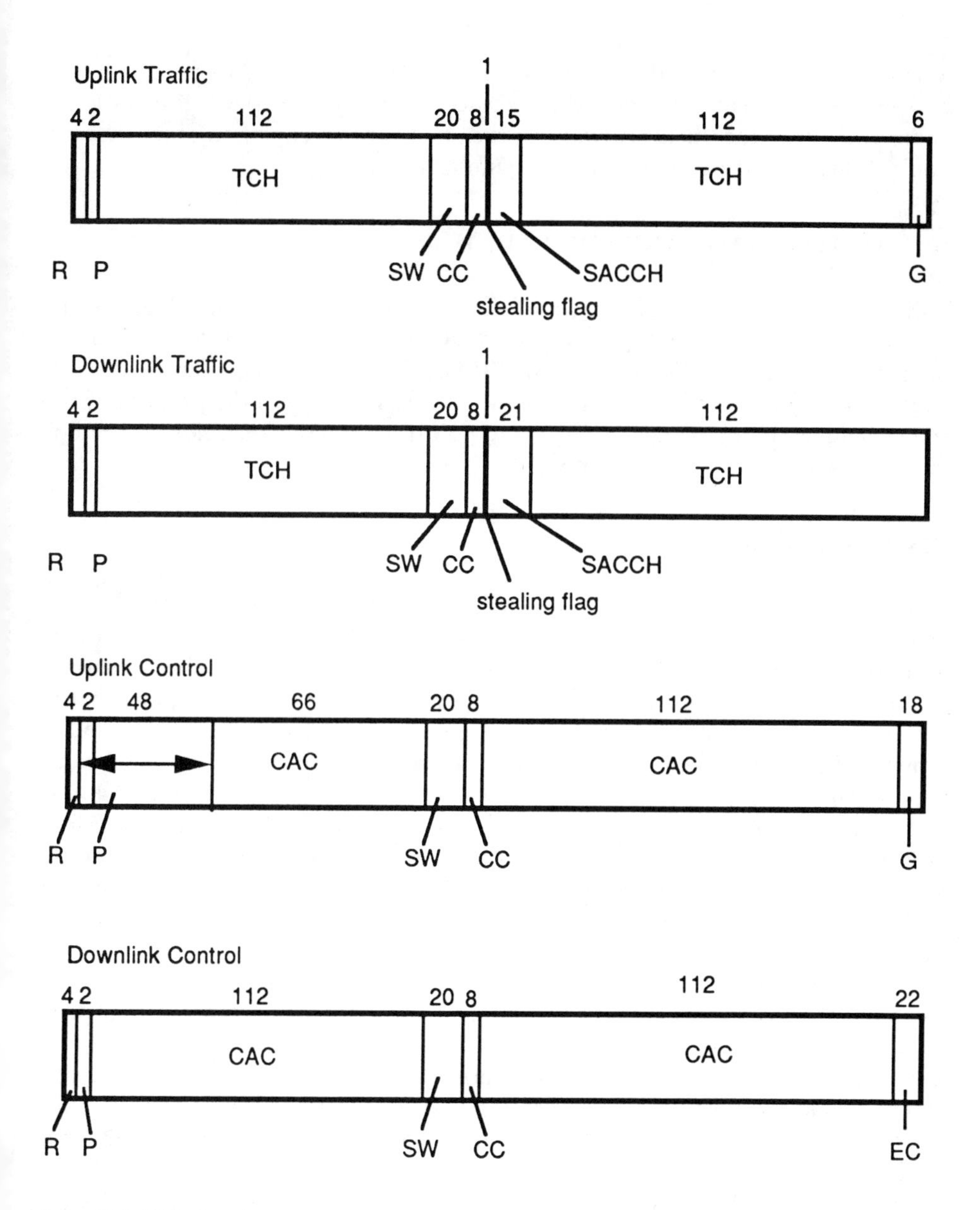

Figure 12.8 PDC time slots.

to time slot collision control, and the mobile's uplink can extend its preamble from 2 to 48 bits. The reader will note that, as was the case with NADC's time slots, the number of bits in each field is always an even number in recognition of the modulation's ability to carry two bits in each symbol.

12.8 TESTING NADC AND PDC RADIOS

With the exception of their unique modulation, and the reduced signaling baggage (particularly in NADC), testing NADC and PDC mobiles and base stations is identical to the GSM techniques discussed in Part III. The DQPSK-type modulation has some interesting display possibilities and has the curious effect, with the proper displays and techniques, of allowing a technician to peer deep into the operation of a radio's transmitter through the antenna port.

12.8.1 Error Vector

Refer to Figure 12.3 and consider the eight points around the periphery of the I/Q plot as the ideal places a carrier's phase should change. Then imagine that, for whatever reasons, a certain phase change point lies elsewhere than where it should, say at exactly 94 deg (just to the left of the top of the vertical Q-axis), and that its power is reduced at that instant, thus moving it toward the center of the plot, as shown in Figure 12.9. Assume, moreover, that at the same instant the rogue point appears on the plot, we are at time $t + 1$ in Figure 12.4 when we need to transmit a 00 symbol through the mobile channel. Under these circumstances, and if we add the normal phase and amplitude distortions that the channel will contribute to the receiver's version of the recovered modulation, it is easy to imagine how a decoder could become confused with the resulting ambiguity of such a point; was a 00 symbol transmitted, or was a 01 symbol intended?

The appropriate parameter against which we judge the transmitted quality of the modulation is the *error vector magnitude*. If we plot the bad phase change point (P) as shown in Figure 12.9, and then draw a vector from the intended point to the actual point, the resultant vector (**E**) is the error vector, which we can scale in percent of the intended vector (**v**). The error vector magnitude is just under 100% for point P in Figure 12.9. Since the receiver's channel decoder can efficiently correct individual symbol errors up to a point, an individual misplacement of a phase change is not significant for us. We therefore consider some average value of a set of such error vector magnitudes taken over an interval of collected symbols plotted in the I/Q plane, and find the value of the average above which the channel decoder will fail to keep up with all the errors under typical fading conditions. We declare the average value to be the *maximum allowed error vector magnitude*. The NADC system considers this value to be about 12.5%.

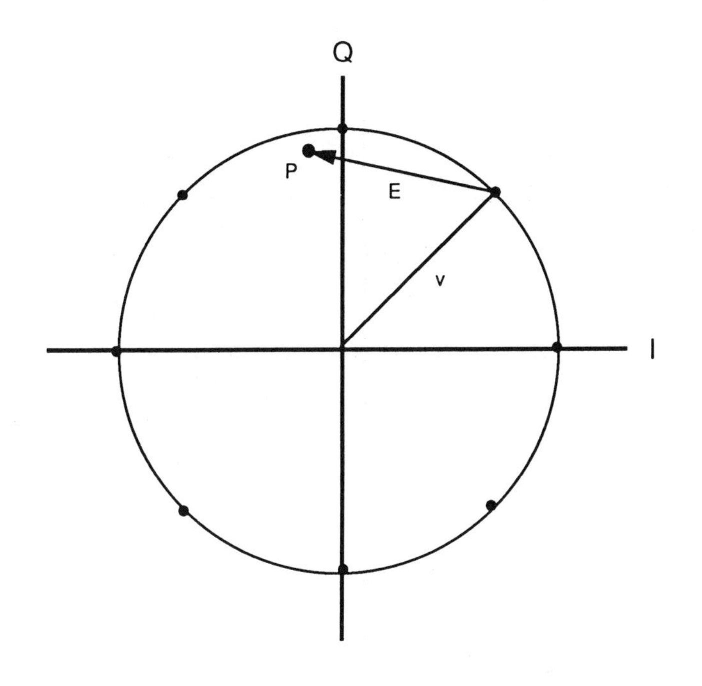

Figure 12.9 Error vector.

Recall from Section 5.15 that we must guard against allowing any parameter of a carrier to change too suddenly, for doing so will cause energy to appear in neighboring channels and the allowed bandwidth for the emissions to be exceeded. This restriction includes phase shifts. The modulating waveforms (I- and Q-waveforms in this case) are then filtered to prevent sudden shifts in phase. The filters cause the carrier's phase to change rather gradually and gently. What would this look like on the I/Q plot?

12.8.2 Constellation Display

Imagine each allowed phase change point to be an airport. If we plot the carrier's phase continuously between points (airports), we would see broad and sweeping trajectories between allowed points such as those depicted between the two points (airport stops) in Figure 12.10. Aircraft make the same gentle turns as they travel between airports in deference to the passengers on board, for to do otherwise, to take direct routes with sudden (unfiltered) sharp turns, would reduce the aircraft cabin to a shambles, injure many passengers, and elevate the serving of drinks and meals to a spectacle of stunt victualing at the phase change points.

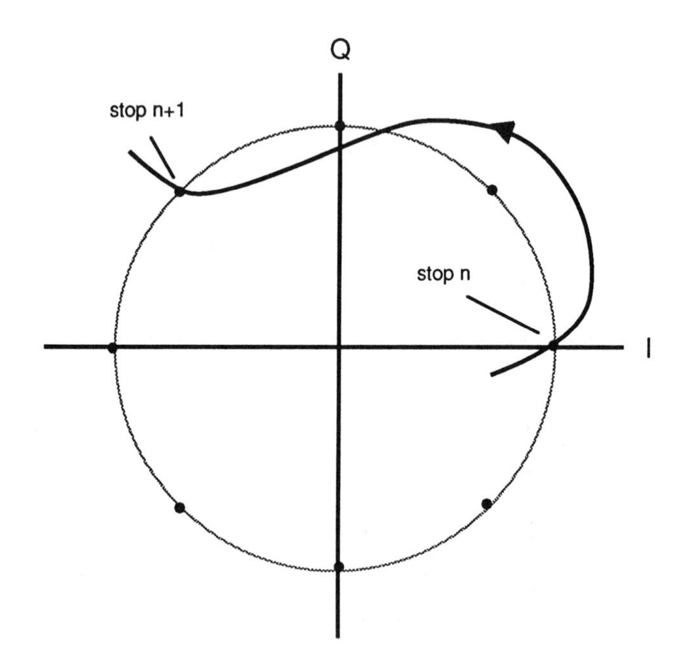

Figure 12.10 Aircraft route and phase change trajectories.

Now, things can go wrong in amplifiers that can, for example, prevent aircraft from following their most gentle routes between airports. This will cause the flights to land outside the airport property. Occasional crash landings can be tolerated, since the landings are always gentle, and each airport has a fleet of buses and drivers to go out and fetch the wayward passengers and their baggage. But if there are too many crash landings in the countryside (the average error vector magnitude exceeds the maximum allowed average), then the airports will run short of buses and drivers, and passengers will become angry as they fail to meet their appointments. A receiver's channel decoder supplies the buses and drivers to recover lost passengers (data) from the darkness and mud between the airports.

Rather than plot the aircraft's routes, it is better and clearer to plot only their landings, wherever they may be. We can do this just as we did in Section 5.15, where we flashed a strobe light at the instant of a phase change. The resulting display (Figure 12.11) will fill the I/Q plot with dots at each point a phase change occurs, which gives the display the appearance of the night sky; hence the name *constellation display*. If the update rate of the display is fast enough, a great deal of information can be gleaned from it. Depending on the details of a particular radio's design and construction, an attending technician can quickly discern the cause of transmitter problems with a great deal of resolution and precision. A fast constellation display

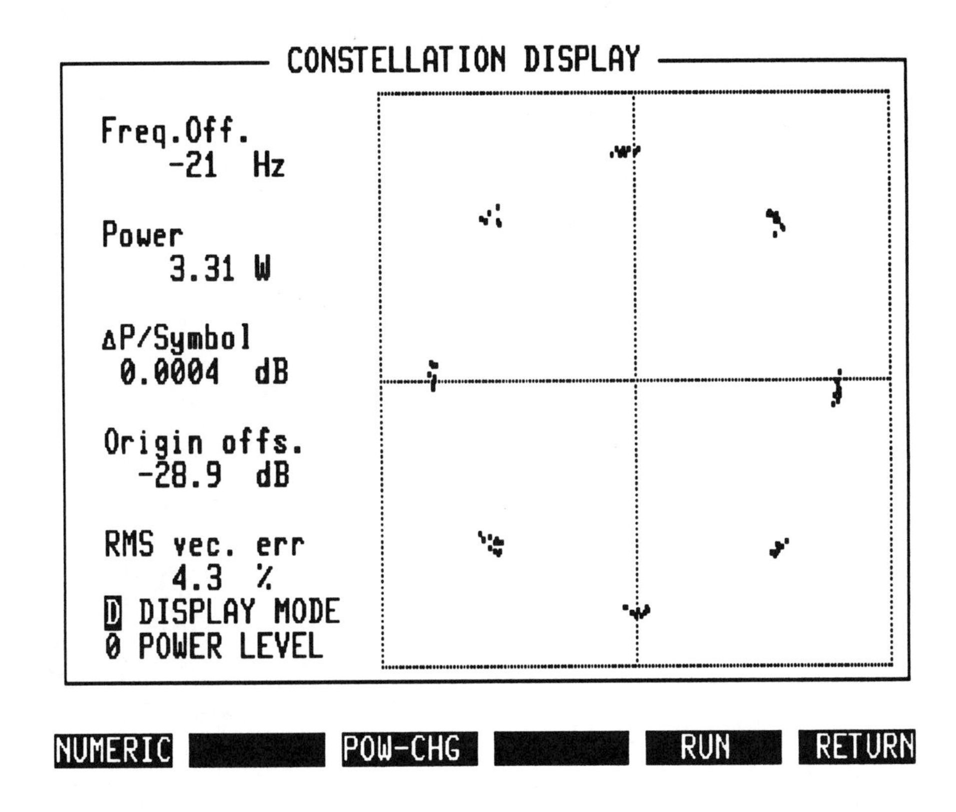

Figure 12.11 Constellation display.

also has the effect of lending an analog feel to the digital radio's operation, which opens up the work to the best technicians, who use all their senses to bring equipment back into proper order. Using the services of good technicians reduces the cost of commissioning and maintaining cell sites and of repairing terminals.

REFERENCE

[1] IS-54, "Cellular System Dual-Mode Mobile Station—Base Station Compatibility Specification," EIA/TIA, May 1990.

CHAPTER 13

Code-Division Multiple Access

13.1 CDMA IN THE UNITED STATES

In sharp contrast to the situation in Europe and Japan, where diverse and incompatible analog cellular systems are being replaced by common regional digital systems, the United States is slowly replacing its nationwide AMPS system with various incompatible analog and digital systems. A popular multibillion-dollar cellular communications industry has thrived in the United States since 1983 on the back of the nationwide AMPS compatibility standard EIA/TIA-553. The standard gave manufacturers easy access to lower costs, and customers bought equipment and services in large quantities, secure in the knowledge that they could enjoy the benefits of their purchases anywhere in North America. Today, there are four separate standards in the United States (two analog and two digital), for which the unifying thread is the dual-mode requirement through the existing AMPS specification. This dual-mode requirement simply means that customers should have a way to roam among the four competing systems with minimum fuss and bother. To realize this, all systems must therefore support the traditional AMPS services as well as any of the enhanced protocols the system operators choose to install. A user who wishes to roam among various systems can expect seamless compatibility between the systems as long as the user's terminal supports the AMPS protocol. Since all systems are dual-mode in operation, owners of AMPS terminals are not denied service in a digital system. Moreover, new mobile terminals are made to be dual-mode, where in addition to whatever enhanced protocol is installed, the traditional AMPS protocol can always be called into service when needed.

13.1.1 Dual-Mode Operation

Two of the new systems (the analog NAMPS and the digital IS-54 systems) accomplish compatibility through *dual-mode* operation by sharing the AMPS control channels and providing a mechanism for handoff to an appropriate enhanced protocol channel in accordance with the particular terminal's capabilities. A dual-mode terminal requesting service indicates its enhanced capabilities to the serving network with some signaling on the RECC when it first enters a system. If the enhanced capabilities of the terminal are compatible with that of the serving network, then traffic is passed on the enhanced resource. If there is no enhanced compatibility in the serving system, then the mobile and the serving system revert to the traditional AMPS resources.

13.1.2 Dual-Mode Operation in CDMA

A second digital system, the CDMA system, does not share any of the existing AMPS resources. The CDMA resources simply exist in the same frequency band with the traditional AMPS system. A dual-mode CDMA terminal has two separate personalities. If the dual-mode CDMA terminal enters a system that supports the CDMA protocol, it gains access to the cellular resources in its CDMA mode. If, on the other hand, the serving system does not support the CDMA protocol, the dual-mode CDMA terminal recognizes the missing CDMA resource and becomes a conventional AMPS phone.

13.1.3 IS-95 Version of CDMA

CDMA means *code-division multiple access*, which implies that users in the system are distinguished from each other by a code rather than a frequency band allotted time slot. There are many CDMA systems in the world, but the CDMA system explained in this book is the one described in IS-95, which is the result of a comprehensive proposal from QUALCOMM, Inc., of San Diego, California, with some cooperative efforts from AT&T, Motorola, and others [1].

A single CDMA channel is 1.23 MHz wide, and a dozen or so subscribers typically share the same wide channel simultaneously. Moreover, because of special measures employed in the CDMA system to reduce the effects of interference from nearby cells, neighboring cells, or sectors in cells, can use the same physical channel. An operator who chooses to offer CDMA service is obliged to clear one or more 1.23-MHz-wide channels of its older 30-kHz AMPS channels and install CDMA resources in as many cells as desired. The operator can install its CDMA resources in all cells on the same 1.23-MHz band if desired. The possibilities for capacity enhancements for such a system are attractive and remain the subject of numerous papers and lively discussions. How can such a system work? How can several subscribers share the same frequency band at the same time? How can neighboring cells share the same band? Why is a discussion of CDMA included in a book on GSM?

To answer the last question, there are two reasons to include such a discussion. First, the IS-95 version of CDMA is an important GSM competitor, and project mangers should be aware of CDMA's key features in order to properly judge events in the changing cellular market. Second, CDMA accomplishes everything GSM does in extraordinarily different ways. Exploring these differences adds to our understanding of radio techniques in general and GSM in particular.

The study of differences and contrasts is an efficient way to enhance our understanding of a system and makes that understanding intuitive.

13.1.4 Basic CDMA

The techniques used in CDMA have been known since the late 1940s have been used in secure military communications since the early 1950s, and some aspects have been shrouded in secrecy until only recently. As explained in Chapter 5, cellular systems allow access to their trunking resources with various radio techniques. One of these techniques is to deliberately spread the spectrum of a radio sharing access in a system with a high-speed code word unique to that radio. This code further modulates the normal payload data, thus spreading the transmitter's output in frequency with the identifying code. The technique is therefore called *spread spectrum radio*. A receiver's correlator in such a system can distinguish the emissions of a particular radio from many others in the same band by regarding the whole occupied spectrum in careful time coincidence with an exact local copy of the desired spreading code in a manner equivalent to that demonstrated in the access part of Chapter 5. The desired signal is recovered by despreading the signal with a copy of the spreading code in the receiving correlator; all other signals remain fully spread and are not subject to demodulation.

Given the scarcity of spectrum, the technique of spreading the spectrum of a signal far in excess of the bandwidth normally required for information transmission seems foolish and wasteful. Spreading the signal bandwidth by some factor, however, lowers the signal power spectral density by the same factor. The digital cellular systems described up to this point (GSM, IS-54, and JDC) are examples of bandwidth-limited systems. The goal in bandwidth-limited systems is to maximize the transmitted information rate within the allowed bandwidth by increasing E_b/N_0, which is the ratio of bit energy to noise power spectral density. For our purposes in this chapter, we can consider E_b/N_0 to be proportional to the SNR.

As explained in Chapter 5, if we confine ourselves to a fixed bandwidth, and as long as we hold the information rate less than the channel capacity, $R < C$, the bandwidth efficiency, R/B, can be increased as long as we increase the SNR simultaneously with whatever measures we use to enhance bandwidth efficiency. The most common way to increase bandwidth efficiency is to increase the agility of the modulation, or the number of bits represented by each symbol. A symbol is a change in a certain aspect of a carrier (amplitude, frequency, or phase), where the change represents one or more bits. We can employ, for example, the method described in Chapter 12 (DQPSK) to increase the number of bits represented by a symbol. Doing so increases the bandwidth efficiency, since multiple bits are moved across the channel with each symbol. The modulation used in the NADC and JDC systems (explained in Chapter 12) carries symbols, each of which represents two bits. The greater the number of bits in each symbol, the greater the bandwidth efficiency.

The price we pay for increasing the number of bits in a symbol is a tendency to increase the receiver's complexity. The GMSK modulation in GSM carries only one bit per symbol, but GMSK is a very elegant and gentle user of spectrum and does not require a particularly complex receiver. GMSK represents an alternative way of increasing bandwidth efficiency by minimizing the required bandwidth rather than increasing the agility of the modulation.

If we keep a constraint on the bandwidth, we can say that the more bits represented by each symbol, the greater must be the SNR, or the more precise a receiver must be in order for a particular modulation method to be viable at such a bandwidth. Since bandwidth-limited systems are limited by their allotted bandwidths, E_b/N_0 becomes the chief factor affecting error rates. Noise becomes the enemy; all our efforts are turned to keeping E_b/N_o as high as possible, even in deep fades.

13.1.5 Spread Spectrum

A spread spectrum system is an example of a power-limited system. We are not constrained by bandwidth in such systems, and the more we spread information in these systems, the less of an enemy noise power becomes for us. There are several kinds of spread spectrum systems: direct sequence, frequency hopping, and chirping types [2]. In a spread spectrum system, the bandwidth of the transmitted signal is much wider than that required for the information to be carried, and the transmitted signal is modulated a second time with a waveform having nothing to do with the payload information. This second modulating waveform determines the final bandwidth of the transmitted signal. Though GSM employs frequency-hopping techniques, it is not generally considered a true spread spectrum system, because the hopping waveform does not cause the resulting transmitted bandwidth to be much wider than that caused by the information being transmitted. Hopping is used in the reverse direction to give low-power GSM mobiles the advantage of a few decibels of diversity gain, particularly at the edge of cells at handoff time.

The spread spectrum technique used in CDMA is of the direct sequence variety. The spreading is accomplished by multiplying the narrowband information by a much wider spreading signal, which is usually a pseudorandom noise (PN) code. If each radio has its own radio-specific PN code, then all radios, except the one we desire to receive from, look like noise to a receiver in such a system.

13.1.6 Why Go to All the Trouble?

An important characteristic of a spread spectrum system is its *processing gain*, G_p, which is proportional to the ratio of the spreading code rate to the data rate.

$$G_p = B_{\text{spread}}/R$$

where B_{spread} is the PN code rate and R is the information data rate. The PN spreading code is often called a *chipping code*, because it "chops up" or "chips" the much slower information bits. We can write the processing gain, therefore, another way:

$$G_p = R_{\text{chip}}/R$$

where R_{chip} is the chip rate. The performance of a spread spectrum system is most dramatic for G_p values greater than 1,000. In such circumstances, the energy per bit is spread out over G_p chips. In the receiver, the energy in each of the G_p chips is accumulated over the much longer data bit's time. For G_p values this high, information can be transmitted at power levels below the ambient noise, thus making the emissions hard to detect by the casual observer.

Consider moving a heavy and fragile piece of valuable furniture up a flight of stairs. If you hire a moving company that employs bandwidth-limited techniques of furniture moving, the furniture will be moved, without a scratch, by a crew of strong and careful men who carry the piece slowly up the stairs. As an alternative, you could hire a competitive moving company which advertises the use of power-limited techniques. In this case, only one small man shows up to do the job, takes the piece of furniture apart, packs each component in a small box, and then throws all the boxes to the top of the stairs. After this rather curious display of furniture moving ability, the small man goes to the top of the stairs to unpack the components and reassemble the piece, which stands before the happy client in perfect condition. As strange as this technique may seem, which method would you use if the furniture was very large and heavy, and the staircase was long, dangerous, and in very bad repair (N_0 was very high)? We do not have to hire a moving company to use power-limited techniques in moving goods and equipment; such techniques are regularly used to smuggle weapons into prisons and contraband across borders.

There is nothing magical about spread spectrum radio; it is not a countermeasure against the realities of the universe, and is, therefore, not a weapon against low

E_b/N_0 at any chipping speed. Spread spectrum techniques are used to diminish the influence of low E_b/N_0 at the expense of the required bandwidth.

To get a better understanding of the effects of adding a second modulating waveform to a carrier, consider the case where the spreading code is much faster than the information rate, and then refer back to Chapter 5 in which we saw the Shannon-Hartley capacity formula:

$$C = B \log_2(1 + S/N)$$

or

$$C/B = \log_2(1 + S/N)$$

If S/N is less than 1, which is not necessarily the case in a cellular system, then B must be large in relation to C in order to maintain a low error rate in the channel. This means that if we are not limited in bandwidth, we can make the error rate in a channel as low as we like by simply increasing B, which we accomplish by increasing the chip rate of the spreading code. We make B for the channel artificially large for decreasing values of S or for increasing values of N. As we will see, the CDMA system discussed here is a power-limited system, which uses a rather low G_p value of only 64. In such a mode, the spreading effect, though it has some effect on lowering the error rate in the presence of noise, is used primarily to separate users desiring access to the CDMA cellular system by making individual users appear to be noise to each other. Noise is transformed from being the enemy, which it always is in a bandwidth-limited system, to being a helpful tool. The noise tool, however, must be used with respect as well as with a keen eye and a steady hand. Bandwidth-limited systems require constant attention to limit the occupied bandwidth in order to avoid interference with adjacent channels and cells. Power-limited systems require that substantial measures be taken to control transmitted power in order to avoid disturbing users in the same band and cell. The measures employed to control power in the CDMA system are interesting, and will be explored later in this chapter.

Even though it does not fit the strict definition of spread spectrum radio, some people consider FM with a high modulation index, as is used in the FM broadcast services, a form of spread spectrum radio. The modulation index m for an FM system is the maximum carrier frequency shift (the frequency deviation) divided by the frequency of the modulating signal (the rate of frequency shift). If the modulation index is high, then there is more carrier deviation than may be absolutely required for listeners at their receiver. However, the greater the modulation index m the less susceptible the signal is to interference and fades. So, even though a high m does not make a spread spectrum radio, the increased robustness of signals with a high modulation index further validates the capacity formula. There is no inherent advantage or disadvantage in using a spread spectrum technique; it is just another way to use the spectrum.

13.2 THE COCKTAIL PARTY EFFECT

As with so many phenomena in the study of radio technology, where we compare a particular process with some aspect of our more common experience, we can equate a power-limited spread spectrum radio system with a common daily occurrence. Such comparisons enhance our understanding and fix the behavior of the system in our minds. This approach is particularly important in this chapter on CDMA, because our brief explanation of the CDMA system comes from this book on GSM, not from a book all about CDMA. The treatment, therefore, is more intuitive than it is rigorous.

A spread spectrum system operates in a manner similar to the dynamics of a cocktail party. Imagine yourself at such a party. Though it is not necessary that cocktails be served in order to make our comparison valid, the imbibing of modest amounts of alcohol enhances and accelerates the power-limiting effects we will observe. As we try to maintain a conversation with a friend at a crowded party, we have to distinguish our friend's voice from everyone else's, as well as listen through the loud band, which some wish were a skilled orchestra and which brutally competes for our attention. We discern the voice of our friend from among all the others in attendance by correlating his voice with our local knowledge of the sound, language, accent, peculiar characteristics, and idiom of his speech. As the night progresses, we observe two things about this power-limited system:

1. The attendees tend to distribute themselves evenly over the party floor as they strive to equalize the level of their voices with each other.
2. As we try to equalize individual voice levels, the average voice level from all in attendance rises slowly throughout the night until we can scarcely hear the band at midnight.

Especially if the band is a bad one (its playing is more like random noise than ordered sounds), its influence on all those making conversation will be to raise the average voice levels of everyone as the band's playing level increases. The band may slowly increase its playing level to counter the effects of alcohol on individual attendees, who, when intoxicated, find it difficult to understand what others are saying. Eventually, the band may become so loud that all conversation will cease, and everyone will be happy when the band finally takes a break from their playing.

The peculiar characteristics of the individual voices at the party are like spreading codes (PN codes), which we use to separate speakers. If each speaker communicates in a completely different language (the G_p is very high), it is easier to separate conversations at our ears. The lower the G_p of the party, the more similar the voices sound, the more carefully we have to listen to a speaker, and the more intolerant we will be of the band's influence. The band is the thermal noise in the background of our power-limited system, and its playing level determines the starting level of all the conversations at the party. As long as the band maintains an acceptable level, conversations can take place. The band, however, can become so loud that no conversations can occur even if the voice characteristics separating one speaker from another are very distinct. There is a soft limit to the number of conversations that can take place simultaneously on the party floor. As the party becomes more crowded, individual voice levels may have to increase. At some point, we will have to move the party to a larger hall or ask the band to leave.

In order to further extend our comparison of the cocktail party to the CDMA system, we must add some curious and strange constraints. If we want to increase the number of guests at the party, which is the same as saying we want to increase the system's capacity, we must add discipline and order to our power-limited party. To accomplish this, we have to do three things.

First, since local law requires we have a band (thermal background noise is an invariant quality of the universe), we hire a good one, which agrees to play at the lowest allowed level. Second, before walking onto the party floor, we agree, in advance, that as the number of attendees increases throughout the night, we fight the natural tendency to talk louder, and decrease our individual voice levels (rather than increase them), thus keeping the average level from all in attendance constant. Third, and very strange indeed, the host, who is a charismatic, popular, and eccentric individual (a cellular base station), has decreed that the only conversations allowed to occur are those between himself and everyone in attendance; attendees cannot converse with each other. The host, you see, owing to some remarkable dexterity of tongue and palate, has the curious ability to carry on multiple conversations at once. The host requires, therefore, that everyone at the party speak at such a level that the voices reach the host's ears at exactly the same level relative to each other; those far from the host must talk louder than those close to him.

The need for strict power control of individual transmitters in a real spread spectrum system has origins that are similar to those we observe at a cocktail party. This is in sharp contrast to bandwidth-limited parties, where people speak as loud as required in order to ensure good communications. A bandwidth-limited party is only required to keep itself confined within its assigned, rented facilities and keep the level of revelry below the level that would disturb the power-limited party next door.

Precise, accurate, and fast power control requires, among other things, that the power in a CDMA channel be adjusted *800 times a second*. Power control is both the most interesting aspect of the CDMA system and its greatest challenge. Only recent advances in device integration have made this type of access feasible in a widely deployed commercial system. Network operators accustomed to frequency planning in their FDMA and TDMA systems can abandon such concerns with CDMA as they learn power planning.

13.3 DIVERSITY IN CDMA

Land mobile radio systems use multiple versions of the same signal, in one or more domains, to counter the destructive effects of fading in the mobile channel. The CDMA system makes use of diversity in three different domains. We say that we use three types of diversity: frequency diversity, spatial diversity, and time diversity.

13.3.1 Frequency Diversity

Since each signal in a CDMA system covers a relatively wide part of the spectrum, frequency diversity is an inherent feature of CDMA. Multipath fading is caused by different delays among the alternative paths between a mobile and a base station. If we regard the CDMA signal in the frequency domain with a spectrum analyzer, it appears as a band of noise about 1.23 MHz wide (see Figure 13.1). A fade will appear to be a notch filter connected to the input port of the spectrum analyzer. The width of the notch is approximately equal to the inverse of the delay between the paths arriving at the receiver. For example, if the delay between paths is 1 μs, then the notch filter will be about 1 MHz wide. If the receiver moves, the notch filter will move across the spectrum occupied by the signal. As the delay between the paths becomes shorter in time, the canceling of the signals from the alternative paths becomes more complete and the notch filter becomes wider. Only delays less

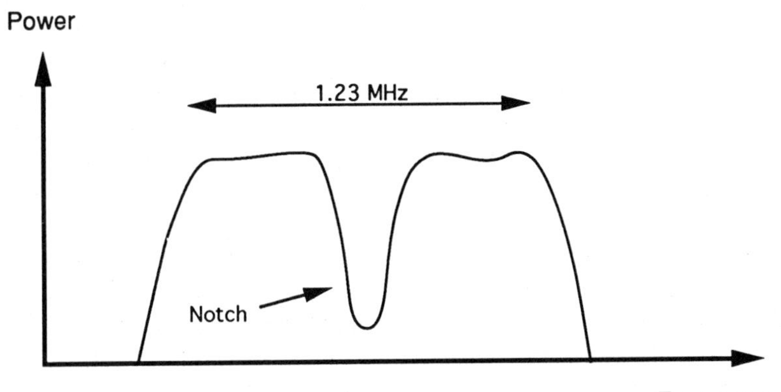

Figure 13.1 Moving notch in a CDMA signal.

than 1 μs will drop a CDMA signal into a deep fade. Delays much longer than 1 μs cause only a power reduction of the received CDMA signal.

13.3.2 Spatial Diversity

The CDMA system uses multiple antennas at the base site's receiver, just as many FDMA and TDMA systems do. If two antennas at a receiver site are far enough apart, then the delays between the paths to these antennas from the mobile will be different, and it will be unlikely that a fade at one antenna will also be experienced at the other antenna at the same instant. An interesting feature of CDMA is that the system allows multiple base stations to transmit to a single mobile, and this allows soft handoffs to be employed. A soft handoff is a *make-before-break* handoff. A more traditional handoff, such as is used in GSM, is a *break-before-make* process. The ability to allow multiple base stations to transmit to a single mobile, with its inherent advantages against multipath fading is, as we will see, made possible by the rake receiver used in CDMA radios (see Section 13.3.4).

13.3.3 Time Diversity

Some of the more common measures traditionally used to employ time diversity in most cellular systems are used in CDMA as well. Typical channel coding schemes are used, where data is convolutionally coded and then it is interleaved before modulating the carrier. In the QUALCOMM system, the forward channel, which is carried by transmissions from the base to the mobile, employs half-rate convolutional coding, which doubles the number of bits representing the original information bits. The reverse channel, which is carried by transmissions from the mobile to the base, employs one-third-rate convolutional coding. This means that two redundancy bits are generated for each information bit; the number of bits representing the original information is multiplied by three. All the bits, both information and redundancy bits, are interleaved, which means they are strewn over a part of the data stream in a pseudorandom order. Both these measures have the effect of separating adjacent bits and filling the space between them with redundant ones. If a sufficiently wide moving notch filter, depicted in the frequency diversity discussion above (Section 13.3.1), moves across the CDMA signal's spectrum, bursts of fade are less likely to obliterate long strings of important data. Only isolated bits will be lost, and these kinds of isolated errors are easier to correct at the receiving end of the channel than the obliteration of long sequences of data. On the receive end of the channel, a Viterbi decoder with soft decision points is used. Viterbi decoding is a fast way to decode a convolutional code. The decoder makes 1 and 0 bit decisions as it works its way through the received code toward the decoded output data. The demodulator ahead of the decoder can pass additional information to the decoder concerning the quality of each recovered symbol from the demodulator. This kind of help from the demodulator is called *making soft decisions on the demodulator's output.* The demodulator marks noisy or questionable symbols as it passes them to the channel decoder. The decoder can choose to ignore bad symbols rather than try to work through them in an effort to recover the original data.

13.3.4 Rake Receiver

Figure 5.7 shows the multipath situation in a mobile environment. The same conditions apply in a CDMA system. Consider a CDMA mobile moving through a cell among buildings, mountains, and other obstacles, which reflect the signal from the host cell site. The different rays arrive at the mobile's antenna at slightly different times (on the order of 1 μs), and will have some arbitrary phase relationship with each other.

As the mobile moves through the field, the rays will usually add destructively, depending on their relative phases at any instant. This, as was also explained in Chapter 5, is the cause of the very destructive fast (Rayleigh) fading phenomenon. All of the diversity schemes explained in this section are designed to counter the extremely destructive effects of this kind of fading. Even though the CDMA system enjoys some frequency diversity gain because of its wide bandwidth, considerable improvement in performance comes from adding the power from several rays in a rake receiver. Among cellular systems, the rake receiver is peculiar to the CDMA system. It adds both spatial diversity and time diversity to the system.

Conceptually, a spread spectrum signal is usually recovered in two steps. First, the fast PN sequence spreading signal is removed, thus collapsing the bandwidth of only the selected signal with the appropriate PN code. All of the other signals remain spread. Second, the remaining narrowband signal is demodulated. Several rays appear at the receiver's antenna, and each ray will have a slightly different phase from the others. The relative phase relationships will change in a random fashion with time. A selected channel from each ray is separately de-spread with an appropriate PN sequence, which is the same PN sequence as that used to spread the original channel at the base station's transmitter. Since each ray arrives at the receiver's antenna at different times, each despreading PN generator in the rake receiver has its phase dynamically adjusted to correlate with different rays containing the same channel information. The recovered and despread data streams appear at the outputs of the mixers in Figure 13.2.

The de-spread signals no longer contain the PN chipping sequence and are sent to an n-branch linear optimal combiner, where $n = 4$ in a CDMA mobile.

Each branch has three stages:

1. The τ stages adaptively cancel delay spread among each of the despread rays. This is called *τ dither tracking.*

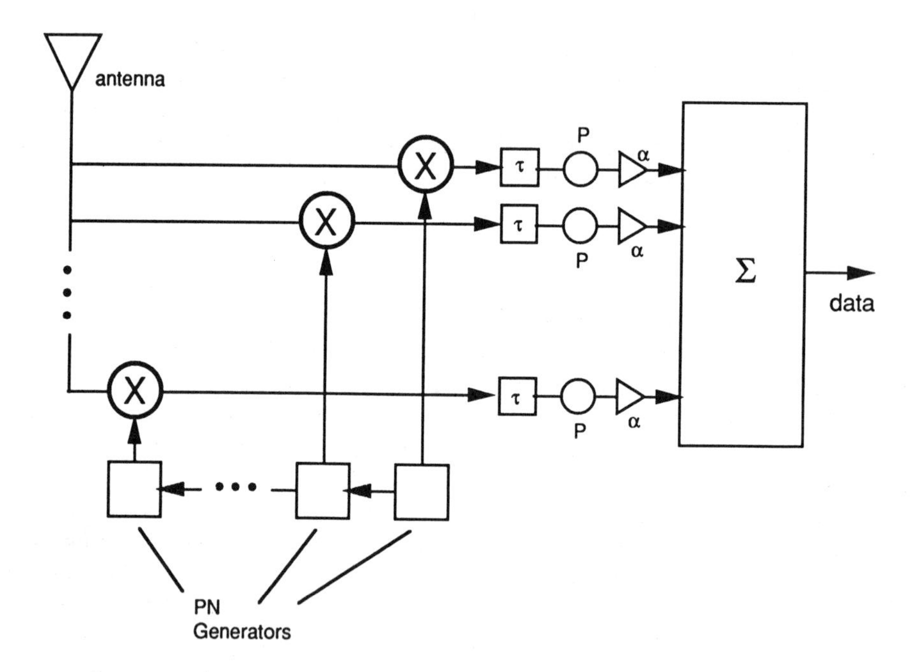

Figure 13.2 Rake receiver.

2. The P stages adaptively adjust the phase of each branch.
3. The α stages adaptively equalize the level out of each branch.

Each of the four branches is called a *finger*, and three of the fingers are optimally combined. The fourth finger is called a *roving finger*, which is used to seek out the next ray to be assigned as one of the three combining fingers.

In addition to time diversity, the rake receiver is used to add spatial diversity and accomplish the soft handoff function. As the mobile moves toward the edge of a cell, the mobile telephone switching office (MTSO) senses that the mobile is running out of RF power range and assigns the mobile's PN sequence to a neighboring cell's host transmitter, which is assigned the same carrier frequency. Two sites are now transmitting traffic to the mobile with the same identifying spreading sequence on the same physical channel. The mobile's rake receiver simply considers one of the rays coming from the newly assigned cell site transmitter as just another ray in need of optimal combining. As the mobile moves further into the new site's coverage area, all its rake fingers finally correlate with rays from the new site, and the old site drops the PN sequence and corresponding traffic from its transmitter.

13.4 CDMA DUAL-MODE BAND ASSIGNMENTS

CDMA channels are allowed to coexist within AMPS systems simply by having the operator clear spectrum space in units of 1.23 MHz to accommodate them. Just as they are in the original AMPS system, the forward and reverse channels are separated by a duplex offset of 45 MHz, where the forward channel is 45 MHz higher than the reverse channel. Figure 12.1 in the previous chapter shows the cellular channel assignments in North America. The primary CDMA physical channel is centered about channel number 283 for the A (non-wire-line) operators, and channel 384 for the B (wire-line) operators. A single CDMA physical channel takes up 41 of the 30-kHz AMPS channels. Operators can, if they wish, assign the same CDMA physical channel to neighboring cells, since frequency planning is not required in the CDMA system. Using a spectrum analyzer to examine the AMPS spectrum in which CDMA resources are present will show 1.23-MHz segments of noise where the CDMA resources are placed. Figure 13.3 is an actual print from the spectrum analyzer in a radio test set and shows one CDMA channel just above some AMPS traffic channels. The figure shows a CDMA channel centered at 892.750 MHz under the center-screen marker. We know the signal is from a base station, because the base station transmits on the high side of the duplex channel. The high side of the duplex channels in the American system lies between 869.01 and 893.97 MHz. The spectrum analyzer's span setting is highlighted and shows the span to be 500 kHz/division. The signal covers slightly more than two divisions in the center of the screen. The print in Figure 13.3 was acquired with the radio test set only a few hundred meters from the CDMA base station. Even at this close proximity, note that the CDMA signal is 30 or 40 dB lower than the accompanying AMPS signals, which were much farther away.

When it is first turned on, a CDMA mobile looks for a physical CDMA channel. If the mobile does not find a CDMA channel, it reverts to traditional narrowband AMPS operation and seeks a control channel among the conventional analog

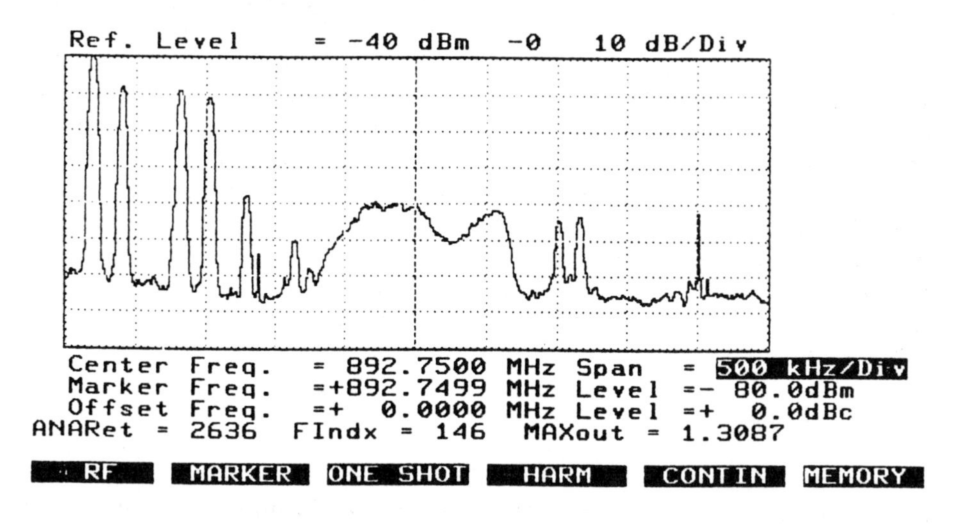

Figure 13.3 A CDMA channel within an AMPS system.

resources. If the mobile does, in fact, find a physical CDMA channel, it looks for and then synchronizes with some logical channels within the forward physical channel. These logical channels carry important control information the mobile needs to continue its successful access into the system. The logical control resources are separated from each other by different spreading codes rather than frequency and time as is done in GSM.

13.5 CDMA FORWARD CHANNEL

The *CDMA forward physical channel* transmitted from the cell site in the 870-MHz band is more accurately called the *forward waveform*. The individual logical channels in the forward waveform are shown in Figure 13.4, and are distinguished from each other by their spreading codes, W0 through W63. It is important to note that when we depict the individual channels as shown in Figure 13.4, we keep in mind that the channels are not separated in frequency or time, as is the case with GSM. The channels are distinguished from each other by their own codes, and are transmitted on top of each other, from the base station to the mobiles, with a common final modulating code identifying them with a particular base station. It is easy to become confused when discussing CDMA waveforms and channels. In this chapter, we call the collection of all the channels on a single carrier, a waveform. We call the individual resources within the waveform *channels*.

In Figure 13.4, we describe the various logical resources in the forward waveform by showing how they are created. The 64 different Walsh functions (W0 through W63), which separate the channels, are shown in boxes across the top of the figure. W0 is always reserved for a resource called the *pilot channel*. W32 is assigned to an optional SCH. W1 through W7 can be assigned to paging channels if needed, and W1 through W63, together with W32 (if W32 is not used for a SCH), can be assigned to traffic channels (T1 through T63). Below each of the boxes are flow diagrams illustrating how each type of channel, within the forward waveform, is created. The signal flow is from the top to the bottom of the figure. The traffic channel's signal flow is not included in the figure, but is thoroughly explained in Section 13.7 of this chapter. Each channel starts with some payload data of some kind. The pilot channel's payload data, for example, is a constant string of zeros. All the other channels, except the pilot channel, experience some channel encoding, and all the channels are spread with a Walsh function specific to each channel. Finally, all of the channels making up the forward base station's output are spread with a common PN code.

13.5.1 Pilot Channel

Since timing is extremely critical for proper despreading and demodulation of CDMA signals, there is always one pilot channel with its own identifying spreading code,

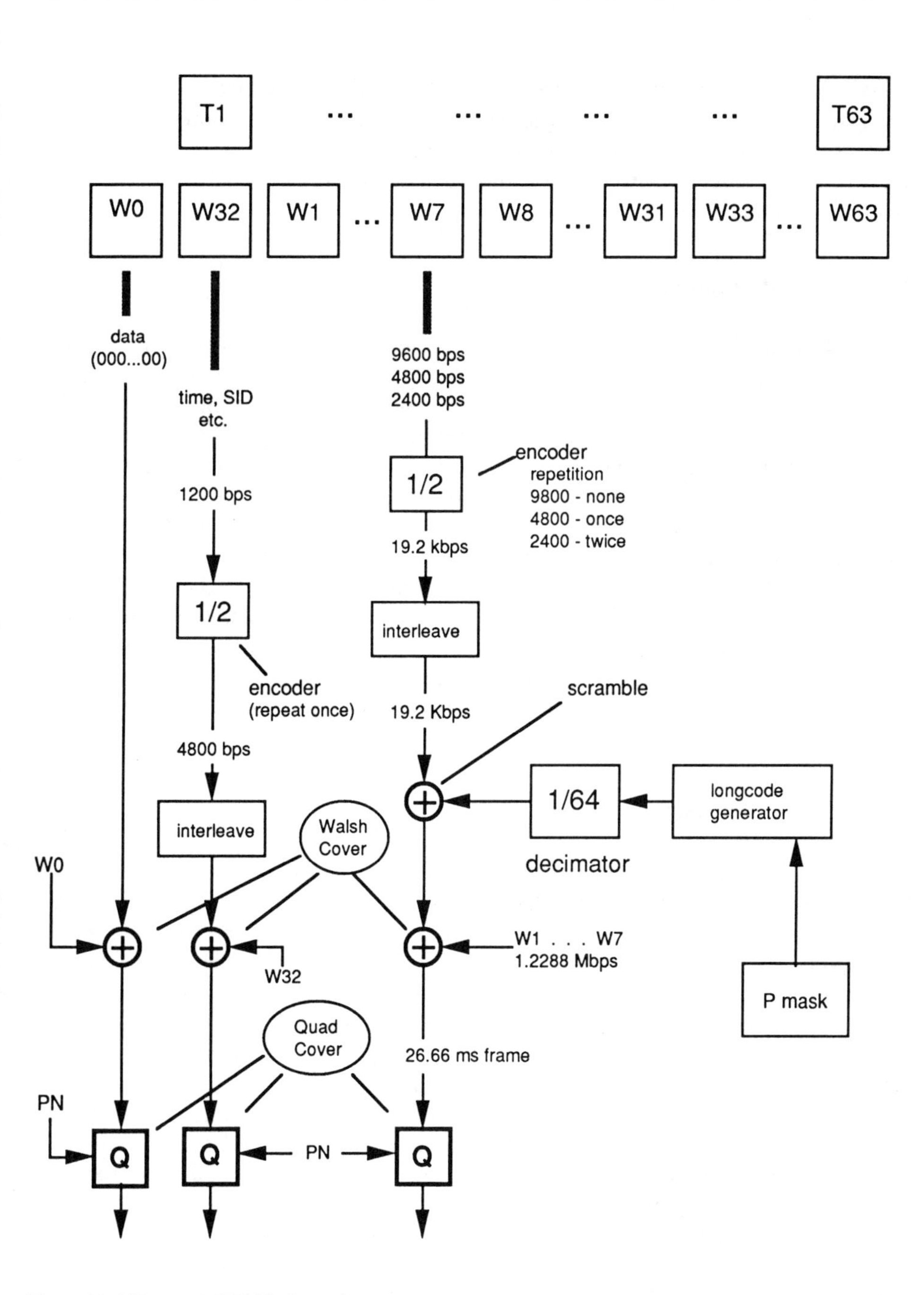

Figure 13.4 Forward CDMA channels.

called *W0*, for the mobiles to refer to. Walsh functions modulate their particular channels at a rate of 1.2288 Mbps. The pilot channel's data content is a trivial sequence of zeros and is not channel-encoded. The mobiles use this channel to synchronize themselves with the base station so they can recognize any of the other channels they may need to.

13.5.2 Synchronization Channel

The SCH is usually present, but may be omitted in very small cells. In the latter case, a mobile gets its required synchronization information from a neighboring cell. When the SCH is present, it is always spread with Walsh function number W32. The SCH's data contain a large amount of information, which all the mobiles in a CDMA system need, and which includes such things as the system identification number (SID), some access procedures, and some precise time-of-day information. The SCH's information is created at rate of 1,200-bps and is half-rate channel-encoded before it is repeated only once to yield a 4,800-bps data rate entering a bit interleaver. The output of the interleaver is modulo-2-added with the SCH's Walsh function W32. Modulating the channel encoded data with a Walsh function is called *giving the data a Walsh cover*. All the logical channels in the forward waveform are spread one more time with a common PN code to generate QPSK. The PN spreading is called *quad covering*, and this process is discussed in more detail throughout this chapter.

13.5.3 Paging Channels

Paging channels are optional, and there are never more than seven of them on a base station's waveform. Paging channels are covered with Walsh functions W1 through W7 at 1.2288 Mbps. PCH data occur at one of three rates (9,600, 4,800, and 2,400 bps), is half-rate-encoded and is repeated a number of times inversely proportional to the original data rate. The output of the interleaver is a constant 19.2 kbps, which is modulo-2-added with a scrambling code. The scrambling code is generated from a decimated version of a very long data stream called a *long code*. The long code is prevaricated with a mask peculiar to paging channels. The mobiles know the long code and the mask that alters it. After Walsh covering, the 26.66-ms frames are quad-covered just as all the other channels in the forward waveform are. The various QPSK outputs are simply added together to make the forward physical channel waveform.

There can be as many as 63 traffic channels, each with its own identifying code, selected from what remains of the set of 64 Walsh codes. If there is a full complement of synchronization and paging channels, then there is a maximum of only 55 traffic channels. System consideration restricts the actual number of assigned traffic channels to a much lower number. Section 13.7 includes a detailed description of how the forward traffic channels are created, and you will see that the traffic channels are merely fancy versions of the paging channels.

13.5.4 Quad Cover

All 64 logical channels are finally modulated, as we have seen, one more time at the base station with a common PN code, which is 2^{15}-1 bits long. The PN code is called a *short code* and has a rate of 1.2288 Mbps. All base stations have the same PN spreading code, but each base station sets its own offset in the short code. There are 512 possible offsets for the short PN spreading code, and the offsets have names 0 through 511. A mobile can easily distinguish transmissions from two base stations by their distinguishing short-code offsets, and then the individual logical channels are distinguished from each other with their Walsh functions.The pilot, synchronization, and paging channels are simplified versions of the forward traffic channels, which we have not examined yet. In contrast to the more complicated traffic channels, the auxiliary channels discussed here (pilot, synchronization, and paging) are simplified versions of the traffic channels, which we explore in Section 13.7.

13.6 CDMA REVERSE CHANNEL

The reverse channel is transmitted from all the mobiles within a particular cell site's coverage area, in the 825-MHz band, to the base station. This means that the mobiles transmit on a band offset 45 MHz below the base station's forward channel output. Figure 13.5 shows how the reverse channel may appear in the base station's receiver. The reverse channel appears as a composite of all the outputs from all the mobiles in the base station's coverage area, and Figure 13.5 depicts each of these outputs as a box numbered 1 through *m* and 1 through *n*. Since they do not all come from the same place, we call the base station's view of all the mobile's outputs the *reverse channel* rather than the reverse waveform. At any time, there may be *m* mobiles engaged in moving traffic and *n* mobiles trying to gain access to the system. Though the reverse channel structure allows up to 62 different traffic channels and up to 32 different access channels, considerably fewer than these numbers of channels will exist at any instant.

Each PCH on the base station's forward waveform can support up to 32 access channels and 62 traffic channels in the reverse direction from the mobiles. In a manner almost identical to the forward channel, all the different mobiles' contributions to the reverse channel are spread twice. First, since each of the logical channels will come from a specific mobile, the mobile spreads its output with a mobile-specific long code at a rate of 1.2288 Mbps with a modulo-2 adder. Second, the adder's output is spread again at a rate of 1.2288 Mbps with the same short-code PN

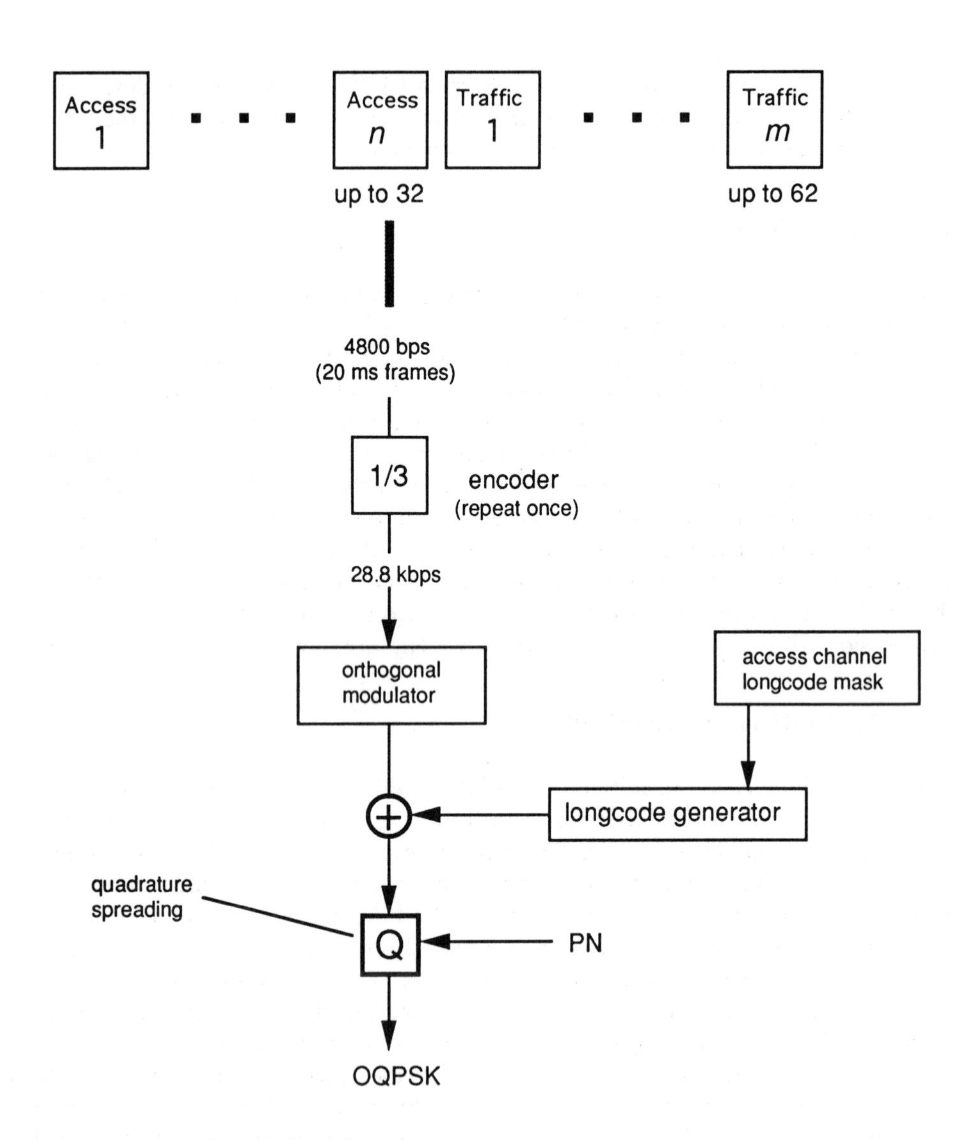

Figure 13.5 Reverse CDMA channel.

sequence the mobile sees on the pilot channel on the base station's forward waveform. The second process is called *quadrature spreading* and gives the mobile a cell-specific code to augment the mobile-specific long code that preceded it. All CDMA mobiles have their own long codes, each of which is a function of its manufacturer and serial number. The long code is, indeed, very long: $2^{42} - 1$ bits!

In the course of acquiring a traffic channel from the base station, the mobile speaks in its own unique voice by sending its ESN to the base station. The base station uses the received ESN to alter one of its many modulating waveforms into a user-specific code, which the mobile recognizes as its own.

13.6.1 Access Channel

There are only two types of channels on the reverse channel, traffic channels and access channels. Section 13.8 describes, in considerable detail, how the reverse traffic channel is constructed. The access channels are almost identical to the traffic channels. Mobiles use the access channel to initiate communications in the system and to respond to PCH messages. The access channel's data occurs at a fixed 4,800-bps rate in 20-ms frames and contains information the network needs to properly log the mobile terminal into service. This information includes, for example, the ESN of the mobile, the MIN, the SCM, and the home system number (SID). The 20-ms frames are one-third-rate channel-encoded and repeated once to yield 28.8-kbps channel-encoded data. A function called a *64-ary orthogonal modulator* follows the interleaver. This function substitutes groups of 6 code bits for all 64 bits of one of the 64 available Walsh symbols. This unusual process is explained in more detail in Section 13.8. The orthogonal modulator's output is covered with a masked long-code sequence, which distinguishes any particular access channel from all the other channels the base station may receive. The long code is altered with a mask that includes such things as the access channel number (*n*), the corresponding PCH in the base waveform, and the current PN code offset. Finally, each access channel is quadrature-spread with a cell-specific PN code to yield OQPSK. This last modulation process is also discussed in more detail in Section 13.8.

13.6.2 RF Power Bursting

As we will see in Section 13.8 of this chapter, once it is assigned to a traffic channel, the mobile's transmitter power is bursted on for a duty cycle inversely proportional to the amount of traffic it has to send to the base station. The more traffic the mobile has to send, the higher the information data rate, and the longer it keeps its RF output intact. The access channel, on the other hand, has a fixed 4,800-bps information rate, and its transmission duty cycle remains at a full 100%. We will see in Section 13.10, however, that even as the access channel's duty cycle is 100%, the mobile does not transmit continuously when it is in an access mode.

13.7 CDMA BASE STATION TRAFFIC CHANNEL

The logical channels in spread spectrum systems are best described, as we have already seen in the previous two sections, by the way they are constructed with the various coding and spreading waveforms that modulate them. In this section, we confine our attention to the traffic channel on the base station's forward waveform. During the discussion, you will note that the forward traffic channel is an elaborate version of the more simple pilot, synchronization, and paging channels.

13.7.1 Payload Data

We start with the payload data in the *variable-rate speech encoder* block on the left side of Figure 13.6. This speech encoder is sensitive to the amount of speech activity present on its input, and its output will appear at one of four rates: 9,600, 4,800, 2,400, or 1,200 bps. These rates change in proportion to how active the speech input may be at any time. The rate is subject to change every 20 ms. The speech encoder's output is convolutionally coded at a half rate, thus doubling the data rate to 19.2 kbps when the input is 9,600 bps. When the speech encoder operates in one of its three reduced rates, the error coder repeats the 20-ms frames as often as required in order to maintain a full 19.2-kbps output to the interleaver. The bits are interleaved (shuffled) before they are modulo-2-added to another 19.2-kbps scrambling code from a 1/64 decimator. The decimator selects every 64th bit from a long-code generator running at 1.2288 Mbps.

13.7.2 Long-Code Generator

The long-code generator creates a very long code, which is $2^{42} - 1$ bits long, from a user-specific mask supplied by a CDMA mobile requesting a traffic channel. Such a code has a period of 995 hours when running at 1.2288 Mbps. The mobile-specific character of the traffic channel's long code comes from the user mask, which alters the long code. The altered long code scrambles the forward traffic in a manner the target mobile can recognize as belonging to itself. For clarification, we review access in cellular systems. Analog cellular systems separate traffic channels by frequency with FDMA techniques. In addition to using FDMA, GSM further separates traffic with TDMA techniques when it separates traffic according to assigned time slots. The CDMA system abandons the TDMA technique and replaces it with a series of codes that "color" the data from a base station's waveform destined for a particular mobile. The "coloring" is the masked long code.

13.7.3 Power Control Subchannel Multiplexer

A second decimator accepts the 19.2-kbps scrambling code from the previous 1/64 decimator (used for the channel's scrambling code), and divides the 19.2-kbps data stream into 1.25-ms frames of 24 bits each. Next, the decimator selects the first 4 bits of each 24-bit frame and sends these selected bits as an address to the multiplexer

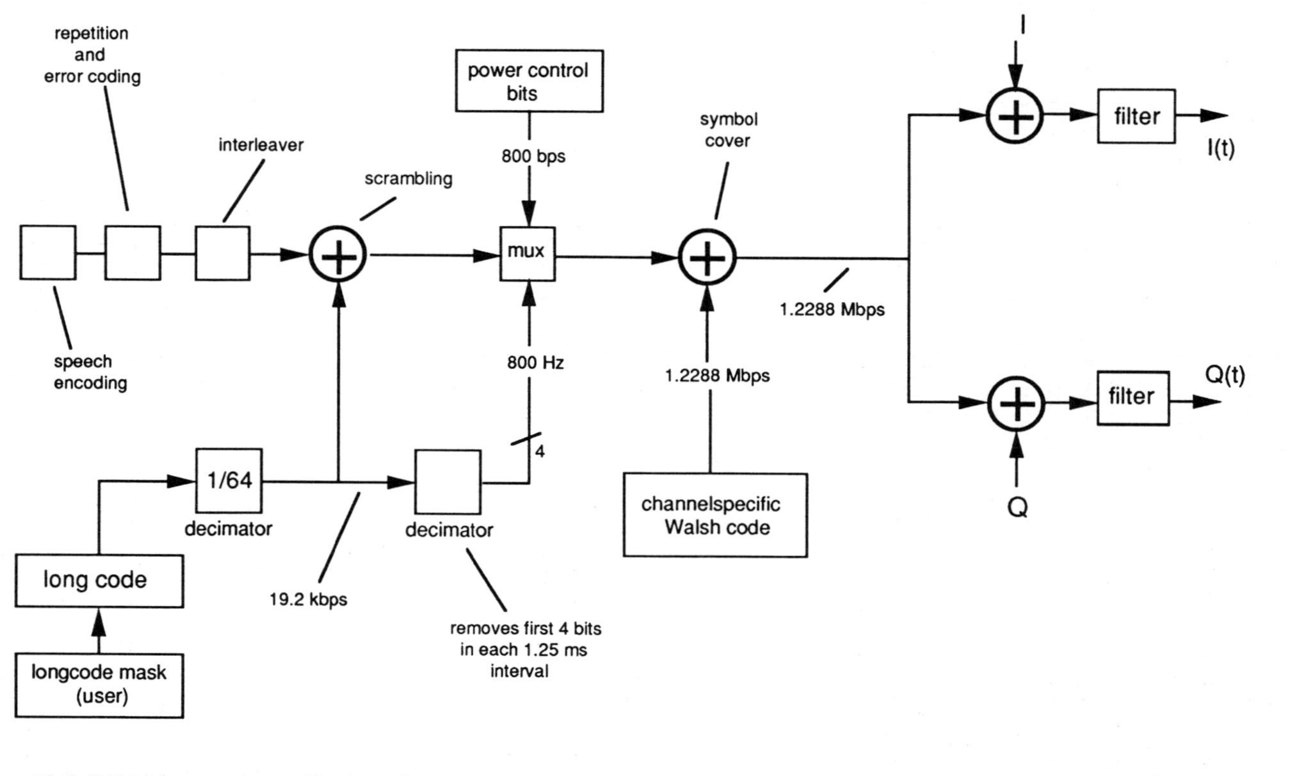

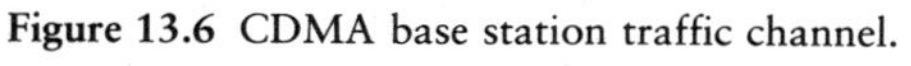

Figure 13.6 CDMA base station traffic channel.

(mux); 4 bits yield 16 possible addresses. The multiplexer builds a power control subchannel by "puncturing" the 19.2-kbps data stream once in every 1.25-ms frame and replacing the punctured bit (together with the one following it) with a single power control bit. The power control bit is two bits long and can be a 1 or a 0, and will replace two bits starting at any of the first 16 bit positions in each 24-bit frame entering the multiplexer from the scrambler. This puncturing occurs at an 800-Hz rate (once in each 1.25-ms frame) with ruthless disregard for how important the sacrificed bits may be. The mobile's decoder is left with the task of separating the power control subchannel from what remains of the traffic data, and then repairing the damage to the remains of the encoded data stream. Once the power control subchannel is recovered, the mobile uses the data therein to make fine adjustments to its RF output power. Section 13.9 more fully explains how RF power is controlled in the whole IS-95 system. The rather desperate technique described here should give the reader an indication of how important precise RF power control is in a CDMA system.

13.7.4 Symbol Covering

The 19.2-kbps data stream, which has been punctured with the power control bits of the power control subchannel, is modulo-2-added to a channel-specific 1.2288-Mbps Walsh code selected from a small catalog of available codes (usually W8 through W63). The selected Walsh function gives the traffic channel an identity (see Figure 13.4). This process is called *Walsh covering*; we "cover" the encoded information bits with a Walsh function. The Walsh functions are the codes the mobile's receiver uses to distinguish between the logical channels it receives from the base station's forward waveform. These Walsh codes are only 64 bits long and repeat every 52 μs.

13.7.5 Quadrature Cover

The forward channel's content is spread one more time with a PN sequence specific to the base station. Actually, all the base stations use the same PN sequence, but each base station selects from among 512 possible offsets for its own identifying spreading code. The PN sequence is 2^{15}-1 bits long (32,768 bits) and exists in two forms: its I-form and its Q-form. The forward channel uses QPSK modulation as shown in Figure 5.41. Such a modulator requires two waveforms in quadrature. The reader should note that this type of modulation carries two bits per symbol, and the symbol information is contained in the absolute phase of the carrier rather than the change in the phase. This is called *quadrature covering* (or quadrature spreading). The PN's data rate is 1.2288 Mbps, which makes its period 26.66 ms.

13.7.6 Signaling

Signaling can occur on the forward traffic channel. A trivial example is the power control subchannel, which is always present in the forward traffic channel. Other types of signaling must periodically occur in order to maintain the integrity of the channel. This type of signaling appears in either blank and burst mode or one of three dim and burst modes. In the blank and burst mode, all of the speech data from the speech encoder are replaced by important signaling data. In the dim and burst modes, only some of the speech data are replaced by signaling data at rates inversely proportional to the accompanying speech data rates. The signaling data substitution is made at the input to the convolutional encoder on the left side of Figure 13.6, at the repetition and error coding block.

13.7.7 Other Forward Channels

Now we will review the other simpler forward channels. The forward pilot, synchronization, and paging channels are simple modifications of the forward traffic channel. All the forward channels are quadrature-spread with a 1.2288-Mbps-delayed PN sequence before being combined in the base station's QPSK modulator. The pilot channel gets a W0 Walsh cover on top of null data (000 . . . 0). The SCH has a fixed 1,200-bps data rate, which contains important site information such as SID (site identification) and a large amount of system timing information. The 1,200-bps data are convolutionally encoded in a half-rate encoder and repeated to bring the data rate to 4,800 bps. the data are interleaved and then modulo-2-added with Walsh function W32. The paging channels are almost identical to the traffic channels, except that the long code's scrambling mask is a public one rather than a confidential user-specific mask. Moreover, the identifying Walsh codes can only be selected from W1 through W7. There is no power control subchannel in any of these overhead channels; there is no bit puncturing.

13.8 CDMA REVERSE TRAFFIC CHANNEL

Just as we did in the previous section with the base station's forward traffic channel, we will consider the mobile's reverse traffic channel in Figure 13.7, starting with the payload data.

13.8.1 Payload Data

The *payload data* come from a variable-rate speech encoder with four possible output data rates: 9,600, 4,800, 2,400, and 1,200 bps. The rate depends on the speech

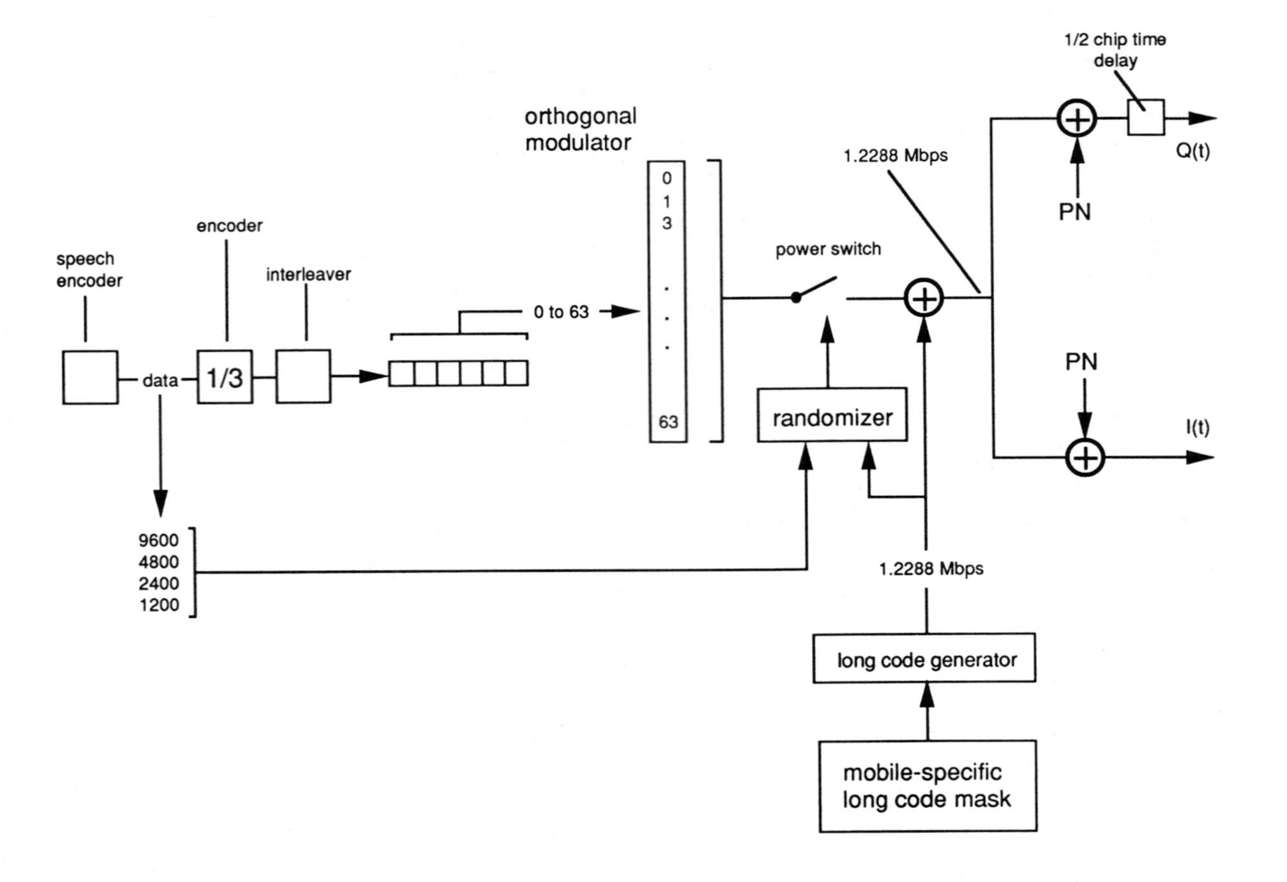

Figure 13.7 Reverse CDMA traffic channel.

activity. Typical speech activity has this speech encoder running at its lowest rate about half of the time. Just as in the base station, the data rate is subject to change every 20 ms and can include signaling information.

The data from the speech encoder are convolutionally encoded in a one-third-rate encoder, which adds two redundancy bits to each data bit, thus multiplying the data rate by three. The data rate out of the one-third-rate encoder is the full 28.8-kbps rate over the usual 20-ms frames. The data are interleaved before they enter the orthogonal modulator. Now, unlike the situation in the base station's forward traffic channel, where the speech data frames are repeated when the speech encoder is running at one of its reduced rates, the mobile gates its output power off, at the power switch in Figure 13.7, just before modulating the data with the mobile's long code. The mobile station turns off its RF output rather than repeat the data.

13.8.2 Orthogonal Modulator

The encoded data from the bit interleave enter a shift register, which is six bits long. Each of the six-bit data segments selects one of 64 possible Walsh codes, W0 through W63. Since there are 64 possible values represented in the six-bit shift register, all 64 Walsh functions are possible. One of the 64 available Walsh codes is selected by the bit pattern in the shift register, then another Walsh code, and another, and so on. Each of the Walsh codes is 64 bits long, and they come from the same little Walsh code catalog with which the base station identifies its logical channels. Each six bits of data is replaced by another 64-bit Walsh function. Walsh chips, then, exit the orthogonal modulator at a 307.2-kbps rate. This is called the *chip rate.* Since the mobile does not have a pilot channel to offer the base station's receiver, it modulates its data with Walsh codes. This has the effect of making the mobile's reverse channel look like a bunch of base station logical channels changing, with time, in accordance with the encoded payload data.

13.8.3 Long-Code Spreading

For now, we will leave the power switch on the output of the orthogonal modulator in the closed position as we see that the 307.2-kbps Walsh chips are modulo-2-added to a 1.2288-Mbps spreading code, which is unique to each mobile station. The spreading code comes from a long-code generator almost identical to the one in the base station, except that the 42-bit long-code mask can be either a public or a private one. The private mask is based on the mobile's ESN. The long code is $2^{42} - 1$ bits long, which makes its period 995 hours at the 1.2288-Mbps spreading rate.

13.8.4 Quadrature Cover

Just as we did in the base station, we cover our transmitted data (burst this time) one more time with a short code, or PN spreading code. The mobile station adopts the PN code it sees on the base station's pilot channel. To conserve battery power, however, the mobile uses OQPSK, which it generates by delaying the Q-adder's output by one-half chip time before applying the code to the modulator. As Figure 5.42 shows, OQPSK avoids the origin and this makes efficient RF power amplifiers easier to design.

13.8.5 Power Bursting

Power is gated on and off in the mobile to reduce the average power in the reverse channel as it appears at the base station's receiver. This has the effect of reducing the interference levels throughout the CDMA system and improves system capacity and performance.

Now we return to the power switch, which we left in the closed position when we looked at the long-code spreading function. If the voice encoder is running at its full 9,600-bps rate, the power switch will remain closed. If, however, the voice encoder is running at one of its reduced rates, the power switch will be opened rather than let repeated payload data frames pass through. The mobile's transmitter is gated on for only one-half, one-fourth, or one-eighth of each of the 20-ms data frames for data rates of 4,800, 2,400, and 1,200 bps, respectively. The power bursting instructions from the speech encoder are sent to a randomizer function, which adds a pseudorandom character to the power bursting. The randomizer selects a quasirandom number of 1.25-ms symbol frames to pass on to the long-code spreader before opening the switch and gating the RF power off. The timing still works in the reverse channel, because the orthogonal modulator has chopped up the original data into much faster Walsh chips, which the base station can still recognize.

13.9 POWER CONTROL LOOPS

For the CDMA system to work properly, RF power in the system must be controlled in two aspects. First, and most important, all the mobile's transmissions must be received at the base station's receiver within 1 dB of each other, even under fast multipath fading conditions. Second, in order to allow as many users as possible to share the use of a cell, only the minimum RF power required for reliable communications is allowed from any base station transmitter. It is normal for a CDMA channel to operate in a constant state of rather high BER when compared to GSM or any other bandwidth-limited systems. We have seen some extraordinary measures installed in the base and mobile units to accomplish these goals.

1. The base station creates a separate RF power control subchannel to adjust the mobile's RF output power 800 times a second in small steps.
2. The mobiles gate their RF power off when transmissions are not absolutely necessary. This would be the case under reduced data rate conditions.

There are three major RF power control mechanisms in the CDMA system.

13.9.1 Open Loop

The mobile begins by estimating the loss in the path between itself and the base station, and then it makes an initial course adjustment in its RF power output based on this estimate. To do this, it despreads and then measures the RF power on one of the forward channels from the base. The mobile combines this reading with some control information the base station sends to the mobile during some initial signaling transactions. This information is a set of operating parameters such as the base station's actual output power, the current traffic load serviced by the base station, and some other details the mobile needs to make its initial RF output power adjustment.

13.9.2 Closed Loop

Path losses vary with frequency in the 800-MHz band. Mobiles make their initial path loss estimates on the base station's forward channel. Since the mobile responds to the base station on the reverse channel, which is 45 MHz lower in frequency than the forward channel is, it is not likely that the initial open loop path loss estimate will be particularly accurate for the reverse channel. The base station, therefore, measures the level at which it receives the mobile's transmissions, and sends a 1 or a 0 bit in its RF power control subchannel back to the mobile. A 0 means an increase in power, and a 1 means a decrease in power. These corrections by the base station continue at an 800-Hz rate as long as the channel between the base and the mobile is established. The mobile responds to the bits it receives on the forward power control subchannel by making small adjustments between 0.5 and 2.0 dB in its RF output power. The size of the adjustment depends on the mobile's data rate at the time the adjustment is needed, as well as the time since an adjustment was last made. Mobiles can make closed-loop adjustments within 24 dB above or below their initial power settings.

13.9.3 Base Station Transmission Power Control

In order to keep the total RF power on the forward channel low, which will allow a maximum number of mobiles to share the use of a cell and not disturb neighboring

may be tested by the brute force method described in the Chapter 12 to an extent inversely proportional to the level of circuit integration in the equipment. The more integrated and modular the radio, the less detailed and tedious the analog functions of the dual-mode CDMA phone need to be tested. As is the case with all dual-mode phones, the tests associated with the digital aspects are dependent on the type of digital service supported. The CDMA functions could include the following:

1. Examining the OQPSK modulation quality from the mobile. The manner in which this is done depends on the level to which repairs to the mobiles are accomplished. High-level testing, which is the simple separation of good mobiles from bad mobiles, requires a tester that gives simple numeric readings of the error vector magnitude. Lower level testing in which mobiles are actually repaired (individual components or modules are exchanged) require graphic displays, such as a fast constellation display, which give an analog character to the radio. The attending technician is given the opportunity to tap on components, for example, and observe the responses.
2. Testing the mobile's receiver. The manner in which the mobile's receiver is tested depends almost entirely on the test modes available in the radio. Regardless of what kind of BER or loopback tests are provided, a base station simulator is needed with accurate RF output levels. The tester could have an interfering forward channel present during receiver tests.
3. The mobile's power control loops. In the absence of an appropriate test mode, the mobile's RF power control loops can be tested by setting up a traffic channel between the tester (the base site simulator) and the mobile in a manner similar to that done in the actual network. This would require simulated pilot and synchronization channels. Once the traffic channel is established, we send the mobile a known and calibrated RF power level on the traffic channel and measure the resulting power from the mobile, adjust the RF power from the tester up and down, and see if the mobile adjusts its own output power in the opposite way.
4. The rake receiver. The rake receiver can be tested by simulating a soft handoff.

There are some important customer satisfaction issues that should be considered when testing CDMA mobile terminals in a field setting. Customers often send in their terminals for service because of what they hear in the terminal's handset. They may, for example, complain of poor voice quality under one condition or another. When a problem is discovered and repaired, the repair has to confirmed to the satisfaction of the customer, not just to the satisfaction of the technician who made the repair. Analog cellular phones have the important quality that their proper repair is easily confirmed by the customer who can witness a simple voice check, for example, with a radio test set. The digital case is different. One can make a strong argument that a BER test of a receiver is sufficient to confirm a repair to the receiver in a CDMA terminal, and that including a voice check merely wastes time checking voice coding and decoding software. Does the customer understand or even care about such technical details? The customer complained about voice quality and would prefer a clear and simple confirmation of its resolution. In this particular case, the proper approach is to provide for voice checks, within the CDMA protocol, at only the highest levels of service, the levels where only the simplest numeric displays are provided on the CDMA test set, where the test set is the most automated and subject to the least control by the technician, and where the customer is most likely to observe the repair process. CDMA test sets in lower level service facilities, facilities further removed from the customer's gaze, should have less automation, faster and more graphic displays, and offer more control over the specific test process by more thoroughly trained and knowledgeable technicians. There is no need for "feel-good" voice checks at such a level. Service managers and test equipment manufacturers have to consider a host of concerns similar to the voice check case above, where test set utility and versatility sometimes contradict the customer's immediate needs.

13.11.2 Base Station Testing

Base station testing is simpler than testing a mobile. We should examine the QPSK modulation from the base transmitter for quality and proper power. We can test the base station receiver's BER in a similar manner as we do with a mobile. The way in which we test the base station's power control loop depends on what kind of test modes are available, and the need to test these loops at all depends on the design of the base station. The RF output of the base station can be examined with a high-quality spectrum analyzer for RF power and spurious spectral components. There is no need to test functions that depend on the proper operation of software such as soft handoffs.

13.11.3 System Considerations

An interesting problem with the IS-95 realization of CDMA is common to all power-limited systems. Since all the users on a physical channel share the same resource, so they also share in any problems an individual user may have. For example, if a mobile terminal's power control loops were to fail such that the phone would transmit at full power all the time, everyone sharing the resource in which the phone is misbehaving would suffer. The phenomenon seems to support the notion that all users sharing a resource must be disciplined in their use. If a phone takes on a rogue characteristic, then everyone suffers. This should be seen in sharp contrast to misbehavior in bandwidth-limited systems such as GSM, where a defective phone usually confines its destructive influences to the assigned channel. Individual users are sensitive to problems in their phones and can be counted on to bring their defective terminal to a service center for repair on their own. Users of a CDMA phone, on the other hand, will generally not be aware of most problems in their terminal as they make the channel difficult to use for everyone.

Servicing phones in the IS-95 system becomes more of a network function than an individual phone repair chore. The opposite is true in GSM; system problems are easily distinguished from individual radio problems. It remains to be seen how Americans and other users of the IS-95 CDMA system will handle this system trouble analysis challenge. Will the network perform periodic measurements on phones over the air? Yes. There are lots of "service flags" in the IS-95 specification. Will there be testing fixtures at public places (bus stops, train stations, airports, shopping centers, phone stores) in which users can place their questionable phones for a few seconds for detailed analysis? Will defective phones be isolated and disabled by the network, perhaps putting a "Bring me to a service center" message on the phone's display?

These kinds of questions should peak the interest of GSM service providers as they consider how they will provide service for their customers. In considering these interesting problems and reviewing the starkly different manner in which CDMA provides mobile phone services compared to GSM, a more intuitive appreciation of all the details of GSM should finally become clear. It is only with an understanding of our history, our neighbors, and our prospects for the future that we can alter and appreciate our own times and circumstances.

REFERENCES

[1] "*Wideband Spread Spectrum Digital Cellular System Dual-Mode Mobile Station-Base Station Compatibility Standard*," TIA/EIA IS-95.

[2] Dixon, R. C., *Spread Spectrum Systems*, 2nd edition, New York: John Wiley & Sons, 1984.

APPENDIX

This appendix provides a list of GSM/DCS official technical specifications and references that are closely related to this book. Any of these, together with a complete list of other technical specifications, can be obtained from the address below.

ETSI Secretariat, 06921 Sophia Antipolis Cedex, Valbonne, France, Tel. +33-92944200, Fax +33-93654716, Telex 470 040F.

This book cites Phase 1 technical specifications exclusively.

GSM/DCS Technical Specification (TS) No.

04.06	MS-BSS—Data Link Layer Specification
04.08	Mobile Radio Interface—Layer 3 Specification
04.21	Rate Adaptation on the MS-BSS Interface
05.02	Multiplexing and Multiple Acces on the Radio Path
05.03	Channel Coding
05.04	Modulation
05.05	Radio Transmission and Reception
05.08	Radio Subsystem Link Control
06.10	Full Rate Speech Transcoding
08.51	BSC-BTS Interface, General Aspects
08.52	BSC-BTS Interface Principles
08.54	BSC-TRX Layer 1
08.56	BSC-BTS Layer 2 Specification
08.58	BSC-BTS Layer 3 Specification
11.10	Mobile Station Conformity Specification
11.20	The GSM Base Station System: Equipment Specification

The following is a list of more general texts and publications applicable to GSM and its supporting technologies.

Balston, D. M., and R. C. V. Macario, eds., *Cellular Radio Systems*, Boston: Artech House, 1993.
Bellamy, J., *Digital Telephony*, 2nd edition, John Wiley & Sons, 1991.
Calhoun, G., *Wireless Access and the Local Telephone Network*, Boston: Artech House, 1992.
Calhoun, G., *Digital Cellular Radio*, Boston: Artech House, 1993.
Dixon, R. C., *Spread Spectrum Systems*, 2nd edition, John Wiley & Sons, 1984.
Harte, L., *Dual Mode Cellular*, P. T. Steiner Publishing, 1992.
Hess, G. C., *Land-Mobile Radio System Engineering*, Boston: Artech House, 1993.
Lee, W. C. Y., *Mobile Communications Engineering*, McGraw-Hill, 1982.
Lee, W. C. Y., *Mobile Cellular Telecommunications Systems*, McGraw-Hill, 1989.
Lee, W. C. Y., *Mobile Communications Design Fundamentals*, Howard W. Sams & Co., 1986.
Linnartz, J.-P., *Narrowband Land-Mobile Radio Networks*, Boston: Artech House, 1993.
Mehrotra, Asha, "*Cellular Radio Analog and Digital Systems*," Boston: Artech House, 1994.
Mouly, M., and M.-B. Pautet, *The GSM System for Mobile Communications*, M. Mouly et Marie-B. Pautet, Palaiseau, France.
Parsons, J. D., and J. G. Gardner, *Mobile Communications Systems*, London: Blackie, 1989.
Peterson, W. W. and E. J. Weldon, *Error Correcting Codes*, 2nd edition, MIT Press, 1972.
Prentiss, S., *Introducing Cellular Communications, the New Mobile Telephone System*, Tab Books, 1984.
Taub, H., and D. L. Schilling, *Principles of Communication Systems*, 2nd edition, McGraw-Hill, 1986.
Tuttlebee, W. H. W., ed., *Cordless Telecommunications in Europe*, London: Springer-Verlag, 1990.
Yacoub, M. D., *Foundations of Mobile Radio Engineering*, CRC Press, 1993.
Wiggert, D., *Codes for Error Control and Synchronization*, Dedham, MA: Artech House, 1988.

GLOSSARY

The following is a list of abbreviations and terms frequently used in the GSM literature. It should serve as a reference or lookup table for the reader who does not know them by heart.

A	interface between MSC and BSC
Abis	interface between BSC and BTS
AC	authentication center
ADC	analog-to-digital converter
AGC	automatic gain control
AGCH	access grant channel, used by BTS to allocate initial signaling channel
AM	amplitude modulation
AMPS	advanced mobile phone system, U.S. cellular standard
ARFCN	absolute radio frequency channel number
ASIC	application-specific integrated circuit
ASK	amplitude-shift keying
BCC	base station color code
BCH	broadcast channels
BCCH	broadcast control channel, base channel from BTS
BCF	base station control function
BDM	bit demodulator
BER	bit error rate
BFI	bad frame indication, used when a speech frame is corrupted
Bm	bearer-mobile channel
bps	bits per second
BPSK	binary phase-shift keying
BS	base station
BSC	base station controller

BSIC	base station identity code, contains identity and TSC
BSS	base station system
BSSMAP	base station system management part
BTS	base transceiver station
CAI	common air interface
CBCH	cell broadcast channel, used for point-to-multipoint SMSs
CC	call control entity
CCCH	common control channel, can be AGCH, PCH, or RACH
CCITT	Commité Consultatif International de Télégraphique et Téléphonique, International Telegraph and Telephone Consultative Committee
CDMA	code-division multiple access, a broadband radio technology
CDVCC	coded digital verification color code
CELP	code excited linear predictive coding
CEPT	Conférence Européenne des Administrations des Postes et des Télécommunications, European Conference of Posts and Telecommunications Administrations
C/I	carrier-to-interference ratio
C/R	command/response
CRC	cyclic redundancy check
CM	connection management
CW	continuous wave
DAC	digital-to-analog converter
DAI	digital audio interface
D-AMPS	Dual-mode AMPS, U.S. analog and digital dual-mode cellular system
DAMPS	another abbreviation for D-AMPS
DCCH	dedicated control channels
DCS 1800	digital cellular system, 1800-MHz band, used for PCN networks
Dm	data-mobile channel
DQPSK	differential quadrature phase-shift keying
DSP	digital signal processing
DTAP	direct transfer application part
DTC	digital traffic channel
DTMF	dual-tone multifrequency, (touch-tone) addressing of numeric keypad
DTX	discontinuous transmission, optional transmission mode of a mobile station, which is used when no voice activity is detected
DUT	device under test
EAMPS	expanded AMPS, a form of AMPS with more RF channels than the 666 channels in AMPS

E-GSM	extended GSM
EIA	Electronics Industries Association
EIR	equipment identity register, network register for mobile terminal equipment
EPI	protocol indicator
EQ	equalizer
ESN	electronic serial number
ETS	European telecommunications standards
ETSI	European Telecommunications Standards Institute
FACCH	fast associated control channel
FCCH	frequency-correction channel, BTS to MS
FDD	frequency-division duplex
FDMA	frequency-division multiple access
FEI	frame erasure indication
FER	frame erasure rate
FN	frame number
FM	frequency modulation
FOCC	forward control channel
FPLMTS	future public land mobile telephone system
FSK	frequency-shift keying
FTA	full type approval, mobile equipment approval
FVC	forward voice channel
GMSC	gateway mobile switching center
GMSK	Gaussian minimum-shift keying, modulation scheme used in the GSM
GSM	Global system for mobile communications (formerly: Groupe Spéciale Mobile, now SMG)
MAHO	mobile-assisted handoff
HLR	home location register, network register for home subscribers
HSN	hopping sequence number, frequency-hopping parameter
HT	hilly terrain
HWC	hardware controller
IE	information element
IEI	information element identifier
I-ETS	Interim European Telecommunications Standards
IF	intermediate frequency
IMEI	international mobile equipment identity
IMSI	international mobile subscriber identity
IMTS	improved mobile telephone service
IS	interim standard

ISDN	integrated services digital network
ISO	International Standards Organization
ITA	interim type approval, mobile equipment approval before FTA was available
JDC	Japanese Digital Cellular, also PDC
JTACS	Japanese TACS
kbps	kilobits per second, a thousand bits in one second
kHz	kilohertz, 1,000 Hertz or cycles per second
LAC	location area code
LAI	location area identity, contains LAC, MCC, and MNC
LAPD	link access protocol digital
LAPDm	link access protocol digital mobile
Lm	low-mobile channel
L2ML	Layer 2 management link
LMT	local maintenance terminal
LPC	linear predictive coding, speech coding function
LPD	link protocol discriminator
LTP	long-term prediction, speech coding function
MAIO	mobile allocation index offset, indicates channels to be used for frequency hopping, channel selector
MAP	mobile application part, network function/entity
MCC	mobile country code, country code of GSM network, three digits
ME	mobile equipment, terminal not fitted with a SIM
MF	mandatory fixed length
MHz	megahertz, 1,000,000 Hertz or cycles per second
MIN	mobile identity number
MM	mobility management
MMI	man-machine interface, user interface
MNC	mobile network code, identifies a GSM-network within a country, two digits
MOC	mobile-originated call
MoU	Memorandum of Understanding, within a group of GSM operators
MS	mobile station, terminal (hardware) equipped with a SIM
MSC	mobile (services) switching center, GSM switch
MSIC	mobile subscriber identification number
MSK	minimum-shift keying
MT	message type
MTC	mobile-terminated call
MTP	message transfer part

MTS	mobile telephone service
MV	mandatory variable length
NADC	North American Digital Cellular
NAMPS	narrowband AMPS
NCC	national color code
N-CDMA	narrowband CDMA
NMT	Nordic mobile telephone system, Scandinavian analog cellular standard for 450 and 900 MHz
NTACS	narrowband TACS
OF	optional fixed length
O&M	operation and maintenance
OMC	operations and maintenance center, network entity for network operation and management
OML	operation and maintenance link
OQPSK	offset quadrature phase-shift keying
OSI	open systems interconnection, functional interconnect model
OV	optional variable length
PAD	packet assembler/disassembler
PCH	paging channel, used to call a mobile station
PCM	pulse code modulation
PCN	personal communications network
PCS	personal communications system, U.S.
PD	protocol discriminator
PDC	personal digital cellular (JDC, Japanese personal communication system), 800/1,500 MHz
PIN	personal identification number, to be presented when using a GSM phone/SIM
P/F	poll/final
PLMN	public land mobile network
PM	phase modulation
PRBS	pseudorandom bit sequence
PSK	phase-shift keying
PSTN	public switched telephone network
PUK	personal unblocking key
RVC	reverse voice channel
QPSK	quadrature phase-shift keying
RA	rural area
RACH	random access channel, used by MS to access a GSM network
RBER	residual bit error rate
RCC	radio common carrier

RECC	reverse control channel
RPE	regular pulse excitation, speech coding function
rms	root mean square
RR	radio resource management
RSL	radio signaling link
RSSI	received-signal strength indication
RX	(radio) receiver
SABM	set asynchronous balanced mode, Layer 2 command
SACCH	slow associated control channel, associated with TCH or SDCCH
SAP	service access point
SAP(I)	service access point (identifier)
SAT	supervisory audio tones
SC	signaling controller
SCCP	signaling connection control part
SCH	synchronization channel, used by BTS to transmit BSIC and time stamps (frame/multiframe counters)
SCM	station class mark
SDCCH	standalone dedicated control channel, signaling channel
SDMA	space-division multiple access
SFH	slow frequency hopping
SID	silence descriptor (frame sent when no voice activity)
SID	system identification number
SIM	subscriber identity module, smart card
SINAD	signal + noise and distortion
SMG	Special Mobile Group, ETSI standardization groups for GSM
SMS	short-messages services, transmission of alphanumeric information, point-to-point (dedicated) or point-to-multipoint (CBCH)
SMT	surface-mounted technology
SNR	signal-to-noise ratio
SS	supplementary services
SSN7	signaling system no. 7, CCITT standard
TACS	total access cellular system, U.K. analog cellular standard
TCH	traffic channel, for speech or data
TCH/FS	traffic channel/full-rate speech
TCH/HS	traffic channel/half-rate speech
TDD	time-division duplex
TDMA	time-division multiple access
TEI	terminal equipment identifier
TI	transaction identifier, Layer 3 parameter
TMSI	temporary mobile subscriber identity

TN	time slot number (0, 1, . . ., 7)
TRAU	transcoder rate adapter unit
TRX	(radio) transceiver
TS	time slot
TSC	training sequence code (0, 1, . . ., 7)
TTL	transistor-transistor logic
TU	typical urban
TX	(radio) transmitter
UI	unnumbered information
Um	GSM air interface (radio interface)
UMTS	universal mobile telephone system
UUT	unit under test
VAD	voice activity detection
VCO	voltage-controlled oscillator
VLR	visitor location register, network register for visiting subscribers, always associated with an MSC
VLSI	very-large-scale integrated circuits
WP	working party

ABOUT THE AUTHORS

Siegmund M. Redl graduated from the Technical University of Munich in 1989 with a degree in communications engineering. He joined Wavetech, Munich, Germany, in early 1990. After some years of providing technical product support for GSM radio test systems and serving for ETSI groups in GSM test standardization, he is now responsible for product management of test equipment for new cellular radio systems and new test equipment products.

Matthias K. Weber graduated from the Technical University of Munich in 1989 with a degree in communications engineering. He joined Wavetech, Munich, Germany, in October 1989. After a short period in R&D of GSM test equipment, he turned to providing technical product support for GSM radio test systems. He is now responsible for the product management of GSM/DCS-1800 test equipment.

Malcolm W. Oliphant graduated from Hawaii Loa College, Kaneohe, Hawaii, in 1972 with a degree in biology and mathematics, and has completed numerous continuing education courses and some graduate work in communications engineering and fine art. He joined Wavetech, Billerica, Massachusetts, in early 1990. He remains responsible for applications support for all the radio test products offered by Wavetech in North America and is involved in some standardization activities.

INDEX

The Artech House Telecommunications Library

Vinton G. Cerf, Series Editor

Advanced Technology for Road Transport: IVHS and ATT, Ian Catling, editor

Advances in Computer Communications and Networking, Wesley W. Chu, editor

Advances in Computer Systems Security, Rein Turn, editor

Advances in Telecommunications Networks, William S. Lee and Derrick C. Brown

Analysis and Synthesis of Logic Systems, Daniel Mange

Asynchronous Transfer Mode Networks: Performance Issues, Raif O. Onvural

ATM Switching Systems, Thomas M. Chen and Stephen S. Liu

A Bibliography of Telecommunications and Socio-Economic Development, Heather E. Hudson

Broadband: Business Services, Technologies, and Strategic Impact, David Wright

Broadband Network Analysis and Design, Daniel Minoli

Broadband Telecommunications Technology, Byeong Lee, Minho Kang, and Jonghee Lee

Cellular Radio: Analog and Digital Systems, Asha Mehrotra

Cellular Radio Systems, D. M. Balston and R. C. V. Macario, editors

Client/Server Computing: Architecture, Applications, and Distributed Systems Management, Bruce Elbert and Bobby Martyna

Codes for Error Control and Synchronization, Djimitri Wiggert

Communications Directory, Manus Egan, editor

The Complete Guide to Buying a Telephone System, Paul Daubitz

Computer Telephone Integration, Rob Walters

The Corporate Cabling Guide, Mark W. McElroy

Corporate Networks: The Strategic Use of Telecommunications, Thomas Valovic

Current Advances in LANs, MANs, and ISDN, B. G. Kim, editor

Digital Cellular Radio, George Calhoun

Digital Hardware Testing: Transistor-Level Fault Modeling and Testing, Rochit Rajsuman, editor

Digital Signal Processing, Murat Kunt

Digital Switching Control Architectures, Giuseppe Fantauzzi

Distributed Multimedia Through Broadband Communications Services, Daniel Minoli and Robert Keinath

Disaster Recovery Planning for Telecommunications, Leo A. Wrobel

Document Imaging Systems: Technology and Applications, Nathan J. Muller

EDI Security, Control, and Audit, Albert J. Marcella and Sally Chen

Electronic Mail, Jacob Palme

Enterprise Networking: Fractional T1 to SONET, Frame Relay to BISDN, Daniel Minoli

Expert Systems Applications in Integrated Network Management, E. C. Ericson, L. T. Ericson, and D. Minoli, editors

FAX: Digital Facsimile Technology and Applications, Second Edition, Dennis Bodson, Kenneth McConnell, and Richard Schaphorst

FDDI and FDDI-II: Architecture, Protocols, and Performance, Bernhard Albert and Anura P. Jayasumana

Fiber Network Service Survivability, Tsong-Ho Wu

Fiber Optics and CATV Business Strategy, Robert K. Yates *et al.*

A Guide to Fractional T1, J. E. Trulove

A Guide to the TCP/IP Protocol Suite, Floyd Wilder

Implementing EDI, Mike Hendry

Implementing X.400 and X.500: The PP and QUIPU Systems, Steve Kille

Inbound Call Centers: Design, Implementation, and Management, Robert A. Gable

Information Superhighways: The Economics of Advanced Public Communication Networks, Bruce Egan

Integrated Broadband Networks, Amit Bhargava

Intelcom '94: The Outlook for Mediterranean Communications, Stephen McClelland, editor

International Telecommunications Management, Bruce R. Elbert

International Telecommunication Standards Organizations, Andrew Macpherson

Internetworking LANs: Operation, Design, and Management, Robert Davidson and Nathan Muller

Introduction to Error-Correcting Codes, Michael Purser

Introduction to Satellite Communication, Bruce R. Elbert

Introduction to T1/T3 Networking, Regis J. (Bud) Bates

Introduction to Telecommunication Electronics, A. Michael Noll

Introduction to Telephones and Telephone Systems, Second Edition, A. Michael Noll

Introduction to X.400, Cemil Betanov

Land-Mobile Radio System Engineering, Garry C. Hess

LAN/WAN Optimization Techniques, Harrell Van Norman

LANs to WANs: Network Management in the 1990s, Nathan J. Muller and Robert P. Davidson

Long Distance Services: A Buyer's Guide, Daniel D. Briere

Measurement of Optical Fibers and Devices, G. Cancellieri and U. Ravaioli

Meteor Burst Communication, Jacob Z. Schanker

Minimum Risk Strategy for Acquiring Communications Equipment and Services, Nathan J. Muller

Mobile Communications in the U.S. and Europe: Regulation, Technology, and Markets, Michael Paetsch

Mobile Information Systems, John Walker

Narrowband Land-Mobile Radio Networks, Jean-Paul Linnartz

Networking Strategies for Information Technology, Bruce Elbert

Numerical Analysis of Linear Networks and Systems, Hermann Kremer *et al.*

Optimization of Digital Transmission Systems, K. Trondle and Gunter Soder

Packet Switching Evolution from Narrowband to Broadband ISDN, M. Smouts

Packet Video: Modeling and Signal Processing, Naohisa Ohta

Personal Communication Systems and Technologies, John Gardiner and Barry West, editors

The PP and QUIPU Implementation of X.400 and X.500, Stephen Kille

Principles of Secure Communication Systems, Second Edition, Don J. Torrieri

Principles of Signaling for Cell Relay and Frame Relay, Daniel Minoli and George Dobrowski

Principles of Signals and Systems: Deterministic Signals, B. Picinbono

Private Telecommunication Networks, Bruce Elbert

Radio-Relay Systems, Anton A. Huurdeman

Radiodetermination Satellite Services and Standards, Martin Rothblatt

Residential Fiber Optic Networks: An Engineering and Economic Analysis, David Reed

Secure Data Networking, Michael Purser

Service Management in Computing and Telecommunications, Richard Hallows

Setting Global Telecommunication Standards: The Stakes, The Players, and The Process, Gerd Wallenstein

Smart Cards, José Manuel Otón and José Luis Zoreda

Television Technology: Fundamentals and Future Prospects, A. Michael Noll

Telecommunications Technology Handbook, Daniel Minoli

Telephone Company and Cable Television Competition, Stuart N. Brotman

Teletraffic Technologies in ATM Networks, Hiroshi Saito

Terrestrial Digital Microwave Communciations, Ferdo Ivanek, editor

Transmission Networking: SONET and the SDH, Mike Sexton and Andy Reid

Transmission Performance of Evolving Telecommunications Networks, John Gruber and Godfrey Williams

Troposcatter Radio Links, G. Roda

UNIX Internetworking, Uday O. Pabrai

Virtual Networks: A Buyer's Guide, Daniel D. Briere

Voice Processing, Second Edition, Walt Tetschner

Voice Teletraffic System Engineering, James R. Boucher

Wireless Access and the Local Telephone Network, George Calhoun

Wireless Data Networking, Nathan J. Muller

Wireless LAN Systems, A. Santamaría and F. J. Lopez-Hernandez

Writing Disaster Recovery Plans for Telecommunications Networks and LANs, Leo A. Wrobel

X Window System User's Guide, Uday O. Pabrai

For further information on these and other Artech House titles, contact:

Artech House
685 Canton Street
Norwood, MA 02062
617-769-9750
Fax: 617-769-6334
Telex: 951-659
email: artech@world.std.com

Artech House
Portland House, Stag Place
London SW1E 5XA England
+44 (0) 171-973-8077
Fax: +44 (0) 171-630-0166
Telex: 951-659
email: bookco@artech.demon.co.uk